BIPOLAR SEMICONDUCTOR DEVICES

McGraw-Hill Series in Electrical Engineering

Consulting Editor

Stephen W. Director, *Carnegie-Mellon University*

Circuits and Systems
Communications and Signal Processing
Control Theory
Electronics and Electronic Circuits
Power and Energy
Electromagnetics
Computer Engineering
Introductory
Radar and Antennas
VLSI

Previous Consulting Editors

Ronald N. Bracewell, Colin Cherry, James F. Gibbons, Willis W. Harman, Hubert Heffner, Edward W. Herold, John G. Linvill, Simon Ramo, Ronald A. Rohrer, Anthony E. Siegman, Charles Susskind, Frederick E. Terman, John G. Truxal, Ernst Weber, and John R. Whinnery

Electronics and Electronic Circuits

Consulting Editor

Stephen W. Director, *Carnegie-Mellon University*

BIPOLAR SEMICONDUCTOR DEVICES

David J. Roulston

Electrical Engineering Department
University of Waterloo
Waterloo, Canada

McGraw-Hill Publishing Company

New York   St. Louis   San Francisco   Auckland   Bogotá
Caracas   Hamburg   Lisbon   London   Madrid   Mexico   Milan
Montreal   New Delhi   Oklahoma City   Paris   San Juan
São Paulo   Singapore   Sydney   Tokyo   Toronto

This book was set in Times Roman.
The editors were Alar E. Elken and John M. Morriss;
the production supervisor was Louise Karam.
The cover was designed by Albert M. Cetta.
Project supervision was done by Harley Editorial Services.
R. R. Donnelley & Sons Company was printer and binder.

BIPOLAR SEMICONDUCTOR DEVICES

1234567890 DOC DOC 89432109

ISBN 0-07-054120-5

Library of Congress Cataloging-in-Publication Data

Roulston, David J.
    Bipolar semiconductor devices/David J. Roulston.
        p.   cm.—(McGraw-Hill series in electrical engineering)
    Includes index.
    ISBN 0-07-054120-5
    1. Diodes, Semiconductor—Design and construction.   I. Title.
II. Series.
TK7871.86.R68   1990
621.381'522—dc20   89-8017

ABOUT THE AUTHOR

David J. Roulston was born in England and received the B.Sc. degree from Queen's University, Belfast, Northern Ireland, in 1957, and the Ph.D. degree from Imperial College, University of London, in 1962.

He spent six years in industry in England and France before joining the Department of Electrical Engineering, University of Waterloo, Ontario, Canada, where he is currently a Professor. He has also held consulting positions with R & D laboratories in Canada, France, Japan, USA, and with UNIDO in India.

His research interests are in modeling physical processes in semiconductor devices: specifically bipolar transistors, microwave diodes, photodiodes, and MESFETs. He has been responsible for development of the BIPOLE computer program for analysis of bipolar devices. He has published over 100 technical papers in the above fields, has had six patents awarded, and has presented invited seminars or conference papers in twelve countries.

Dr. Roulston is a Fellow of the Institution of Electrical Engineers (England). He has been Associate Editor for Bipolar Devices of the IEEE Transactions on Electron Devices for two years and is currently on the Honorary Editorial Advisory Board of Solid State Electronics. For the academic year 1988–1989 he was elected a Visiting Fellow at Wolfson College, Oxford, where he has spent a sabbatical leave writing and researching in the Department of Metallurgy and Materials Science of Oxford University.

To
Christine, Hélène, Philippe,
AND
Jessica

CONTENTS

3 PN Junction Diodes   55

4 Transient and High Frequency Behavior of Diodes   91

PREFACE

This book is intended for graduate level students in Electrical Engineering and practicing semiconductor device engineers in the industry. Parts may also be used in final year undergraduate courses in semiconductor devices. Users of the book will normally have already had an introductory course in semiconductor devices, however, the material is accessible to any student or practicing engineer with a general electronics or physics background.

This text is the outcome of over twelve years of teaching a graduate level course on bipolar devices and corresponding time working closely with industrial device design engineers as a consultant. Interaction with industrial users of the BIPOLE computer program has also served as an education experience, through which much has been learned about bipolar device operation. This program was developed as a spinoff from the author's research and has been used to generate many of the computed curves in this text.

The material is presented in such a way that it is pedagogical but also contains much practical information of direct use for design purposes. The approach is to keep complex theory to a minimum, compatible with relating adequately the device structure parameters (doping profiles, region thickness, mask geometries, etc.) to the device terminal characteristics (I-V laws, high frequency cutoff, switching times, etc.). To facilitate reading, sections of less importance are set in smaller print and may be omitted at a first reading. In many cases, data from computer simulations is included both to assist in the understanding of the device and also in choosing device design parameters for a specific terminal specification. The pedagogical approach is thus combined throughout the text with a strong emphasis on engineering practicality.

After a review chapter on the basic physical properties of semiconductors necessary for treatment of bipolar devices (including such essential material as an overview of heavy doping effects such as bandgap narrowing and Auger recombination), the first part of the book, up to and including chapter 6, is devoted to PN junctions and diodes. This material is useful in itself for understanding the behavior of diodes and for relating terminal characteristics to the physical structure (impurity profiles, junction depths, etc.); it is also useful as an introduction to phenomena of importance in bipolar transistor theory which forms the subject matter in the second part of the book. For example, the reader is introduced in chapter 2 to the various types of PN junctions: abrupt symmetrical, abrupt P^+N, linearly graded, diffused (gaussian). The importance of the NN^+ low-high junction is also discussed and the concept of a retarding field for minority carriers is introduced. The significance of free carriers inside the space-charge region is highlighted since this is very important in low current transistor operation.

Chapter 3 presents the theory for a range of diode structures: the P^+N narrow base diode (equivalent to the diode-connected transistor found in many integrated circuits), the wide base diode, the important P^+NN^+ and PIN diodes under low, medium, and high current conditions. This chapter terminates with a discussion of avalanche multiplication and contains some useful design information obtained from computer simulations.

Chapter 4 introduces the reader to the transient behavior of diodes. This includes some interesting and important nonlinear effects observed in PIN diode switching at both 50/60 Hz and at microwave frequencies, and a brief discussion of the *open circuit voltage decay* behavior sometimes used for lifetime evaluation.

Chapter 5 gives an overview of some practical aspects of diodes for various applications. It includes operation of the hyperabrupt varactor diode, the microwave step recovery diode, and the low frequency rectifier structure.

Chapter 6 covers the PIN photodetector diode and the solar cell. The PIN photodetector diode is an important bipolar optical device because of its use in fiber optic systems. The solar cell is a fascinating bipolar structure which, although pertaining to a very specialized field, introduces the reader to some fundamental concepts. The treatment includes a clear explanation of why the power conversion efficiency of solar cells is restricted.

The basic operation of the bipolar transistor is reviewed in chapter 7. This is in sufficient detail to make the book completely self-contained, even if the reader has not been previously exposed to bipolar transistor theory (in fact the author has used this chapter, with some success, as an introduction to transistors in a third-year course). Only the constant doping approximation to the practical double-diffused device is considered here. The treatment includes a discussion of all important characteristics including base resistance and its exact theoretical value, current gain, plus cutoff and maximum oscillation frequencies. The chapter terminates with a discussion of collector-base and collector-emitter breakdown and includes a set of computer generated design curves for a wide range of collector doping levels and thicknesses. These curves are used subsequently in later parts of the book and in problems. By the end of this chapter, the reader will have a complete grasp of the essential features of a bipolar transitor and will be ready to proceed to the next three chapters where the author expands upon the basic theory.

In chapter 8, the diffused emitter region is considered in some detail. This is the only part of the transistor which independently can be tailored for a specific gain (in contrast to the base region whose parameters determine also cutoff and maximum oscillation frequency). The treatment is based on approximating the emitter effective impurity profile by an exponential and a constant doping region. This leads to a clearly visible explanation of the differences between deep and shallow emitters (in terms of their recombination behavior). The effects of bandgap narrowing and Auger recombination are included in the analysis. The modern polysilicon-emitter structure is also treated, including the important difference in behavior between devices with and without a thin interfacial oxide layer.

Chapter 9 continues in the vertical direction by studying the emitter-base space-charge region, using the linear graded junction approximation to evaluate both junction capacitance and low current gain falloff. The neutral base region is then studied by approximating the impurity profile to two exponential regions: a retarding field followed by an accelerating drift field. The theory is kept quite simple and yet enables a useful engineering accuracy and understanding of base transport to be obtained, both as it affects current gain and high frequency cutoff. The Gummel integral (the basis of most widely used CAD circuit simulation models) is then introduced; this leads naturally to an examination of *Early voltage* and its dependence on operating bias (a point missed in many other treatments and yet of fundamental importance in thin base VLSI devices when using circuit models for computer simulations).

A complete discussion of all terms contributing to common emitter current gain bandwidth product f_t is given. This includes delays due to free carriers within the base-collector and emitter-base space-charge layers and in the emitter neutral region. This leads to a reexamination of the maximum oscillation frequency and a discussion of the "optimum design"—trading low base resistance versus high f_t.

Chapter 10 examines the various phenomena creating high current falloff in performance. These include emitter current crowding, high level injection, base widening, and quasi-saturation. This latter effect is particularly important for a clear understanding of both logic inverters and high voltage switches.

Chapter 11 covers transient switching behavior including a discussion of high voltage inductive load switching. Thermal behavior is treated briefly in this chapter.

Chapters 12 and 13 cover topics specific to various practical structures. Firstly discrete devices, where topics relating to RF power and high voltage power switching devices are discussed; the Darlington pair and the thyristor (SCR) are also presented. Secondly, in chapter 13, various aspects of the integrated circuit transistors are analyzed; this includes sections on base and collector resistance (both of which necessitate additional material to that dealt with for discrete transistors), a treatment of the widely used lateral PNP structure, and (by extension of the PNP work) a study of *integrated injection logic* or IIL. Aspects of ECL and the increasingly important BICMOS structures are also covered here.

Chapter 14 on "Advanced Technology Devices" is important for setting the scene in terms of performance of modern bipolar transistors. The reader is introduced to the benefits of polysilicon technology, oxide isolation, and the effect of the extra extrinsic base step. Some examples are included here on experimental devices, including V-groove bipolar structures. An introduction is given to GaAlAs-GaAs heterojunction bipolar transistors, in which the essential theoretical aspects are emphasized.

The final chapter surveys the field of numerical analysis and attempts to assist the reader in determining just what can be achieved with existing software. This includes an introduction to some of the equations used in the BIPOLE program. The problems in this chapter involve the writing of some computer code. Some of the limitations of circuit CAD models for bipolar transistors are discussed and methods of overcoming them are presented.

The author has used the contents of this book in teaching graduate courses and also in some aspects of undergraduate courses. As a first graduate course in semiconductor bipolar devices, the material of chapters 1 through 4, followed by chapters 7 through 10 has been found ideal for a one-term course of 26 to 30 lecture hours. If it is desired to focus only on bipolar transistors, sections 3.3, 3.4.3, and 3.5.2 on the wide base, P^+NN^+ and PIN diodes plus all of chapter 4 could be omitted and a substantial part of chapter 13 included instead. A follow-on course has then consisted of a brief review (mainly of some material from chapters 1, 2, and 3), followed by the material of chapters 5 and 6, followed by chapters 11 through 15. A final-year undergraduate course on the design of semiconductor devices has made use of chapter 7, selected parts of chapters 9 and 10, followed by parts of chapters 12 and 13. Finally, as mentioned above, chapter 7 has been used in a third-year course and has met with a good response from the students. The treatment in chapter 7 is simpler, and yet closer to reality, than that found in many textbooks.

The contents of this book have resulted from the author's interaction with students and colleagues at the University of Waterloo over many years and also from discussions with colleagues in industrial R & D laboratories. It is difficult to single out individuals through fear of omitting the names of particular friends and colleagues who have undoubtedly helped me at one time or another, to gain a deeper understanding of semiconductor devices. Nevertheless, among colleagues at the University of Waterloo, Professor C. R. Selvakumar deserves special thanks for contributing his expert knowledge and time at various stages of preparing this text. I have had many stimulating discussions with Professor S. C. Jain (in Waterloo, Delhi, Oxford, and at IMEC, Leuven) and it is thanks to him that the section on bandgap narrowing has been brought up to date. Among past students, Dr. François Hébert and Dr. Alan Solheim have contributed particularly through their enthusiasm, originality, and hard work. On the many occasions when the computer just would not work, and in the large task of keeping the BIPOLE computer program working and expanding during the course of using it while writing this book, grateful thanks are extended to Heidi Strayer. Roger Grant, Manager of the Silicon Device and Integrated Circuits Laboratory at Waterloo, has succeeded in keeping our fabrication facility running over many years; this has provided an invaluable educational experience for both the author and his students—it is only when examining real devices (some of which are described in this text) that the limits of theoretical work become apparent. Of my many industrial colleagues, Messieurs J. B. Quoirin, M. Depey, and R. Henry all of Thomson—CSF (now SGS Thomson) in France are to be thanked for the many fascinating problems we have solved together. Dr. Kokhle and his staff at the CEERI laboratory, in Rajasthan, India have provided me on several occasions with stimulating conversations and an environment in which to learn more about semiconductors and to contemplate the early stages of the manuscript.

Without the initial encouragement of Alar Elken at McGraw-Hill, this book would not have progressed. Thanks to his efforts, a number of invaluable comments and suggestions were received from reviewers at various stages of writing the manuscript. Grateful thanks are due to the following: Dorothea E. Burk, University of Florida; D. W. Greve, Carnegie Mellon University; Dimitris Pavlidis, University of Michigan; Gregory E. Stillman, University of Illinois at Urbana–Champaign; and Earl Swartzlander, TRW Defense Systems Group.

The book has been largely written while the author was on sabbatical leave with the Department of Metallurgy and Materials Science at the University of Oxford, and as a Visiting Fellow at Wolfson College. For arranging my visit and the facilities provided for this task, special thanks are due to Dr. Roger Booker, Professor Sir Peter Hirsch, and Sir Raymond Hoffenberg.

Finally, in addition to all the professional assistance which I have received for this work, I thank my daughter Hélène, who through a very conscientious effort at word-processing, proof reading, and general invaluable help, enabled the typescript to meet the publisher's deadline.

David J. Roulston

BIPOLAR SEMICONDUCTOR DEVICES

CHAPTER
1

OVERVIEW
OF
BASIC
SEMICONDUCTOR
PROPERTIES

1.1   INTRODUCTION

The purpose of this chapter is to give an overview on the most important semi-conductor properties relevant to the ensuing treatment on bipolar devices. There are many excellent textbooks on basic semiconductor properties [1, 2, 3] and the interested reader is referred to these for a more detailed treatment of the topics to be mentioned in this chapter. Our intent in the following is to give only the essential information which will be used in subsequent chapters. After a very brief background discussion of the energy band diagrams for pure and doped semicon-ductors, we proceed to discuss current flow and the concept of mobility and its dependence on doping level. This is directly related to diffusion constant, through the Einstein relation. The well established Shockley-Read-Hall recombination model is then discussed. Some temperature-dependent effects of the above param-eters are considered where relevant. The chapter concludes with a discussion on heavy doping effects, which are very important in the emitters of most bipolar transistors. These effects concern bandgap narrowing, Auger recombination, and minority carrier mobility. It is emphasized that the purpose of this chapter is not to explore the physics of these effects but rather to introduce them in such a way that the material is available for use in subsequent parts of this text.

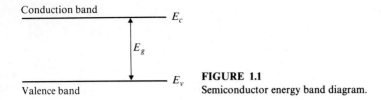

Conduction band

Valence band

FIGURE 1.1
Semiconductor energy band diagram.

1.2 PROPERTIES RELATED TO THE ENERGY GAP

The main semiconductor used today, i.e., silicon, has a diamond lattice crystal structure, as does germanium, the original material used in early semiconductor devices. These are both elements from column IV of the periodic table and are covalent bond structures. One of the main compound semiconductors used today, a III/V compound, gallium-arsenide, has a zinc blende lattice structure with a small component of ionic bonding [4].

The value of the bandgap, separating the conduction band from the valence band, is the single most important parameter characterizing the properties of the semiconductor crystal, see Fig. 1.1. Table 1.1 shows that silicon has a bandgap intermediate between that of germanium and gallium-arsenide. The bandgap, E_g, has a slight temperature-dependence given by

$$E_g(T) = E_g(0) - \frac{\alpha T^2}{T + \beta} \tag{1.1}$$

where T is the absolute temperature. The values of the constants are given in Table 1.1 [5] for the three common semiconductors. Figure 1.2 shows the density of available states, the Fermi probability distribution, and the electron concentration versus energy for a semiconductor. By considering the kinetic energy of an electron in the conduction band $E - E_c$ and its momentum, and by applying the concept of minimum momentum volume (from Heisenberg's uncertainty principle) it can be shown [6, 7] that the density of states in the conduction band $N(E)$ is proportional to the square root of the energy above the bottom of the conduction band:

$$N(E) \propto (E - E_c)^{1/2} \tag{1.2}$$

TABLE 1.1
Values of constants for Eq. (1.1) after Thurmond
[5], and of $(N_c N_v)^{1/2}$ at 27°C

	$E_g(0)$	$\alpha \times 10^{-4}$	β	$(N_c N_v)^{1/2} \times 10^{19}$
GaAs	1.519	5.405	204	0.18
Si	1.170	4.73	636	1.71
Ge	0.7437	4.774	235	0.8

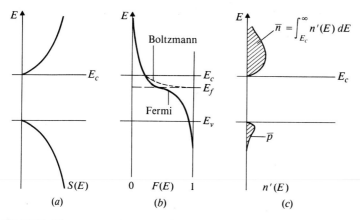

FIGURE 1.2
(a) Density of available states versus energy; (b) Fermi probability function, with Boltzmann probability shown by broken curve; (c) electron concentration versus energy.

where E_c is the conduction band edge (Fig. 1.1). The probability of occupancy of a state is given by the Fermi-Dirac distribution function $F(E)$:

$$F(E) = \frac{1}{1 + \exp\left[(E - E_F)/kT\right]} \tag{1.3}$$

where E_F is the Fermi level and k is Boltzmann's constant. For most semiconductor problems, $F(E)$ can be approximated by the Boltzmann distribution in the conduction band by omitting the '1' in the denominator. By integrating over energy above E_c the product of the density of available states times the probability of occupancy, the electron concentration in the conduction band is obtained (see App. 1). This is usually written in the form

$$\bar{n} = N_c \exp\left(\frac{E_F - E_c}{kT}\right) \tag{1.4}$$

where N_c, the effective density of states function, is given by

$$N_c = \frac{4\sqrt{2}}{h^3}(\pi m_e^* kT)^{3/2} \tag{1.5}$$

where h is Planck's constant and m_e^* the effective electron mass. A similar relationship applies for holes in the valence band (energies below E_V, Fig. 1.1):

$$\bar{p} = N_V \exp\left(\frac{E_V - E_F}{kT}\right) \tag{1.6}$$

where N_V differs from N_c only in the effective mass of holes m_h^* which, in general, is not equal to m_e^*, although in silicon and germanium the difference is only about a factor of two. It may be shown that N_c and N_V have values of 2.8×10^{19} and 1.0×10^{19} per cm^3, respectively, at 300 K for silicon.

For calculations, it may be noted that at room temperature (27°C) the value of kT is 0.0259 eV. Combining Eqs. (1.4) and (1.6) gives the very basic and useful relationship

$$\overline{np} = N_c \, N_V \, \exp\left(-\frac{E_g}{kT} \right) \tag{1.7}$$

where $E_g = E_c - E_V$ is the bandgap of the semiconductor. Equation (1.7) is often written in the form

$$\overline{np} = n_i^2 \propto \exp\left(-\frac{E_g}{kT} \right) \tag{1.8}$$

where n_i is the intrinsic carrier concentration, i.e., the concentration of electrons (or holes) in a pure (nondoped) semiconductor, and is strongly temperature-dependent. This result is known as the *mass action law*. Note that since $kT = 0.0259$ eV at 300 K, n_i for silicon is 1.6×10^{10} cm^{-3}.

Taking the ratio of Eqs. (1.4) and (1.6) and rearranging terms gives

$$E_F = \frac{E_c + E_V}{2} + \frac{kT}{2} \ln \frac{N_V}{N_c} + \frac{kT}{2} \ln \frac{\bar{n}}{\bar{p}} \tag{1.9}$$

For doped semiconductors the Fermi level shifts either upward for N type (Fig. 1.2), or downward for P type, approaching the band edges for highly doped material. If the doping level is so high that the Fermi level passes the edge of the bandgap, the material is referred to as degenerate and the above relations strictly no longer apply [since they have been based on the assumption that the "tail" of the Fermi probability distribution, Fig. 1.2(b), can be replaced by the Boltzmann distribution, within the conduction (or valence) band].

Donor atoms, typically phosphorus or arsenic from column V of the periodic table, can be assumed in most cases to have given up their fifth electron. A simple calculation of the ionization of the donor atom can be made using the hydrogen atom model. The ionization energy for the hydrogen atom is

$$E_H = \frac{m_0 \, q^4}{8 \varepsilon_0^2 \, h^2} = 13.6 \text{ eV} \tag{1.10}$$

where m_0 is the mass of the free electron, q the electronic charge, ε_0 the permittivity of free space. The ionization energy for the donor at an energy level E_d below the conduction band edge is obtained using Eq. (1.10) but with the dielectric constant of the semiconductor, and replacing the value of m_0 by the effective mass of electrons. This gives

$$E_d = \left(\frac{\varepsilon_0}{\varepsilon_0 \, \varepsilon_r} \right)^2 \left(\frac{m_e^*}{m_e} \right) E_H \tag{1.11}$$

where ε_r is the relative dielectric constant of the semiconductor (11.9 for Si, 13.1 for GaAs, 16 for Ge). This formula gives values of the ionization energy of order 0.01 eV for Ge, Si, and GaAs. The corresponding band diagram information is given in Fig. 1.3. For charge neutrality the total number of free electrons, n, is given by the total number of ionized donors N_D^+ plus the number of free holes:

$$n = N_D^+ + p \qquad (1.12)$$

The concentration of ionized donors is now given by:

$$N_D^+ = N_D\left[1 - \frac{1}{1 + (1/2) \exp\left[(E_D - E_F)/kT\right]}\right] \qquad (1.13)$$

where N_D is the total donor concentration and the factor 1/2 is due to the fact that the donor level can have one electron with either spin, or no electrons. Note that we use $E_D = E_C - E_L$ as shown in Fig. 1.3. For semiconductors doped with acceptor atoms such as boron from column III of the periodic table, the free hole concentration is given by

$$p = N_A^- + n$$

and the concentration of ionized acceptors is

$$N_A^- = \frac{N_A}{1 + 4 \exp\left[(E_A - E_F)/kT\right]} \qquad (1.14)$$

where E_A is the acceptor energy level close to E_V.

At normal operating temperatures of bipolar devices and over the range of doping levels encountered in practice it is customary to assume 100 percent ionization of donor and acceptor atoms. Errors introduced when this is not the case

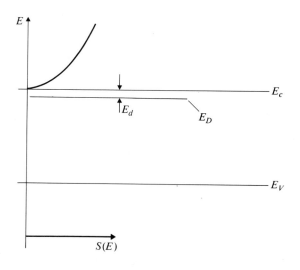

FIGURE 1.3
Energy band diagram for ionization of donor.

will normally be masked by other more important phenomena, such as heavy doping effects, to be discussed later in this chapter. Henceforth we will not distinguish between N_D, N_D^+ nor between N_A, N_A^-.

1.3   CURRENT FLOW

Current in a semiconductor may be carried by two independent mechanisms: drift and diffusion. Drift current due to an applied electric field E is determined by the mobility μ of the carriers as in the following equations:

$$J_{e \text{ drift}} = q\mu_n nE \qquad (1.15a)$$

$$J_{h \text{ drift}} = q\mu_p pE \qquad (1.15b)$$

The conductivity σ of the semiconductor is given by ($\sigma = J/E$)

$$\sigma = q(\mu_n n + \mu_p p) \qquad (1.16)$$

Diffusion current on the other hand is due to the gradient of the carrier concentrations and is given by

$$J_{e \text{ diff}} = qD_n \frac{dn}{dx} \qquad (1.17)$$

$$J_{h \text{ diff}} = -qD_p \frac{dp}{dx} \qquad (1.18)$$

where D_n, D_p are the electron and hole diffusion coefficients, respectively. For nondegenerate semiconductors the mobility and diffusion coefficients are related by the Einstein relation (see App. 2)

$$\frac{D_n}{\mu_n} = \frac{D_p}{\mu_p} = \frac{kT}{q} = V_t \qquad (1.19)$$

where V_t is sometimes referred to as "thermal voltage." Its value is 0.0259 volts at 300 K. For degenerate semiconductors more complicated expressions are available [6].

Because mobility and diffusion coefficients are fundamental parameters determining transport in all semiconductor devices, we shall expand our discussion on mobility, noting that the diffusion coefficient can always be obtained through the Einstein relation. Mobility is the parameter which determines the way in which the carrier moves in response to an electric field while suffering multiple collisions. It is given in terms of the mean free time between collisions by

$$\mu = \frac{q\bar{t}}{m} \qquad (1.20)$$

where $\bar{t}$ is the mean free time between collisions of particles moving under random thermal motion averaged over all energy levels. $\bar{t}$ is determined by various scattering mechanisms [8]. At low values of electric field and low doping level, the mobility is determined by lattice scattering and has a $T^{-3/2}$ dependence. For high doping levels, ionized impurity scattering will dominate. In this case the temperature-dependence is such that the mobility has a $T^{3/2}$ dependence. At high carrier concentrations, carrier-carrier scattering will further reduce the mobility.

Since temperature effects can be important it is desirable to have a formula for mobility versus doping level and temperature for use in predicting device characteristics. Figure 1.4 shows electron mobility in phosphorus-doped silicon as a function of doping from a combination of experimental and theoretical data [9]. The dotted curves are obtained using the following expression:

$$\mu_n = 88T_n^{-0.57} + \frac{7.4 \times 10^8 T^{-2.33}}{1 + [N/(1.26 \times 10^{17} T_n^{2.4})]0.88 T_n^{-0.146}} \tag{1.21}$$

where $T_n = T/300$. The hole mobility for P type material is given by

$$\mu_p = 54.3T_n^{-0.57} + \frac{1.36 \times 10^8 T^{-2.23}}{1 + [N/(2.35 \times 10^{17} T_n^{2.4})]0.88 T_n^{-0.146}} \tag{1.22}$$

Less extensive data is available for germanium and gallium-arsenide but Table 1.2 summarizes the basic mobility values for these two materials along with those

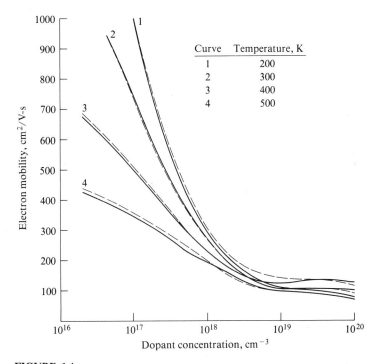

Curve	Temperature, K
1	200
2	300
3	400
4	500

FIGURE 1.4
Electron mobility in phosphorus doped silicon; continuous curves are "experimental," dotted curves are from Eq. (1.21). (*After Arora et al.* [9]. *Reproduced with permission* © *1982 IEEE.*)

TABLE 1.2
Mobility values at room temperature. The minimum values are for a doping level of 10^{20} cm^{-3} using data from Prince [44], Sze and Irvin [48], and Yang [49].

	$\mu_{e\,max}$	$\mu_{e\,min}$	$\mu_{p\,max}$	$\mu_{p\,min}$
Ge	4806	900	2000	110
GaAs	8000	1300	380	70
Si	1500	100	480	45

for silicon. Figure 1.5 shows the variation of mobility with doping level for silicon at room temperature.

At higher values of electric field, the mobility ceases to be a constant. Figure 1.6 shows the drift velocity versus electric field for the three common semiconductor materials. In the presence of an electric field, the carriers acquire energy from the field and lose it to phonons; at sufficiently high fields the carriers acquire more energy than they have at thermal equilibrium. The net result is a saturation in the drift velocity given by [6]

$$v_d = \sqrt{\frac{8E_p}{3\pi m_0}} \sim 10^7 \text{ cm/s} \qquad (1.23)$$

where E_p is the optical phonon energy (0.037 eV for Ge, 0.063 eV for Si, 0.035 eV for GaAs). The double valued or negative differential resistance region for GaAs is related to the existence of two "valleys" in the band structure (electron energy versus wave vector) with different mobilities and different effective masses for electrons. As we shall see subsequently, there are many situations in bipolar devices where carriers actually acquire their saturated drift velocity and in this

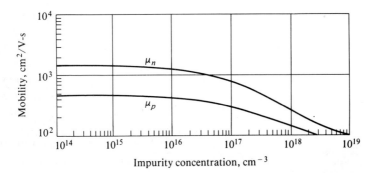

FIGURE 1.5
Electron and hole mobility versus doping level for silicon. (*After Sze* [6], *reproduced with permission, John Wiley* © *1981 and Sze and Irvin* [48], *reproduced with permission from Solid State Electronics* © *1968 Pergamon Press PLC.*)

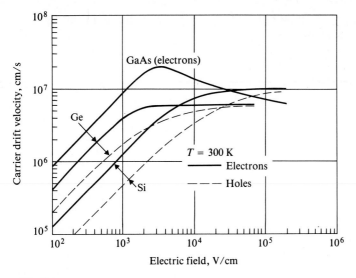

FIGURE 1.6
Carrier drift velocity versus electric field. (*After Sze* [6], *reproduced with permission John Wiley* © *1981, and using data from Jacobini et al.* [45], *Smith et al.* [46], *and Ruch et al.* [47].)

case, of course, the concept of mobility no longer applies. A convenient analytic expression representing the variation of velocity or mobility with electric field is the Caughey-Thomas relationship [10]:

$$v = \frac{\mu_0 E}{[1 + (E/E_c)^\beta]^{1/\beta}} \qquad (1.24)$$

The values for electrons and holes in silicon are given in Table 1.3.

1.4 RECOMBINATION

In an ideal "pure" semiconductor if the equilibrium relation $np = n_i^2$ is disturbed (for example, raising n and p by optical excitation, producing hole-electron pairs), there will be a natural tendency for the "excess" free carriers to "recombine." In

TABLE 1.3
Parameter values for low doping level drift velocity in silicon (1.24) at 300 K, after Caughey and Thomas [10]

	μ_0	β	E_c
Electrons	1375	2	10^4
Holes	500	1	10^4

the simplest case, this recombination would occur across the bandgap as radiative recombination, as shown in Fig. 1.7(a), with emission of a photon or transfer of energy to another particle.

In general, however, the recombination of excess carriers is accelerated by the presence of recombination levels within the forbidden energy gap, E_g. The effect of such recombination centers is greatest when they are located near the middle of the bandgap (obviously, recombination centers at either band edge, to take an extreme case, will have no effect on band-to-band recombination). The intermediate energy levels may occur due to the presence of certain types of "impurity" atoms (not to be confused with the P-type or N-type impurities usually present in much larger concentration). Gold is a good example of such an "impurity" in silicon. Sometimes it is actually introduced on purpose to increase the recombination rate. The intermediate energy levels may also occur due to dislocations in the crystal lattice (again this may occur due to imperfect fabrication techniques or may be brought about on purpose by high energy particle irradiation). The basic recombination process is illustrated in Fig. 1.7(b). We can have electron capture c_n and emission e_n, hole capture c_p and emission e_p at the intermediate energy level recombination centers.

For the study of device characteristics to follow, we shall refer to the well-known Shockley-Read [11, 12] recombination model. The recombination rate U is given by (see App. 3 for derivation)

$$U = \frac{np - n_i^2}{\tau_{p0}(n + n_1) + \tau_{n0}(p + p_1)} \tag{1.25}$$

Here n_1, p_1 are the carrier concentrations when the Fermi level E_F is at E_t (the energy level of the recombination centers), τ_{p0}, τ_{n0} are the minority carrier life-

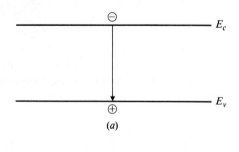

(a)

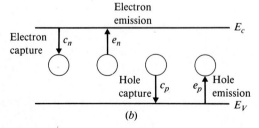

(b)

FIGURE 1.7
Recombination processes. (a) Direct band to band recombination; (b) recombination via intermediate levels.

times in extrinsic (doped) N-type and P-type material, respectively. U is the net recombination $(np > n_i^2)$ or generation $(np < n_i^2)$ rate of carriers. Mathematical treatments may be simplified by neglecting the n_1 and p_1 concentrations as being small compared to n and p, respectively.

The time-dependent relationship between current density, recombination rate, and distance is expressed by the continuity equations which may be derived by taking a nonuniform excess carrier concentration distribution and considering the current (of one type of carrier) entering and leaving an incremental strip of width dx:

$$\frac{\partial n}{\partial t} = G_n - U_n + \frac{1}{q}\frac{\partial J_n}{\partial x} \qquad (p\text{-type}) \tag{1.26}$$

$$\frac{\partial p}{\partial t} = G_p - U_p - \frac{1}{q}\frac{\partial J_p}{\partial x} \qquad (n\text{-type}) \tag{1.27}$$

G_n and G_p are the generation rates due to external effects (e.g., optical excitation).

1.5 SUMMARY OF BASIC SEMICONDUCTOR EQUATIONS

Combining the drift and diffusion equations [Eqs. (1.15) to (1.18)] gives the following two equations

$$J_n = qD_n\frac{dn}{dx} + q\mu_n nE \tag{1.28}$$

$$J_p = -qD_p\frac{dp}{dx} + q\mu_p pE \tag{1.29}$$

These are perfectly general and valid under all static (i.e., time-independent) situations, provided the appropriate values for D and μ are used. It is worth noting at this stage that in the special case of thermal equilibrium (which may frequently be approximated even when current is flowing) these two equations with J_n and J_p set to zero give the Boltzmann relations

$$n = n_i\,e^{qV(x)/kT} \tag{1.30}$$

$$p = n_i\,e^{-qV(x)/kT} \tag{1.31}$$

where V is defined to be zero at the point in the material where $n = n_i$. In addition, where space-charge neutrality does not apply, Poisson's equation must be invoked:

$$\frac{dE}{dx} = \frac{q}{\varepsilon}(N_D - N_A + p - n) \tag{1.32}$$

where N_D, N_A, p, and n are all, in general, position-dependent. $\varepsilon = \varepsilon_0 \varepsilon_r$ is the permittivity of the semiconductor and has a value 1.05×10^{-12} F/cm for silicon.

It must not be forgotten that in time-dependent solutions it is sometimes necessary to include displacement current. There are only a few occasions in the device characteristics to be considered in this text where this is necessary; it is, of course, equivalent to capacitive current flow. Finally, it may be noted that Eqs. (1.28) and (1.29) may also be expressed in terms of quasi-Fermi levels as explained in App. 4.

1.6   HEAVY DOPING EFFECTS

In nearly all bipolar devices the doping level in at least one region will be of order 10^{20} to 10^{21} cm^{-3}. When the doping level exceeds about 10^{18} cm^{-3}, a number of high doping effects come into play. The phenomena to be considered in this section are bandgap narrowing (BGN), Auger recombination, minority carrier mobility.

1.6.1   Bandgap Narrowing (BGN)

In the past ten years there has been a wealth of literature on the significance of bandgap narrowing or bandgap reduction in heavily doped regions of bipolar devices, in particular concerning the emitter of vertical NPN transistors and the surface diffused layer of solar cells. Early experimental observations were related to optical measurements by Vol'fson and Subashiev [13]; the first extensive experimental and theoretical results relating specifically to bipolar transistors were those of De Man [14], Van Overstraeten et al. [15], Mertens et al. [16], and Kleppinger and Lindholm [33].

A more comprehensive theory for high impurity densities was published recently by Berggren and Sernelius [34]. This theory applies only when the average interelectron distance is smaller than the effective Bohr radius. Ghazali and Serre [39] published a multiple scattering theory, which is applicable at moderate and low doping levels also. The results of these calculations [39] are shown in Fig. 1.8(a). As the doping level increases, the impurity band seen in the figure becomes skewed and moves up, the conduction band edge moves down, and finally the two bands merge with each other. In the high impurity density theory [34] only a shift of the conduction band edge with nearly parabolic density of states occurs due to electron donor interaction. The valence band edge moves up by approximately the same amount. Further, nearly equal and larger shifts of the two band edges occur due to many-body effects, i.e., carrier-carrier interactions [34]. In addition, local fluctuation of electrostatic potential due to nonuniform spatial distribution of impurity atoms introduces band tails [17] as shown in Fig. 1.8(b). However, the effect of band tails on the position of the Fermi level or on the value of pn product is much smaller than that calculated in [17], as the classical theory of band tails used in [17] overestimates the tails considerably.

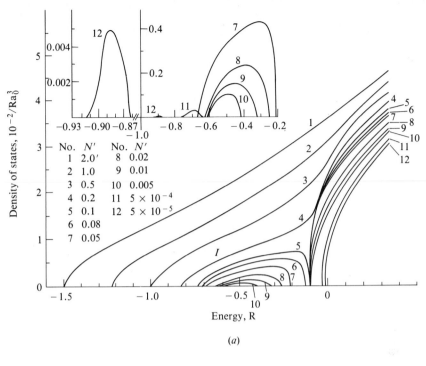

(a)

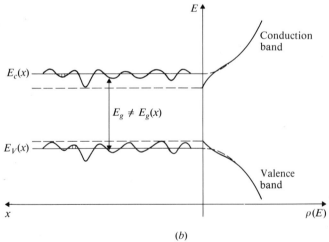

(b)

FIGURE 1.8
(a) Normalized density of states curves versus energy (normalized with respect to effective Rydberg energy R). The curves are given for various doping levels, normalized with respect to Mott's critical concentration $\pi/3(4a_0)^3$, a_0 is the effective Bohr radius. (*From Ghazali and Serre* [39] *reprinted with permission from Solid State Electronics* © *1985 Pergamon Press PLC.*)
(b) Formation of band tails due to fluctuation of local electrostatic potential. (*After Lee and Fossum* [17], *reproduced with permission* © *1983 IEEE.*)

The BGN determined from optical absorption measurements (the latest paper being that of Schmid [37]) are considerably smaller than those obtained by electrical or luminescence measurements. It has been shown in [38] that the values obtained from the absorption measurements are unreliable because of the difficulty in evaluating correctly the free carrier absorption and subtracting it from the observed absorption of the heavily doped silicon.

Until recently there was no universal agreement about the experimental values of BGN or which method gives the most reliable values for BGN. It has also been shown that the earlier values of BGN derived from electrical measurements were too high at doping levels above 3×10^{19} cm^{-3} [18, 19], see also [35]. It has been shown in [36] that the theoretical results of [34] agree with the values of BGN obtained from the luminescence data [32] as well as the electrical data as evaluated in [18] and [19] and shown in Fig. 1.9.

It is necessary to use Fermi-Dirac statistics to calculate the Fermi level when the semiconductor is heavily doped. It can be shown that the modified intrinsic carrier concentration is given by

$$n_{ie}^2 = \exp\left(\frac{\Delta E_g'}{kT}\right)FC = n_i^2 \exp\left(\frac{\Delta E_g}{kT}\right) \tag{1.33}$$

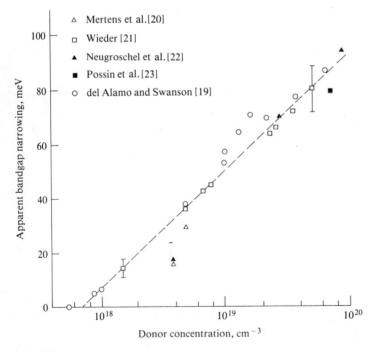

FIGURE 1.9
Summary of bandgap narrowing data from different sources. (*From Ref.* [19], *after Del Alamo, Swirhun, and Swanson. Reproduced with permission © 1983, 1985 IEEE.*)

where *FC* takes into account the effect of Fermi-Dirac statistics (it reduces the value of n_i), $\Delta E'_g$ is the actual bandgap narrowing, and ΔE_g is the apparent BGN value.

The values of apparent BGN calculated from the theory of [34] are not very different from those given by Slotboom and DeGraaf [24], which we shall use in this text.

$$\Delta E_g = 0.009\left\{\ln\left(\frac{N}{10^{17}}\right) + \sqrt{\left[\ln\left(\frac{N}{10^{17}}\right)\right]^2 + 0.5}\right\} \qquad (1.34)$$

where *N* is the ionized impurity concentration.

In order to facilitate writing of the transport equations, use of an effective impurity concentration N_{eff} is frequently invoked.

The effect of the bandgap shift is to introduce an additional electric field term in the equation for minority carrier current. We will consider hole current in heavily doped N-type material in the following study, although it should be realized that the equations apply similarly to electron current in P-type material.

$$J_p = -qD_p\frac{dp}{dx} + q\mu_p p E_{\text{tot }p}$$

where the electric field is made up of the conventional $-dV/dx$ term plus an additional term involving the bandgap narrowing

$$E_{\text{tot }p} = -\frac{dV}{dx} + \frac{1}{q}\frac{d\,\Delta E_g}{dx}$$

Using the above, assuming zero electron current density (this will be considered further in Sec. 2.7.2), and substituting from Eqs. (1.28) and (1.33) for n_{ie}^2 gives the equation for hole current density as:

$$J_p = -qD_p\frac{dp}{dx} + q\mu_p\frac{kT}{q}\left[\frac{1}{n_{ie}^2}\frac{dn_{ie}^2}{dx} - \frac{1}{N}\frac{dN}{dx}\right]$$

If we now define the effective impurity concentration N_{eff} for minority carrier transport as

$$N_{\text{eff}} = N\exp\left(-\frac{\Delta E_g}{kT}\right) \qquad (1.35)$$

we see that the equation for hole current may be written in the form

$$J_p = -qD_p\frac{dp}{dx} + q\mu_p p\frac{kT/q}{N_{\text{eff}}}\frac{dN_{\text{eff}}}{dx}$$

This form is identical in appearance to the normal transport equation with N_{eff} substituted for the actual doping level *N*, so we shall use this in subsequent treatments in this text in all cases *where minority carrier transport is involved.*

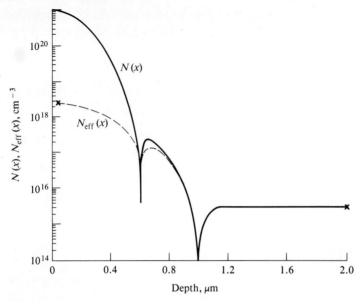

FIGURE 1.10
Impurity profile $N(x)$ for a typical bipolar transistor with the effective doping level $N_{eff}(x)$ superimposed.

Figure 1.10 shows a typical impurity profile $N(x)$ of a double-diffused bipolar transistor with the effective doping profile $N_{eff}(x)$ superimposed. It is seen that the bandgap narrowing causes a two-orders-of-magnitude difference between the actual and effective doping levels in the region near the surface of the emitter. It is therefore clear that any attempt at a quantitative analysis of minority carrier transport in the heavily doped region (of particular importance in emitters of transistors and solar cells) must include this very important effect.

It should be noted that wherever the space-charge region is calculated using Poisson's equation, and for all *majority carrier* current calculations, it is the actual ionized doping level N, as opposed to the effective doping level, which must be used. The logic of this may be readily appreciated if it is realized that bandgap narrowing changes the effective intrinsic carrier concentration from n_i to n_{ie} and hence through the mass-action law [Eq. (1.8)] the minority carrier concentration is changed. The majority carrier concentration is, for all practical purposes (assuming complete ionization), equal to the doping level.

1.6.2   Auger Recombination

A second very important effect which occurs at high doping levels is due to Auger recombination. This is due to the direct band-to-band recombination between an electron and a hole across the forbidden gap, accompanied by the transfer of energy to another free electron or hole, as shown in Fig. 1.7. The recombination lifetime for the Auger process is given by:

$$\tau_A = \frac{1}{C_N N^2} \qquad (1.36)$$

where the Auger recombination coefficient C_N has been measured for electrons in silicon to be in the range 0.4×10^{-31} to 6×10^{-31} cm^6 s^{-1} [25, 26].

In general, it is necessary to include both Auger and SRH recombination. The latter is also dependent on doping level and an experimentally obtained empirical expression for the Shockley-Read-Hall lifetime is given by [26]:

$$\tau_{\text{SRH}} = \frac{\tau_0}{1 + N/N_{\text{ref}}} \qquad (1.37)$$

The combined effect of Auger and SRH recombination is given by the following equation:

$$\frac{1}{\tau_{\text{eff}}} = \frac{1}{\tau_A} + \frac{1}{\tau_{\text{SRH}}} \qquad (1.38)$$

Figure 1.11 shows a collection of measured minority lifetime data in silicon-versus-doping level [27]. An alternative formulation for lifetime due to Selvakumar [27] is

$$\frac{1}{\tau_p} = \frac{C_s^2 N^2}{D_p(N) n_{ie}^4} \qquad (1.39)$$

where $D_p(N)$ is the hole diffusion coefficient and C_s is a fitting constant in the range 1.4×10^6 to 5.4×10^6 cm^{-2} s^{-1}. Use of this lifetime formula will be discussed further in Chap. 8. It should be noted that there is a considerable spread in lifetime data for a given doping level due to process variations, as noted in Arora et al. [26]. It may be assumed that the upper bounds for the data shown in Fig. 1.11 represent the highest achievable lifetimes to date; however, the origins of the recombination centers are not thoroughly understood.

1.6.3   Minority Carrier Mobility

Normally it is assumed that the minority carrier mobility (for example, hole mobility in N-doped material) is the same as that given for the same carrier type when it is a majority carrier (for example, hole mobility in P-type material of the same doping level). Recent results [19, 30] indicate that this may not be a valid assumption in heavily doped regions. It is clear that it is very difficult to separate out the effects of bandgap narrowing, reduced lifetime due to Auger recombination, and minority carrier mobility. Most measurements, for example, of emitter injection efficiency will only yield results which must assume two of these parameters in order to determine the third one.

In [27] it is suggested that consistent results for injection into heavily doped regions are obtained by suitable choice of mobility. If moderate BGN is

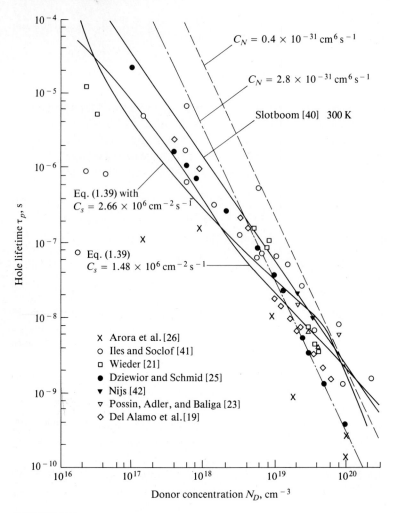

$C_N = 0.4 \times 10^{-31}\,\mathrm{cm^6\,s^{-1}}$

$C_N = 2.8 \times 10^{-31}\,\mathrm{cm^6\,s^{-1}}$

Slotboom [40]   300 K

Eq. (1.39) with
$C_s = 2.66 \times 10^6\,\mathrm{cm^{-2}\,s^{-1}}$

Eq. (1.39)
$C_s = 1.48 \times 10^6\,\mathrm{cm^{-2}\,s^{-1}}$

X   Arora et al. [26]
o   Iles and Soclof [41]
□   Wieder [21]
●   Dziewior and Schmid [25]
▼   Nijs [42]
▽   Possin, Adler, and Baliga [23]
◇   Del Alamo et al. [19]

Hole lifetime τ_p, s

Donor concentration N_D, cm^{-3}

FIGURE 1.11
Summary of hole lifetime data from different sources. (*From Refs.* [27] *and* [26], *after Selvakumar and Roulston* [27]. *Reprinted with permission from Solid State Electronics* © *1987 Pergamon Press PLC.*)

assumed as in the models of Slotboom and deGraaf [24] or Dumke [32], the minority carrier mobility appears to be fairly close to majority carrier mobility values, e.g., [9]. If a strong BGN model is used like that of Lanyon and Tuft [31] or Neugroschel et al. [22], a mobility smaller than the majority carrier value is required in order to obtain consistent results. Use of a weak BGN model, on the other hand, such as that due to Bennett [30], requires the use of a value higher than that for majority carriers.

The results reported by Del Alamo et al. [19] and reproduced in Fig. 1.12 indicate that the hole mobility in heavily doped N-type silicon is approximately twice the value for P-type silicon. However, it must be recognized that the mobil-

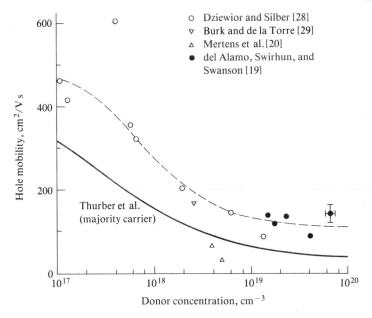

FIGURE 1.12
Hole mobility in heavily doped N-type and P-type silicon. (*After Del Alamo, Swirhun and Swanson* [19]. *Reproduced with permission © 1985 IEEE.*)

ity difference depends on the model used for bandgap narrowing; and it is difficult to assert, with certainty, given the spread in both theoretical and experimental data existing at the time of writing, exactly what values should be used. The agreement between the electrically measured results is, however, sufficient for most practical engineering purposes, specially when the inevitable spread in lifetime values is also taken into account.

1.7 CONCLUSIONS

We have presented the reader in this chapter with an overview of the basic semiconductor properties necessary for the treatments of bipolar devices which follow in this text. The chapter included a review of heavy doping effects which become particularly important in the determination of the properties of the emitter (to be studied in Chap. 8).

In the remainder of this text we shall use only the one-dimensional time-independent form of the transport and continuity equations, firstly because this is normally adequate for bipolar devices, and secondly because simplifying the problems in this manner generally leads to a clearer understanding of the underlying physical behavior. Indeed, in many cases, we make a further number of simplifying approximations (such as neglect of electric field in neutral regions in many cases, zero current assumption in the space-charge layer, infinite lifetimes in some devices, and so on) in order to obtain concise and readily usable results.

It should be clear to the reader from the information given in this introductory chapter that lack of precise quantitative knowledge about some important mechanisms (such as recombination rates and bandgap narrowing, which even for silicon, are not fully resolved) renders "exact" analyses of rather dubious practical significance. Moreover, there are a number of computer simulation programs available which perform full two-dimensional (and time-dependent) analyses for various semiconductor devices, so hand analyses of too great complexity are of little interest. Rather, we shall emphasize in the rest of this text the underlying principles of operation of the various devices; we shall include analytic results which in most cases are adequate for engineering design purposes and for "roughing out" preliminary designs before proceeding with detailed computer simulations.

For the many approximations involved in such analytic treatment, we shall endeavor to indicate their limits of validity so that care can be taken not to apply equations outside these limits. Most of the numerical examples will be for silicon devices. However, from time to time we shall give examples of devices made from other material, including the introduction to heterojunctions in Chap. 14.

PROBLEMS

1.1. Using the data from Table 1.1, calculate E_g and the intrinsic carrier concentration for silicon at $T = -50°C$, $+27°C$, $+150°C$, and for GaAs and Ge at $27°C$.

1.2. When using Eq. (1.12), it is almost always sufficient to set $p = n_i^2/N_D$ and $n = N_D$. Use Eqs. (1.8) and (1.12) to find an exact expression for electron concentration, n, in terms of N_D and n_i, and evaluate the error involved in using the approximate form for the conditions: (i) $N_D = 10n_i$, (ii) $N_D = 3n_i$. What conclusions can you draw for practical doping levels?

1.3. Using Eq. (1.9), calculate the doping level for which the Fermi level is coincident with the conduction band edge at $27°C$ for (i) silicon, (ii) germanium.

1.4. Show that Eq. (1.15) is equivalent to Ohm's law.

1.5. Table 1.2 gives values of maximum and minimum mobility. Using Eqs. (1.21) and (1.22), find the electron and hole mobility in silicon for intermediate doping levels of 10^{14}, 10^{17}, and 10^{20} cm^{-3}, at 300 K.

1.6. From the recombination rate expression [Eq. (1.25)], derive simplified equations for N-type material for two cases: (i) low level injection, $p \ll N_D$, $N_D \gg n_i$, (ii) high level injection, $p \sim n \gg N_D$.

1.7. Derive the Boltzmann relation [Eq. (1.30)] from Eq. (1.28).

1.8. This question concerns the relative magnitudes of drift and diffusion current, using Eqs. (1.15) and (1.17). Calculate the diffusion current density which results if an electron concentration $n(0) = 10^{16}$ cm^{-3} is created at a point $x = 0$ and falls to zero at $x = 1$ μm. Calculate the electric field that would be required to produce the same value of J_n in material doped $N_d = 10^{16}$ cm^{-3}, and the voltage drop over a one-micron distance.

1.9. In order to see the important difference between effective doping (as it affects minority carrier transport) and actual (ionized) doping level, use Eqs. (1.34) and (1.35) and plot on a log-log graph, effective doping versus N, for $T = 100°C$, $27°C$, and $-50°C$.

1.10. Measured values for typical transistors indicate recombination parameters given by: $N_{\text{ref}} = 10^{17}$ cm^{-3}, $\tau_0 = 0.1$ μs, [Eq. (1.37)], and $C_N = 10^{-31}$ cm^6 s^{-1} Eq. (1.36). Using Eq. (1.38) calculate and plot the carrier lifetime τ_{eff} versus doping level in the range found in typical emitters from 10^{16} to 10^{21} cm^{-3}. Find the doping level at which SRH and Auger recombination contributions are approximately equal.

1.11. Use the results of Prob. 1.10 with the mobility formula [Eq. (1.22)], $T = 300$ K, and calculate the values of hole diffusion length in an N-type material doped 10^{17}, 10^{19}, and 10^{21} cm^{-3}. Comment on the probable significance of the result for the emitter of a transistor where the above doping range is covered in a depth of order one micron, given that the peak doping level at the surface is about 10^{21} cm^{-3}.

REFERENCES

1. W. C. Dunlap, *An Introduction to Semiconductors*, Wiley, New York (1975).
2. J. C. Moll, *Physics of Semiconductors*, McGraw-Hill, New York (1964).
3. R. A. Smith, *Semiconductors*, 2d ed., Cambridge University Press, London (1979).
4. R. K. Willardson and A. C. Beer (eds.), *Semiconductors and Semi Metals, Vol. 2, Physics of III-V Compounds*, Academic, New York (1966).
5. C. D. Thurmond, "The standard thermodynamic function of the formation of electrons and holes in Ge, Si, GaAs, and GaP," *J. Electrochem. Soc.*, vol. 122, p. 1133 (1975).
6. S. M. Sze, *Physics of Semiconductor Devices*, 2d ed., Wiley-Interscience, New York (1981).
7. J. Lindmayer and C. Y. Wrigley, *Fundamentals of Semiconductor Devices*, Van Nostrand-Reinhold, New York (1965).
8. M. S. Adler et al., *Introduction to Semiconductor Physics*, SEEC, vol. 1, Wiley (1964).
9. N. D. Arora, J. R. Hauser, and D. J. Roulston, "Electron and hole mobilities in silicon as a function of concentration and temperature," *IEEE Trans. Electron Devices*, vol. ED-29, pp. 284–291 (February 1982).
10. D. M. Caughey and R. E. Thomas, "Carrier mobilities in silicon empirically related to doping and field," *Proc. IEEE (Lett.)*, vol. 55, pp. 2192–2193 (December 1967).
11. W. Shockley and W. T. Read, "Statistics of recombination of holes and electrons," *Phys. Rev.*, vol. 87, pp 835–842 (September 1952).
12. R. N. Hall, "Electron-hole recombination in Germanium," *Phys. Rev.*, vol. 87, p. 387 (1952).
13. A. A. Vol'fson and V. K. Subashiev, "Fundamental absorption edge of silicon heavily doped with donor and acceptor impurities," *Soviet Physics—Semiconductors*, vol. 1, p. 327 (1967).
14. H. J. De Man, "The influence of heavy doping on the emitter efficiency of a bipolar transistor," *IEEE Trans. Electron Devices*, ED-18, pp. 833–835 (October 1971).
15. R. J. Van Overstraeten, H. J. De Man, and R. P. Mertens, "Transport equations in heavily doped silicon," *IEEE Trans. Electron Devices*, ED-20, pp. 290–298 (March 1973).
16. R. P. Mertens, H. J. De Man, and R. J. Van Overstraeten, "Calculation of the emitter efficiency of bipolar transistors," *IEEE Trans. Electron Devices*, ED-20, pp. 772–778 (September 1973).
17. D. S. Lee, J. G. Fossum, "Energy-band distortion in highly doped silicon," *IEEE Trans. Electron Devices*, ED-30, pp. 626–634 (June 1983).
18. J. del Alamo and R. M. Swanson, "The physics and modeling of heavy doped emitters," *IEEE Trans. Electron Devices*, ED-31, pp. 1878–1888 (December 1984).
19. J. del Alamo, S. Swirhun, and R. M. Swanson, "Simultaneous measurement of hole lifetime, hole mobility, and bandgap narrowing in heavily doped n-type silicon," *IEDM Tech. Dig.*, pp. 290–293 (December 1985).
20. R. P. Mertens, J. L. Van Meerbergen, J. F. Nijs, and R. J. Van Overstraeten, "Measurement of the

minority carrier transport parameters in heavily doped silicon," *IEEE Trans. Electron Devices*, ED-27, pp. 949–955 (May 1980).

21. A. W. Wieder, "Emitter effects in shallow bipolar devices: measurements and consequences," *IEEE Trans. Electron Devices*, ED-27, p. 1402 (August 1980).

22. A. Neugroschel, S. C. Pao, and F. A. Lindholm, "A method for determining energy-gap narrowing in highly doped semiconductors," *IEEE Trans. Electron Devices*, ED-29, pp. 894–902 (May 1982).

23. G. E. Possin, M. S. Adler, and B. J. Baliga, "Measurement of the PN product in heavily doped epitaxial emitters," *IEEE Trans. Electron Devices*, ED-31, pp. 3–17 (January 1984).

24. J. W. Slotboom and H. C. de Graaff, "Measurements of bandgap narrowing in Si bipolar transistors," *Solid State Electronics*, vol. 19, pp. 857–862 (1976).

25. J. Dziewior and W. Schmid, "Auger coefficients for highly doped and highly excited silicon," *Appl. Phys. Lett.*, vol. 31, pp. 346–348 (September 1977).

26. D. J. Roulston, N. D. Arora, and S. G. Chamberlain, "Modeling and measurement of minority carrier lifetime versus doping in diffused layer of n^+p silicon diodes," *IEEE Trans. Electron Devices*, ED-29, pp. 284–291 (February 1982).

27. C. R. Selvakumar and D. J. Roulston, "A new simple analytic emitter model for bipolar transistors," *Solid State Electronics*, vol. 30, pp. 723–728 (July 1987).

28. J. Dziewior and D. Silber, "Minority carrier diffusion coefficients in highly doped silicon," *Appl. Phys. Lett.*, vol. 35, pp. 170–172 (July 1979).

29. D. E. Burk and V. de la Torre, "An empirical fit to minority hole mobilities," *IEEE Electron Device Lett.*, EDL-5, pp. 231–232 (1984).

30. H. S. Bennett, "Hole and electron mobilities in heavily doped silicon: comparison of theory and experiment," *Solid State Electronics*, vol. 26, pp. 1157–1166 (December 1983).

31. H. P. D. Lanyon and R. A. Tuft, "Bandgap narrowing in moderately to heavily doped silicon," *IEEE Trans. Electron Devices*, ED-26, pp. 1014–1018 (July 1979).

32. W. P. Dumke, "Comparison of bandgap shrinkage observed in luminescence from N^+Si with that from transport and optical absorption measurements," *Appl. Phys. Lett.*, vol. 42, pp. 196–198 (January 1983).

33. D. D. Kleppinger and F. A. Lindholm, "Impurity-concentration-dependent density of states and resulting Fermi level for silicon," *Solid State Electronics*, vol. 14, pp. 407–416 (May 1971).

34. K.-F. Berggren and B. E. Sernelius, "Very heavily doped semiconductors as a 'nearly-free-electron-gas' system," *Solid State Electronics*, vol. 28, pp. 11–15 (January/February 1985).

35. S. C. Jain, E. L. Heasell, and D. J. Roulston, "Recent advances in the physics of silicon *p-n* junction solar cells including their transient response," *Progress in Quantum Electronics*, vol. 11, pp. 105–204 (1987).

36. J. Wagner and J. A. del Alamo, "Bandgap narrowing in heavily doped silicon: a comparison of optical and electrical data," *Jnl. Appl. Phys.*, vol. 63, pp. 425–429 (January 1988).

37. P. E. Schmid, "Optical absorption in heavily doped silicon," *Phys. Rev.*, B23, pp. 5531–5536 (May 1981).

38. S. T. Pantelides, A. Selloni, and R. Car, "Energy-gap reduction in heavily doped silicon: causes and consequences," *Solid State Electronics*, vol. 28, pp. 17–24 (January/February 1985).

39. A. Ghazali and J. Serre, "Disorder, fluctuations, and electron interactions in doped semiconductors: a multiple scattering approach," *Solid State Electronics*, vol. 28, pp. 145–149 (January/February 1985).

40. J. W. Slotboom, "Minority carrier injection into heavily doped silicon," *Solid State Electronics*, vol. 20, pp. 167–170 (1977).

41. P. A. Iles and S. F. Socloff, in Photovoltaic Specialists Conf. (1975).

42. J. F. Nijs, PhD. Thesis, Katholieke Universiteit, Leuven, Belgium (1982).

43. H. C. Casey and M. B. Panish, *Heterojunction Lasers*, Academic Press, New York (1978).

44. M. B. Prince, "Drift mobility in semiconductors I—germanium," *Phys. Rev.*, vol. 92, pp. 681–687 (November 1953).

45. C. Jacobini, C. Canali, G. O. Ottaviani, and A. A. Quaranta, "A review of some charge transport properties of silicon," *Solid State Electronics*, vol. 20, pp. 77–89 (February 1977).

46. P. Smith, M. Inoue, and J. Frey, "Electron velocity in Si and GaAs at very high electric fields," *Appl. Phys. Lett.*, vol. 37, pp. 797–798 (November 1980).

47. J. G. Ruch and G. S. Kino, "Measurement of the velocity-field characteristics of gallium arsenide," *Appl. Phys. Lett.*, vol. 10, pp. 40–42 (January 1967).

48. S. M. Sze and J. C. Irvin, "Resistivity, mobility, and impurity levels in GaAs, Ge, and Si at 300 K," *Solid State Electronics*, vol. 11, pp. 599–602 (June 1968).

49. E. S. Yang, *Fundamentals of Semiconductor Devices*, McGraw-Hill, New York (1978).

CHAPTER
2

THE PN
AND NN$^+$
JUNCTION

2.1 INTRODUCTION

Most semiconductor devices can be studied by making the assumption that space-charge neutrality exists far from any discontinuities or rapid changes in impurity concentration versus distance, and applying Poisson's law only in the region of such discontinuities. In this chapter we shall first of all examine the "space-charge region" of a symmetric PN junction, and clarify the meaning of *depletion-layer approximation*. The built-in barrier potential is defined and bias is introduced. The presence of "free carriers" inside the "depletion layer" is then studied. Next we examine the nonsymmetric P$^+$N and the linearly graded junctions before considering the important practical case of a double diffused impurity profile. Depletion-layer capacitance is discussed before proceeding to a discussion of the injection properties of the P$^+$N junction and the properties of the NN$^+$ or retarding field junction. This latter topic is particularly important because it occurs in many different structures. The tools acquired as a result of the analyses in this chapter will be used to derive the I-V characteristics of some real diodes in the following chapter.

2.2 THE SPACE-CHARGE LAYER IN THE SYMMETRIC PN JUNCTION

2.2.1 The Edge of the Space-Charge Layer

Figure 2.1 shows the diagrams for a PN junction which we need for the following discussion. Diagram (a) shows the P and N regions on either side of the junction $x = 0$. There is clearly some transition region between the P and N regions since we know from the mass action law [Eq. (1.8)] that on the N side, the minority carrier hole concentration has a value (making the usual valid approximation that $p_n \ll N_D$)

$$p_n = \frac{n_i^2}{N_D} \tag{2.1}$$

whereas on the P side sufficiently far from the junction, the hole concentration must have a value practically equal to the acceptor concentration N_A. A similar argument applies to the transition of electron concentration from a value N_D on the N side to a value n_i^2/N_A on the P side, sufficiently far from the junction. The second diagram shows this transition of electron and hole concentrations from their high values on one side to their quite low values on the other side of the *transition region*, a name sometimes given to this central region. It is now quite clear from diagram (b) that a space-charge must exist in this transition region since from Poisson's equation we have

$$\frac{dE}{dx} = \frac{q}{\varepsilon} [N(x) - n(x) + p(x)] \tag{2.2}$$

where E is the electric field, q is the electronic charge, ε the permittivity, and $N(x)$ the net doping level ($N_D - N_A$).

Diagram (c) shows the net space-charge. Notice that diagram (b) is drawn on a log scale in order to accommodate the large differences in carrier concentrations, whereas the net space charge shown in diagram (c) uses a linear scale. The reader may easily verify that for all practical purposes this space-charge diagram consists of nearly horizontal regions followed by a sharp decrease toward zero as the edge of the transition region is reached, assuming that diagram (b) is roughly correct (this will be verified subsequently in Sec. 2.2.4).

Diagram (d) shows the electric field, obtained by integration of the space-charge of diagram (c), in other words a nearly triangular distribution, because of the nearly constant space-charge on either side.

The last two diagrams show the potential distribution (e) resulting from the integration of the electric field and the energy band diagram (f). Recall that these two are related since the energy E may be represented by qV, where V is the voltage.

Before examining the solution for electric field and potential, let us focus our attention on the "tail" of the transition or space-charge region, where the structure is nearly space-charge neutral ($dE/dx = 0$) [1].

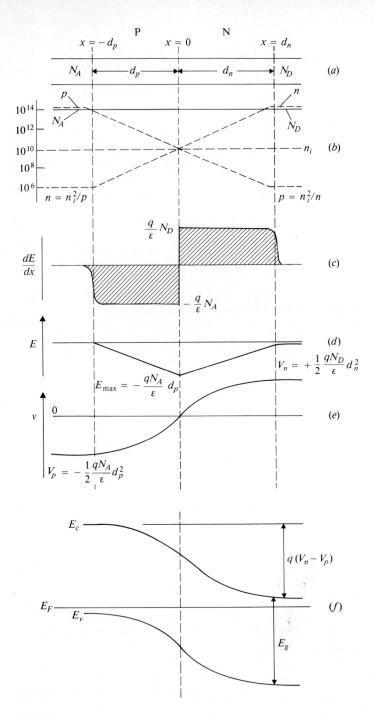

FIGURE 2.1

The symmetric PN junction. (a) The structure; (b) the free carrier concentration distribution on a log vertical scale; (c) the space-charge diagram on a linear vertical scale; (d) the electric field; (e) the potential distribution; (f) the energy band diagram.

The depletion layer "tail"

Restricting the analysis to the case of the symmetric junction used in the preceding discussion, we can use the Boltzmann relations from Eqs. (1.30) and (1.31) to replace the $p(x)$ and $n(x)$ terms in Eq. (2.2) by the corresponding potential change and the intrinsic carrier concentration n_i to obtain

$$\frac{dE}{dx} = \frac{q}{\varepsilon}\left[N(x) - 2n_i \sinh\left(\frac{qV(x)}{kT}\right)\right] \qquad (2.3)$$

Now let us rearrange this result by defining

$$\frac{N}{2n_i} = \sinh\left(\frac{qV''_{bi}}{kT}\right) \qquad (2.4)$$

Here V''_{bi} is one part of the built-in barrier of the PN junction (one-half in this case of a symmetric junction). We may thus rewrite Eq. (2.3) in the form

$$\frac{dE}{dx} = \left[\frac{4qn_i}{\varepsilon}\right]\cosh\left(\frac{q}{kT}\frac{V''_{bi}+V(x)}{2}\right)\sinh\left(\frac{q}{kT}\frac{V''_{bi}-V(x)}{2}\right) \qquad (2.5)$$

As the "edge" of the space-charge layer is approached, the potential $V(x)$ tends to the value V''_{bi}, so that Eq. (2.5) can be approximated by

$$\frac{dE}{dx} = \frac{q^2 N}{\varepsilon kT}\,[V''_{bi} - V(x)] \qquad (2.6)$$

Replacing $-E$ by dV/dx and substituting the extrinsic Debye length

$$L_D = \left[\frac{kT\varepsilon}{q^2 N}\right]^{1/2} \qquad (2.7)$$

we obtain

$$\frac{d^2V}{dx^2} = \frac{V(x) - V''_{bi}}{L_D^2} \qquad (2.8)$$

This equation has the solution

$$V(x) - V''_{bi} = A\,\exp\left(-\frac{x}{L_D}\right) \qquad (2.9)$$

It is thus seen that the potential varies exponentially as a function of distance at the edge of the space-charge region, with a characteristic length L_D. We would thus expect the change from neutrality (where n is close to N_D on the N side) to "full" space charge ($n \ll N_D$ on the N side) to occur within a few Debye lengths. A simple calculation shows that for a doping level of 10^{16} cm^{-3}, the value of L_D is of the order 0.03 μm. This result, first observed by Shockley, is crucial to our ability to handle most semiconductor device structures using analytic methods (as opposed to numerical simulation), since the Debye length in question is usually significantly smaller than the widths of the neutral regions and space-charge layers. It is worth observing at this point that the Debye length may also be expressed in the form:

$$L_D = \sqrt{D_n \tau_{\text{rel}}} \tag{2.10}$$

where τ_{rel} is the dielectric relaxation time of the material defined by

$$\tau_{\text{rel}} = \frac{\varepsilon}{q \mu_n N_D} \tag{2.11}$$

$$= \frac{\varepsilon}{\sigma_N} \tag{2.12}$$

where σ_N is the conductivity of the N doped region. Similar relations apply to the P material.

Physically, the dielectric relaxation time τ_{rel} is the time required for space-charge neutrality to be reestablished after a disturbance in free carrier concentration. Since resistivities are of order ohm cm at medium doping levels, τ_{rel} is typically of order picoseconds. For most practical cases, it may be assumed that neutrality is reestablished instantaneously. In fact this assumption is even made in many numerical analysis programs (see Chap. 15).

Now that we have established the important fact (to be substantiated later on with typical values) that the transition region consists of a space-charge layer which is a function only of the doping level over most of its thickness (or width), we can proceed to solve Poisson's equation in a very simple manner within this space-charge region. Since this region is almost completely depleted of free carriers, it is also frequently called the *depletion layer* of the PN junction. In fact, the three terms: transition region, space-charge layer, and depletion layer are generally used synonymously.

2.2.2 Depletion-Layer Approximation

Restricting our attention to the transition region on the N side from $x = 0$ to $x = $ some value d_n, we can say that to a good approximation, the free carrier concentrations may be neglected in comparison with the doping level N_D, as can be seen from Fig. 2.1(b). This is certainly true for the hole concentration $p(x)$; based on our preceding discussion, it is also true for the electron concentration $n(x)$ up to a few Debye lengths from the "edge" of the region in question. We therefore now solve Poisson's equation using the "depletion approximation"

$$\frac{dE}{dx} = \frac{q}{\varepsilon} N_D \tag{2.13}$$

valid for $0 < x < d_n$. The value d_n is referred to as the depletion-layer thickness (or width) on the N side of the junction. A similar equation applies on the P side. Integration of Eq. (2.13) gives

$$E(x) = \frac{q}{\varepsilon} N_D(x - d_n) \tag{2.14}$$

where we have used the condition that the field must be zero at the depletion-layer boundary. [Note that this depletion-layer approximation forces us to assume an abrupt change from the space-charge region $(x < d_n)$ to the neutral region $(x > d_n)$.] A second integration yields the potential as a function of distance

$$V(x) = -\frac{qN_D}{\varepsilon}\left(\frac{x^2}{2} - d_n x\right) \tag{2.15}$$

where we have arbitrarily (for convenience) defined the voltage to be zero at the junction $x = 0$.

Of particular interest in determining some of the electrical terminal characteristics of the structure are the quantities peak or maximum field $(x = 0)$ and total voltage $(x = d_n)$ on the N side:

$$E_{max} = -\frac{q}{\varepsilon}N_D d_n \tag{2.16}$$

$$V_{tot\ n} = \frac{q}{2\varepsilon}N_D d_n^2 \tag{2.17}$$

Note that the sign attached to the electric field has no practical significance, since drawing the junction with the N side on the left and the P side on the right will alter this sign. On the other hand, it is important to note that where potential is concerned, the N side of any PN junction is always positive with respect to the P side.

The total voltage between the P and N sides is given by writing a similar expression to the above for $V_{tot\ p}$ to obtain

$$V_{tot} = \frac{q}{2\varepsilon}(N_D d_n^2 + N_A d_p^2)$$

$$= \frac{q}{\varepsilon}N_D d_n^2 \tag{2.18}$$

since we are considering here the case N_D (N side) $= N_A$ (P side).

Finally, by combining Eqs. (2.16) and (2.17), we obtain the relation between peak electric field and total N side voltage

$$E_{max} = 2\frac{V_{tot\ n}}{d_n} \tag{2.19}$$

The peak value of field is seen to be double what it would be for the same voltage applied across a perfect dielectric (where the field is constant with respect to distance).

2.2.3   The Barrier Voltage and External Applied Bias

The value of the total potential between the P and N sides of the junction *outside the space-charge layer* may be obtained from the Boltzmann relations discussed in Chap. 1. Since

$$n(x) = n_i \exp\left[\frac{V(x)}{V_t}\right] \tag{2.20}$$

and

$$p(x) = n_i \exp\left[-\frac{V(x)}{V_t}\right] \tag{2.21}$$

where $V_t = kT/q$, the ratio of the carrier concentrations in thermal equilibrium on the two sides is

$$\frac{n_{n0}}{n_{p0}} = \exp\left[\frac{V_n - V_p}{V_t}\right] \tag{2.22}$$

Now since the voltage difference is equal to the built-in barrier potential or contact potential, and since n_{n0} is equal to N_D and

$$n_{p0} = \frac{n_i^2}{N_A} \tag{2.23}$$

for all practical purposes, we obtain directly the result

$$V_{bi} = V_t \ln\left(\frac{N_A N_D}{n_i^2}\right) \tag{2.24}$$

For silicon at room temperature with doping levels of 10^{16} cm^{-3} on both sides, this value is 0.7 V.

Reverse bias increases the potential difference between the P and N sides (making the N side more positive with respect to the P side). Forward bias has the opposite effect. Adopting the convention that reverse bias is negative and vice versa enables us to write the total voltage across the junction as

$$V_{tot} = V_{bi} - V_a \tag{2.25}$$

where V_a is the value of applied bias.

It is clear from the fact that the depletion-layer thickness increases with increasing total voltage, that the approximations involved in the use of the depletion approximation (Debye length small compared to space-charge and neutral-region widths) will be even better for reverse bias, but worse for forward bias (as far as use of the results for determination of the properties of the junction related to the depletion layer itself are concerned).

2.2.4 Free Carrier Concentration Distribution

We are now in a position to determine the shape of the free carrier concentration within the space-charge layer of the junction. In order to maintain generality and to derive a result which will come in useful subsequently, we will assume arbitrary bias. Furthermore, we will assume for the moment that the Boltzmann relations are still valid (this will subsequently be justified as a good approximation for many cases of practical interest). Restricting our attention to the electron concentration near the junction, $x \ll d_n$, we may approximate the potential as a linear function of distance from Eq. (2.15):

$$V(x) = \frac{qN_D}{\varepsilon} d_n x \qquad (2.26)$$

The Boltzmann relation for electrons is:

$$n(x) = n_i \exp\left[\frac{V(x)}{V_t}\right] \qquad (2.27)$$

Combining these two equations gives:

$$n(x) = n_i \exp\left(\frac{x}{x_c}\right) \qquad (2.28)$$

where

$$x_c = \frac{\varepsilon V_t}{qN_D d_n} \qquad (2.29)$$

Or, by substituting for d_n from Eq. (2.18),

$$\frac{x_c}{d_n} = \frac{V_t}{V_{bi} - V_a} \qquad (2.30)$$

At zero bias, this ratio is of order 1/30. We thus see that referring to Fig. 2.1, for a distance approaching one-half d_n, the electron concentration is indeed a straight line on the semilog plot. This characteristic will be used subsequently in deriving low current dc behavior of forward biased diodes and emitter-base junctions of transistors.

2.3 THE P⁺N ABRUPT JUNCTION SPACE-CHARGE LAYER

A simple extension of the above results may be applied to the case where one side of the junction is more heavily doped than the other side. Figure 2.2 shows the situation that exists in such a structure where the P side is much more heavily doped than the N side. From Eq. (2.16), since the electric field $E_{max} = (q/\varepsilon)N_D d_n$ must also be equal to $(q/\varepsilon)N_A d_p$ it is easy to see that the ratio of depletion-layer width on the P side to that on the N side is simply

$$\frac{d_p}{d_n} = \frac{N_D}{N_A} \qquad (2.31)$$

Also, by comparing this result with Eq. (2.17) it is clear that the ratio of the voltage on the P side to that on the N side is given by

$$\frac{V_P}{V_N} = \frac{N_D}{N_A} \qquad (2.32)$$

In other words, on the heavily doped side of a junction, the depletion layer is narrow and the voltage drop is low, compared to the values on the lightly doped side. The general trends shown in Fig. 2.2 are thus seen to be true. Note also that, given the above conditions on depletion-layer widths and assuming the previous results concerning the approximate shape of the free carrier distribution inside

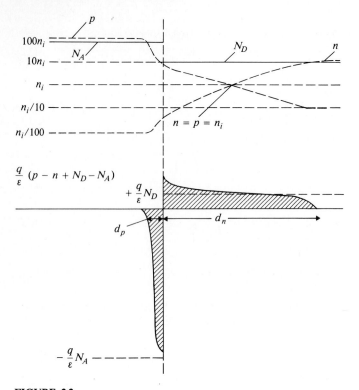

FIGURE 2.2
The P^+N junction showing depletion-layer widths and free carrier concentration distributions.

the depletion layer, the position at which the electron concentration becomes equal to the hole concentration lies on the lightly doped N side and not at the metallurgical junction.

There are many practical cases where the "one-sided abrupt junction" approximation can be used to simplify analyses of diodes. If the P side is doped more than 10 times the level of the N side, then for most engineering purposes, the characteristics may be determined by considering only the depletion layer on the lightly doped N side. This means that we may neglect most of the equations which involve the doping on the heavily doped side. This is particularly useful for depletion-layer capacitance and breakdown voltage calculations. An exception is the calculation of built-in barrier voltage [Eq. (2.24)] where both N_A and N_D must be used.

2.4   THE LINEARLY GRADED PN JUNCTION SPACE-CHARGE LAYER

Figure 2.3 shows a linearly graded PN junction. In this case the structure is symmetrical but with a doping level which is a linear function of distance

$$N(x) = ax \tag{2.33}$$

The simplified form of Poisson's equation (neglecting free carriers) now becomes

$$\frac{dE}{dx} = \frac{q}{\varepsilon} ax \tag{2.34}$$

and the reader may readily verify that the solutions for electric field and potential versus distance on the N side are now given by

$$E(x) = \frac{qa}{\varepsilon}\left[\frac{x^2}{2} - \frac{d_n^2}{2}\right] \tag{2.35}$$

$$V(x) = -\frac{qa}{\varepsilon}\left[\frac{x^3}{6} - \frac{d_n^2 x}{2}\right] \tag{2.36}$$

where, as in the previous analysis, we use the fact that the field is zero at the edge of the depletion layer and we set (arbitrarily) the potential to zero at the metallurgical junction.

The peak value of field and the total voltage on the N side are now given by

$$E_{max} = \frac{q}{\varepsilon}\frac{ad_n^2}{2} \tag{2.37}$$

$$V_{tot\,n} = \frac{q}{\varepsilon}\frac{ad_n^3}{3} \tag{2.38}$$

Since it is usually the total depletion-layer thickness d_{tot} which is the important quantity, these equations may be rewritten as

$$E_{max} = \frac{q}{\varepsilon}\frac{ad_{tot}^2}{8} \tag{2.39}$$

$$V_{tot\,n} + V_{tot\,p} = V_{tot} = \frac{q}{12\varepsilon} ad_{tot}^3 \tag{2.40}$$

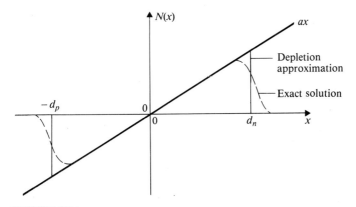

FIGURE 2.3
The linearly graded PN junction and ideal space-charge layer. The actual space-charge layer is shown by the broken curve.

The zero applied bias position is determined by using the general expression for voltage in thermal equilibrium obtained in Sec. 2.2.3 and letting the doping levels N_A, N_D be position-dependent

$$V_i(x) = V_t \ln \left(\frac{N_A(x)N_D(x)}{n_i^2} \right) \tag{2.41}$$

where $N_D(x) = ax$ for negative x and $N_A(x) = ax$ for positive x for an NP junction.

By equating this voltage with that obtained from the Poisson equation for the linear graded junction and comparing the two values, it can be seen that they coincide at only one position on either side of the junction, as shown in Fig. 2.4.

By considering the solution for various gradients, the results shown in Fig. 2.5 are obtained for built-in barrier potential and zero bias depletion-layer width versus a.

It should be noted that since the doping level continues to rise in value outside the space-charge region, there is an electrostatic potential in this region. This fact, combined with the difference between the "ideal" depletion-layer solution and the "exact" Poisson's equation solution means that different interpretations can be put on the meaning of built-in barrier potential for this type of

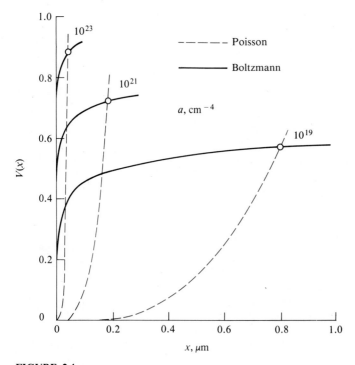

FIGURE 2.4
Illustration of potential variation for a linearly graded junction: (i) from Poisson's equation; (ii) from the Boltzmann equations.

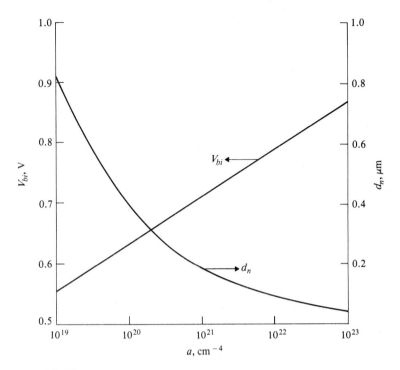

FIGURE 2.5
Computed built-in barrier potential and corresponding depletion-layer thickness for various linear graded junctions of gradient "*a*."

impurity profile [2]. In particular, the value determined from capacitance versus voltage data (sometimes referred to as the off-set voltage) is $2V_t$ lower than the above values; this is discussed further in Sec. 2.6. The differences are of mainly academic interest, since the linearly graded profile is in practice only an approximation to some gaussian-type function, as will be seen in the following section. It should also be noted that for small values of gradient a, the depletion-layer approximation ceases to be valid and exact solutions [i.e., including $n(x)$, $p(x)$, in Poisson's equation] must be used [3, 4, 5].

2.4.1 Shape of Free Carrier Concentration Distribution

The shape of the free carrier concentration distribution inside the depletion layer may be determined by a similar analysis to that used above in Sec. 2.2.4. From Eq. (2.36) it is seen that the potential is once again a linear function of distance for values of x near the center of the region (i.e., x close to 0), hence

$$v(x) = c_1 x \tag{2.42}$$

where

$$c_1 = \frac{qad_n^2}{2\varepsilon} \tag{2.43}$$

This leads to the result for hole concentration on the N side

$$p(x) = n_i \exp\left(-\frac{x}{x_c}\right) \tag{2.44}$$

where

$$x_c = \frac{kT}{q} \frac{2\varepsilon}{qad_n^2} \tag{2.45}$$

or, allowing for a forward bias V_a

$$\frac{x_c}{d_n} = \frac{4}{3} \frac{kT/q}{V_{bi} - V_a} \tag{2.46}$$

This result differs only by the factor 4/3 compared to the result for a symmetrical abrupt PN junction [Eq. (2.30)].

2.4.2   Effect of Neglecting Free Carriers

The lowest value of junction gradient, a, for which the simple analytic results for the depletion-layer width, etc., are valid may be found by including the free carrier concentration distribution in the solution to Poisson's equation as in [3, 4, 5]. Figure 2.6 shows some computed zero bias solutions for two different values of gradient. It is seen that the deviation from the straight line space-charge distribution is better for the case of the larger gradient. In both cases, the total charge on either side of the junction is approximated quite closely by the depletion approximation (although the shape of the charge distribution differs significantly).

The critical factor in determining the validity of the straight line "depletion approximation" is the ratio of the doping level at the depletion-layer edge $N_e = ad_n$ divided by the carrier concentration at the junction n_c. For a ratio N_e/n_c equal to 1000, the error in calculating the total charge on one side of the junction is about 30 percent. This may be demonstrated quite simply using graphical methods with a linear potential distribution inside the space-charge region. For ratios greater than this value of 1000, the simple analytical results may be considered valid for most engineering purposes. This corresponds to gradients of about 10^{16} cm^{-4} at zero bias or about 10^{23} cm^{-4} at a forward bias of 0.6 V at room temperature. Since practical diffused junctions have gradients in the range 10^{19} to 10^{23}, the simple analytic results for Poisson's equation (neglecting the free carriers) are quite useful up to moderate bias levels, but will always fail at high forward bias. This is not in fact too serious a problem, since the principal electrical terminal characteristic associated with this space-charge region is the junction capacitance, and at even moderate forward bias other capacitive effects tend to dominate, as will be discussed in Chap. 4.

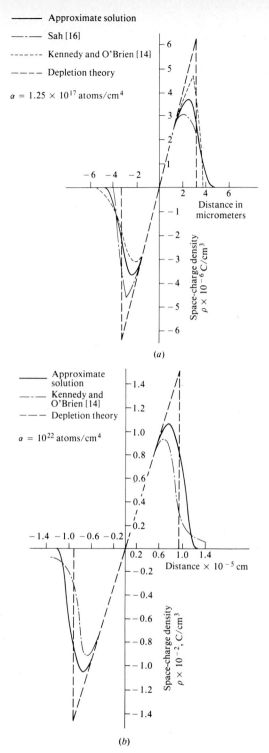

FIGURE 2.6

Computed space-charge solution for linearly graded junction at zero bias: (a) $a = 1.25 \times 10^{17}$ cm^{-4}; (b) $a = 10^{22}$ cm^{-4}. (*After Sirsi and Boothroyd* [6] *1976. Reproduced with permission © 1976 IEEE.*)

2.5 THE DIFFUSED IMPURITY PROFILE

Figure 2.7 shows the impurity profiles for a single-diffused P^+N diode and for a double-diffused diode or transistor. The diode structure (b) corresponds to a hyperabrupt varactor (Chap. 5) or avalanche photodiode structure (Chap. 6). The

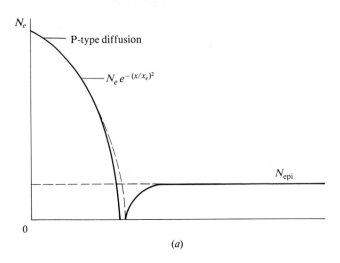

(a)

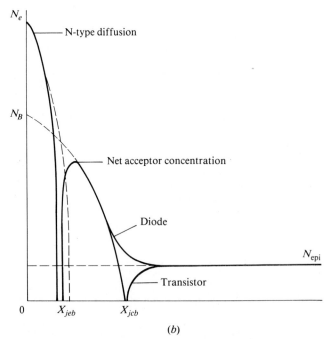

(b)

FIGURE 2.7
Impurity profile for (a) single diffused P^+N diode, (b) double-diffused diode or transistor.

transistor profile is found in nearly all NPN structures to be discussed in Chap. 7 onward.

For the simple diffused structure (a), the one-sided abrupt junction solution to Poisson's equation works well for reverse bias, since the depletion layer extends mainly into the lightly doped N region. In forward bias the junction tends toward the linearly graded case. It is worth noting at this point that an excellent engineering solution to the space-charge layer thickness may be obtained by very simple numerical analysis for any shape of impurity profile. Using the full depletion approximation, i.e., Eq. (2.2) with $n(x)$ and $p(x)$ set to zero, and integrating numerically, one may solve for both electric field and potential. This technique will be discussed in more detail in Chap. 15.

As an example for the simple diffused junction of Fig. 2.7(a) we reproduce the solution obtained by exact numerical analysis and using the depletion approximation in Fig. 2.8(a) and (b) for moderate forward bias. Notice that although the depletion layer "tail" is a significant fraction of the total space-charge layer width, the boundaries obtained by the depletion approximation in Fig. 2.8(b) agree well with the "exact" solution.

For the junction formed by superposition of the two gaussian diffusions, Fig. 2.7(b), the doping level is decreasing on both sides as the junction is approached. If the profile is represented by a double gaussian function

$$N(x) = N_e \exp - \left(\frac{x}{x_e}\right)^2 - N_b \exp - \left(\frac{x}{x_b}\right)^2 + N_{epi} \tag{2.47}$$

and if it is recognized that the location of the junction is given (neglecting the contribution of N_{epi} in the vicinity of the junction) by the condition

$$N_e \exp - \left(\frac{X_{jeb}}{x_e}\right)^2 = N_b \exp - \left(\frac{X_{jeb}}{x_b}\right)^2 \tag{2.48}$$

and also that

$$N_b \exp - \left(\frac{X_{jbc}}{x_b}\right)^2 = N_{epi} \tag{2.49}$$

where X_{jbc} is the position at which the N-type diffusion becomes equal to the constant value N_{epi}, then a simple Taylor series expansion for the doping versus distance in the vicinity of $x = X_{jeb}$ enables the value of the gradient a to be expressed as

$$a = \frac{dN(x)}{dx}\bigg|_{x=X_{jeb}} = N_b \frac{2X_{jeb}}{x_b^2} \exp\left[-\left(\frac{X_{jeb}}{x_b}\right)^2\right]\left[1 - \left(\frac{x_b}{x_e}\right)^2\right] \tag{2.50}$$

Figure 2.9 shows a linear impurity profile superimposed on the double diffused guassian in the vicinity of the junction X_{jeb}. It is clear that the approximation can be excellent over a considerable range of x, specially for forward bias values, as can be seen from the computed position of the depletion layer boundaries at zero bias for these typical cases. However, as the reverse bias breakdown

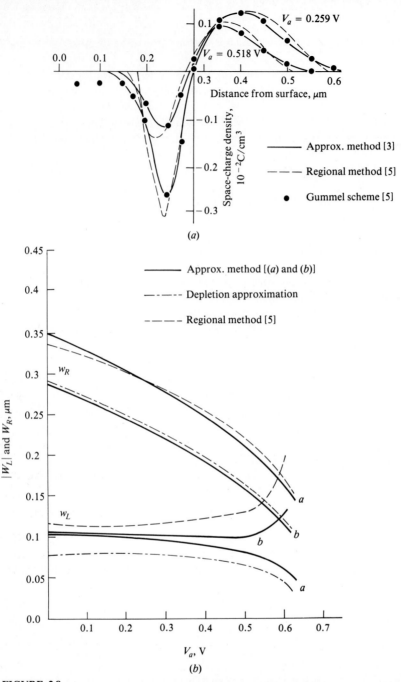

FIGURE 2.8

(a) Space-charge density for a diffused junction as in Fig. 2.7(a) for $N_e = 10^{18}$, $N_{\text{epi}} = 8 \times 10^{15}$ cm^{-3}, $x_e = 0.137$ μm (the numerical solutions are given for two values of forward bias); (b) space-charge layer widths using numerical analyses [3, 5] and using the depletion approximation (*after Eltoukhy and Roulston* [3]). Curves "*a*" use termination of space-charge layer definition given by Kennedy et al. [14]; curves "*b*" use termination of space-charge definition given by Engl et al. [15].

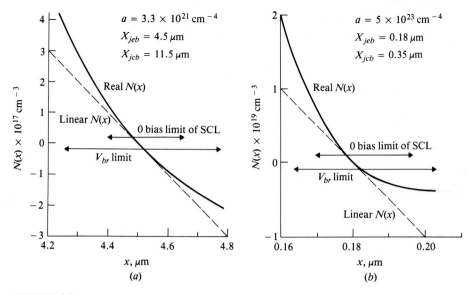

FIGURE 2.9
Linear gradient approximation to a double-diffused doping profile, showing depletion-layer boundaries at zero applied bias and breakdown. (a) $a = 3.3 \times 10^{21}$ cm^{-4}; (b) $a = 5 \times 10^{23}$ cm^{-4}.

voltage is approached (Sec. 3.8), the linearly graded approximation is usually very poor.

2.6 DEPLETION-LAYER CAPACITANCE

The small-signal capacitance associated with the dipole of charge in the depletion layer of a PN junction is defined by

$$C_j = \frac{dQ}{dV} \tag{2.51}$$

This is the capacitance that would be measured by any standard radio frequency bridge method, where the junction has a small sinusoidal signal applied to the terminals (this may be superimposed on an applied dc bias). Figure 2.10 shows the way in which the charge is added at the edges of the depletion layer for a uniformly doped junction. Note that the overall junction remains electrically neutral and that therefore charge of equal magnitude but opposite sign gets added at each of the two depletion-layer boundaries. It is clear that the resulting capacitance will be identical to that of a parallel plate capacitor with plates separated by a distance equal to the total depletion-layer thickness $d_p + d_n$:

$$C_j = \frac{\varepsilon A}{d_p + d_n} \tag{2.52}$$

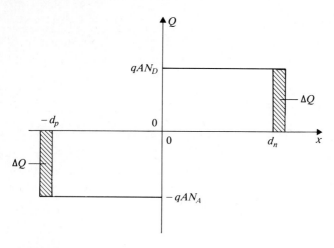

FIGURE 2.10
Illustration of incremental charge added at depletion-layer edges when voltage is altered.

where A is the cross-sectional area of the junction. Of course, exactly the same result will be obtained by substituting the expression for charge $Q = qAd_n N_D$ for either of the abrupt junctions already considered, and then substituting for the widths in terms of the junction voltage and differentiating. This is left as an exercise to the reader.

Since we already obtained expressions for total depletion-layer thickness for each of the three types of junctions considered (abrupt symmetrical, abrupt one-sided, linear graded), it is a simple matter to derive expressions for depletion-layer capacitance versus total junction voltage:

$$C_j = \frac{\varepsilon A}{K_c V^\gamma} \tag{2.53}$$

where the total voltage is $V = V'_{bi} - V_a$ and V'_{bi} is sometimes referred to as the offset voltage [7]. For most practical purposes $V'_{bi} = V_{bi} - 2V_t$ where the $2V_t$ term arises from the effect of the majority carrier distribution at the depletion-layer tail [7], as shown in Figs. 2.2, 2.3, and 2.7. The value of the offset voltage V'_{bi}, may be obtained experimentally from a plot of $C_j^{-\gamma}$ versus V. This is not as simple as may appear, since γ must first be obtained from a plot of log (C) versus log (V), using a guessed or default value for V'_{bi}.

For the symmetrical abrupt junction

$$K_c = \sqrt{\frac{4\varepsilon}{qN_D}} \tag{2.54}$$

$$\gamma = \frac{1}{2} \tag{2.55}$$

and for the one-sided abrupt P^+N junction,

$$K_c = \sqrt{\frac{2\varepsilon}{qN_D}} \qquad (2.56)$$

$$\gamma = \frac{1}{2} \qquad (2.57)$$

For the linearly graded junction, the constants are

$$K_c = \left(\frac{12\varepsilon}{qa}\right)^{1/3} \qquad (2.58)$$

$$\gamma = \frac{1}{3} \qquad (2.59)$$

The two values of the exponent γ are well known to engineers working on device modeling, such as SPICE CAD models, where for a collector-base junction, γ close to 0.5 is normally obtained, whereas for an emitter-base junction a value of γ close to 0.33 is usually measured [9].

A detailed treatment of the linearly graded junction by Morgan and Smits [4] shows that for very shallow gradients, the junction region tends to be almost neutral and the above simple depletion approximation is no longer valid. Furthermore, the validity is only approximate even for quite steep gradients. This is mainly of academic interest only since, as mentioned in Sec. 2.4, most practical junctions are formed by double gaussian functions (as in Figs. 2.7 and 2.8) and even at zero bias there is enough "error" in using the linearly graded approximation that the exact linearly graded solution is not particularly relevant. An extensive treatment is given in [13] of various solutions to the depletion layer.

An interesting situation arises in the case of a very nonsymmetrical P^+N abrupt junction. Gummel [7] showed that in this case the offset voltage V'_{bi} is not a constant. The depletion-layer approximation becomes very bad on the heavily doped side. This fact can be inferred from Fig. 2.2 where the "spillover" of holes into the lightly doped N side means that for zero or small forward bias the heavy doping of the P side is not "seen" by the depletion layer. Gummel showed that the offset voltage becomes nearly independent of the doping on the heavily doped side for the very asymmetric junction at zero or small forward bias. Of course, for sufficient reverse bias, the depletion layer can be forced more into the heavily doped side and its doping-level dependence eventually appears as a contribution to V'_{bi}. At zero bias, the departure from the full depletion approximation occurs for a doping level ratio equal to or greater than 100 to 1 for a lightly doped region of 10^{15} cm^{-3} [7].

This spillover effect is not usually of much practical consequence but will appear again in the treatment of the microwave diode limiter in Chap. 5.

There are two additional sources of deviation from the ideal depletion-layer theory for the capacitance-voltage characteristics of PN junctions. Firstly, it should be recognized that if the doping level at the depletion-layer boundaries is very high, heavy doping bandgap narrowing effects may influence the value of potential (the built-in barrier potential depends on the intrinsic carrier concentration) [10] and thus the offset voltage. This effect is unlikely to be noticeable in most practical devices; under reverse bias, the small BGN voltage effect will be negligible compared to the total bias voltage, while under forward bias, corresponding to narrow depletion layers, the doping level will normally be below that at which BGN effects manifest themselves (since most practical junctions approach the graded junction for forward bias).

The second source of deviation concerns large forward bias. Since, as we saw in Eqs. (2.28) and (2.44), the free carrier concentration inside the space-charge region increases with forward bias, there is a capacitive effect associated with this charge. As pointed out by Lindholm [11], the "free charge" capacitance becomes significant when the applied forward bias becomes comparable to the built-in barrier voltage. In [12] it is shown that the effect of the free carriers is to approximately double the capacitance value when the voltage is within about $2V_t$ of the built-in barrier voltage. Once again, this effect is not of too much practical significance; this is because under these high forward bias conditions, other capacitive effects associated with free carriers in the neutral regions tend to dominate, as we shall see in later chapters.

2.7 FORWARD BIASED PN JUNCTIONS

2.7.1 Injection and the Generalized Law of the Junction

In order to determine the carrier distributions outside the depletion layer under forward bias (necessary for the calculation of the terminal currents) we must first find the relation between the applied bias voltage V_a and the values of "injected" carrier concentrations at both depletion-layer boundaries. It is reasonable to assume that if the bias is not too high, the conditions inside the depletion layer will be approximately those pertaining in thermal equilibrium, and that the Boltzmann relations can be used. We will justify this assumption subsequently.

From Eq. (2.20) we can write the relation between electron concentration and voltage, both position-dependent:

$$\frac{n_n(0)}{n(x)} = \exp\left[\frac{V_N - V(x)}{V_t}\right] \tag{2.60}$$

where V_N is the voltage and $n_n(0)$ the electron concentration at $x = 0$, the depletion-layer boundary on the N side of the junction. Taking this point as reference, $V_N = 0$, gives

$$n(x) = n_n(0) \exp\left[\frac{V(x)}{V_t}\right] \tag{2.61}$$

From the corresponding Boltzmann relation for holes [Eq. (2.21)], we can write

$$p_p(W) = p(x) \exp\left[- \frac{[V_p - V(x)]}{V_t} \right] \tag{2.62}$$

where $p_p(W)$ and V_p are the hole concentration and voltage at $x = W$, the edge of the depletion layer on the P side. Since V_p is made up of the internal barrier potential V_{bi} and the applied (terminal) voltage V_a, where V_a opposes the built-in barrier voltage for forward bias, we obtain

$$p(x) = p_p(W) \left[\exp\left(- \frac{V_{bi}}{V_t} \right) \right]\left[\exp\left(\frac{V_a}{V_t} \right) \right]\left[\exp\left(- \frac{V(x)}{V_t} \right) \right] \tag{2.63}$$

Combining the above two equations (2.61) and (2.63) we obtain

$$p(x)n(x) = p_p(W)n_n(0) \exp\left(- \frac{V_{bi}}{V_t} \right) \exp\left(\frac{V_a}{V_t} \right) \tag{2.64}$$

It is most important to note that this result is a perfectly general relationship between the free carrier concentration $p(x)$, $n(x)$ at any point x within the space-charge layer in terms of the applied voltage V_a, the internal barrier potential V_{bi}, and the majority carrier concentrations $p_p(W)$ and $n_n(0)$ at either edge of the depletion layer.

Two particular forms of Eq. (2.64) are immediately apparent. Writing the total voltage V_{jt} across the depletion layer as

$$V_{jt} = V_{bi} - V_a \tag{2.65}$$

we have, for $x = 0$,

$$p_n(0) = p_p(W) \exp\left(- \frac{V_{jt}}{V_t} \right) \tag{2.66}$$

and for $x = W$,

$$n_p(W) = n_n(0) \exp\left(- \frac{V_{jt}}{V_t} \right) \tag{2.67}$$

Figure 2.11 illustrates the situation under discussion. These results may be expressed thus: "the ratio of carrier concentrations of one type between one space-charge boundary and the other boundary is equal to the exponential of the total normalized voltage between the two boundaries"; on this and other occasions it will be convenient to refer to voltages normalized with respect to $V_t = kT/q$ [0.0259 V at 300 K (27°C), or 0.025 V at 16°C, a more convenient number for hand calculations, corresponding to a rather cool room temperature]. Note once again that the total voltage is always positive on the N side with respect to the P side of a junction.

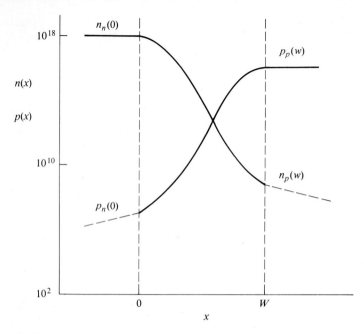

FIGURE 2.11
Carrier concentration variation through a PN junction.

In the common, but nevertheless special, case of low level injection [injected carrier concentration much less than the impurity concentration (net donor or acceptor doping level)], the following simplified relations are valid:

$$p_p(W) = N_A \tag{2.68}$$

$$n_n(0) = N_D \tag{2.69}$$

Under these conditions, Eq. (2.64) simplifies to

$$n(x)p(x) = n_i^2 \, \exp\left(\frac{V_a}{V_t}\right) \tag{2.70}$$

and the injected carrier concentrations at either depletion-layer edge are now given by:

$$p_n(0) = p_{n0} \, \exp\left(\frac{V_a}{V_t}\right) \tag{2.71}$$

$$n_p(W) = n_{p0} \, \exp\left(\frac{V_a}{V_t}\right) \tag{2.72}$$

We recall that p_{n0}, n_{p0} are the thermal equilibrium values of hole and electron concentrations on, respectively, the N and P sides of the junction and are

given by

$$p_{n0} = \frac{n_i^2}{N_D} \tag{2.73}$$

$$n_{p0} = \frac{n_i^2}{N_A} \tag{2.74}$$

It is interesting to study the limit of validity of the above result. The Boltzmann relations are based on zero current flow, but the main applications of Eqs. (2.70) to (2.72) are for forward biased junctions. In App. 5 we derive an expression for the ratio r_{cur} of hole diffusion current $J_{diff\ scr}$ at the center of the depletion layer of a forward biased junction to the hole diffusion current $J_{diff\ neut}$ flowing in the neutral region (i.e., outside the depletion layer boundary $x = W$ in the above analysis). Although we shall be studying current flow only in the next chapter, it is worth noting at this point the applicability of the above result. Appendix 5 shows that the ratio $r_{cur} = J_{diff\ scr}/J_{diff\ neut}$ is of order 10^8 at zero bias and decreases as exp $-(V_a/2V_t)$.

The ratio therefore remains much greater than unity up to very high forward bias for typical devices. Since, in the absence of recombination, the total hole current must be the same at every point, this is equivalent to saying that inside the space-charge layer (at least near the center), the hole diffusion current is much greater than the total hole current—in other words, the hole diffusion current within the space-charge layer must be almost canceled by an equal and opposite hole drift current. This is precisely the condition implied by use of the Boltzmann equations in writing Eqs. (2.60) and (2.62) at the start of the above analysis. We thus conclude that Eqs. (2.66) and (2.67) based on use of the Boltzmann relations are valid over a wide range of applied bias. In fact, at the high bias values where the approximation starts to become less valid, other effects come into play, related to high level injection in the neutral regions, and the error in *total applied (terminal) voltage* for a given current will be normally a very small fraction of the total. Of course, the special formulas for low level injection [Eqs. (2.68) to (2.72)] are not universally valid and we will explore high level injection in considerable detail when dealing with diodes in Chap. 3 and transistors particularly in Chaps. 10 and 15.

It is also of interest at this point to note the quasi-Fermi level concept for a forward biased junction. Figure 2.12 shows the quasi-Fermi level diagram for a typical NP diode. We recall from App. 4 that zero current corresponds to a zero gradient of the corresponding quasi-Fermi level. Thus, for the Boltzmann equation to be valid across the space-charge layer, the electron and hole quasi-Fermi levels must be constant in this region. The gradient outside the space-charge layer corresponds to the fact that minority carrier current flows in the neutral region, as we shall see in the next chapter.

It is also worth noting that from equation (A4.2) and the corresponding equation for holes, plus Eq. (2.70) above, the separation of the quasi-Fermi levels within the space-charge layer is qV_a.

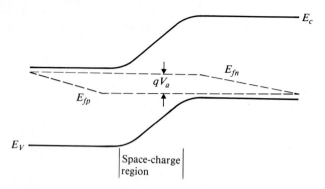

FIGURE 2.12
Diagram showing quasi-Fermi levels E_{fn} and E_{fp} in forward biased junction. Within the space-charge layer the separation is qV_a.

2.7.2 The NN^+ low-high junction—the retarding field

Figure 2.13 shows a region of a semiconductor device in which there is a transition from a moderately doped N_{epi} region to a heavily doped N^+ region. Let us examine the free carrier distribution for this low-high junction, firstly in thermal equilibrium then for the more important case where the minority carrier concentration is raised above thermal equilibrium. Initially, we will assume that the doping level in the N^+ region is low enough that bandgap narrowing can be neglected. The results of this study will then be used in the analysis of the P^+NN^+ diode. The Boltzmann equations may be used directly to deduce the variation of potential and carrier concentration across the low-high junction. In particular, we obtain the ratios of carrier concentrations at either side of the junction as

$$\frac{n(W^+)}{n(W^-)} = \exp\left(\frac{V_{lh}}{V_t}\right) \tag{2.75}$$

and

$$\frac{p(W^+)}{p(W^-)} = \exp\left(-\frac{V_{lh}}{V_t}\right) \tag{2.76}$$

where V_{lh} is the total voltage across the low-high transition region, the positions W^- and W^+ are the x values at the start and end of the low-high transition region as shown in the figure. Note that using the majority carrier (electron) concentration and equating it to the doping level gives an expression for the total voltage across the junction:

$$V_{lh} = V_t \ln\left(\frac{N_D^+}{N_{epi}}\right) \tag{2.77}$$

This must be the same as the result in Eq. (2.76), which leads to the conclusion that

$$\frac{p(W^+)}{p(W^-)} = \frac{N_{epi}}{N_D^+} \tag{2.78}$$

For a moderately doped epitaxial layer, N_{epi} is typically 10^{15} cm^{-3}. Let us use a value of doping for the heavily doped N$^+$ region of 10^{18} cm^{-3}. The corresponding transition region voltage V_{th} is thus about 0.2 V. If the transition occurs over a distance of a few microns, it is clear that the peak value of electric field will be only of order 1000 V/cm. The space-charge associated with the field changing from zero to 1000 V/cm and back to zero over a distance of order 1 micron can be estimated crudely as $(dE/dx)_{av} = 10^7$. This can be compared to the corresponding value $(q/\varepsilon)N_{epi}$ of about 10^8. It is clear that for this typical case the amount of space-charge is small; in fact, it would be quite reasonable to treat this low-high junction as approximately space-charge neutral throughout. This is the approximation that we shall make both here and subsequently when treating low-high transition regions (note that this is a case where the term transition region is not synonymous with space-charge or depletion region).

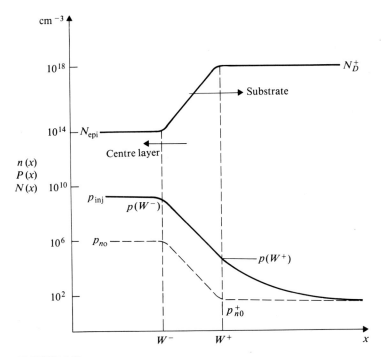

FIGURE 2.13
Impurity profile and minority carrier distribution in the NN$^+$ junction. The hole concentration distribution is shown for both thermal equilibrium and nonthermal equilibrium conditions. For the latter case, it is assumed that the hole concentration p_{inj} is produced by injection at some position to the left, as would occur in a P$^+$NN$^+$ diode.

Now let us consider the non-thermal equilibrium situation. If we accept the argument used in examining the forward biased PN junction, we can state the result that the minority carrier concentration must still decrease in the ratio fixed by the value of the low-high junction voltage V_{lh}, that is, the inverse ratio of the increase in doping level. Any excess minority carriers at the position W^- will be reduced by a factor N_{epi}/N_D^+ by the time they have traveled to the position W^+. It is important to note that this decrease has nothing to do with loss by recombination. In general, it may be expected that the minority carrier current at $x = W^+$ will be the same as the minority carrier current at $x = W^-$. The phenomenon is often explained by the "retarding field" presented by the increase in majority carrier (doping) concentration. In fact if we consider the case of zero electron current, the electric field may be found.

$$J_n = 0 = qD_n \frac{dn}{dx} + q\mu_n nE \tag{2.79}$$

hence we find

$$E = -\frac{D_n}{\mu_n} \frac{1}{n} \frac{dn}{dx} \tag{2.80}$$

or

$$E = -V_t \frac{1}{n} \frac{dn}{dx} \tag{2.81}$$

That this is a retarding field, acting against the normal flow of minority carriers (i.e., away from the point of high concentration created by injection) can be seen by writing the equation for hole current density

$$J_p = -qD_p \frac{dp}{dx} + q\mu_p pE \tag{2.82}$$

and substituting for the electric field, E

$$J_p = -qD_p \frac{dp}{dx} - q\mu_p pV_t \frac{1}{n} \frac{dn}{dx} \tag{2.83}$$

Since dp/dx in the situation under discussion is negative, and since the gradient dn/dx is positive, it is clear that the hole current is *reduced* from the (diffusion current) value it would have in the absence of the built-in retarding field.

This is a general property which we shall come across frequently in studying bipolar devices. It may be summarized by the statement: if the doping level increases moving away from the point where carriers are injected, then a built-in field is created which acts against minority carrier current flow.

The above treatment applies for moderate doping levels on both sides of the low-high junction. However, since the N^+ side will normally be heavily doped, the bandgap will be narrower and the intrinsic concentration n_i will increase to a value n_{ie}. As we mentioned in Chap. 1, this can conveniently be taken care of by using the effective doping level $N_{D\,\text{eff}}^{+}$ instead of N_{D}^{+}, when considering the effect on minority carrier transport. The value of the low-high voltage may thus be written as

$$V_{lh} = V_{t} \ln \left(\frac{N_{D\,\text{eff}}^{+}}{N_{\text{epi}}} \right) \qquad (2.84)$$

and the change in hole concentration is given by

$$\frac{p(W^{+})}{p(W^{-})} = \frac{N_{\text{epi}}}{N_{D\,\text{eff}}^{+}} = \frac{N_{\text{epi}}}{N_{D}^{+} \exp\left(-\Delta E_{g}/kT\right)} \qquad (2.85)$$

The thermal equilibrium value of hole concentration just inside the N⁺ region is given by

$$p_{no}(W^{+}) = \frac{n_{ie}^{2}}{N_{D}^{+}} = \frac{n_{i}^{2}}{N_{D\,\text{eff}}^{+}}$$

Figure 2.14 shows the situation for the case of injection for an N⁺ doping level of 10^{20} cm⁻³ for which $\Delta E_{g} = 0.1$ eV is assumed.

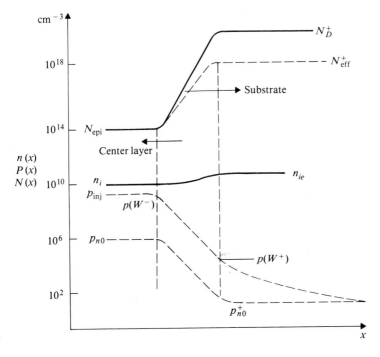

FIGURE 2.14
Impurity profile and effective impurity profile for an NN⁺ junction with heavily doped regions. The hole concentration distributions are given as in Fig. 2.13, and the intrinsic carrier concentration distribution is also included.

2.8   CONCLUSIONS

In this chapter, we have examined in some detail the PN junction, specifically the space-charge layer. We have seen that for most practical purposes, the depletion approximation may be used to simplify Poisson's equation to calculate the peak electric field for given total junction voltage. Use of the Boltzmann relations enabled us to determine the built-in barrier potential of the junction and to examine behavior under forward bias. The low-high junction was then studied and shown to introduce a retarding electric field on the minority carriers, thus decreasing the value of the normal diffusion current. It was seen that the effect of bandgap narrowing will normally have an effect on the reflecting properties of this junction. The various results will be used in subsequent chapters to examine the electrical terminal characteristics of various practical PN junction diodes.

PROBLEMS

2.1. Consider a symmetric PN junction with N_D (N side) $= N_A$ (P side) $= 10^{14}$ cm^{-3}. Calculate the depletion-layer thickness d_n for the total voltage on the N side equal to (i) 0.5 V, (ii) 5 V; compare d_n to the extrinsic Debye length in each case. Repeat the calculations for doping levels of 10^{16}, 10^{18}, 10^{20} cm^{-3}, and observe the extent of the depletion layer "tail" in each case. Sketch the space-charge diagrams.

2.2. For each of the four PN junctions considered in Prob. 2.1, calculate the built-in barrier voltage V_{bi} and the peak value of electric field at zero bias. Breakdown occurs for electric fields between 10^5 and 10^6 V/cm (this will be studied in detail in the next chapter); what are the implications of this fact for the case of a junction with high doping levels on both sides?

2.3. For a symmetrical abrupt junction with doping levels on each side equal to 10^{16} cm^{-3}, calculate the characteristic length of the free carrier concentration near the junction and compare x_c to the depletion-layer width d_n for an applied bias V_a equal to (i) 0 V, (ii) 0.5 V. Sketch the free carrier concentration on a logarithmic vertical scale versus distance and the space-charge (linear scale) versus distance. Use the extrinsic Debye length to estimate the shape at the depletion layer edges and use Eq. (2.70) to find the magnitude of the carrier concentration at the junction.

2.4. Consider a nonsymmetric PN junction as in Fig. 2.2, with $N_A = 10^{18}$ cm^{-3}, $N_D = 10^{16}$ cm^{-3}. Use the solution of Poisson's equation to find the depletion-layer widths d_n, d_p at zero bias and estimate from a sketch the position at which $n = p = n_i$.

2.5. Use Eq. (2.34) and derive the results contained in Eqs. (2.35), (2.36), and (2.37).

2.6. Consider a linearly graded PN junction with an extrinsic Debye length L_{Dn} associated with the doping level at the depletion-layer boundary $x = d_n$. Show that the ratio L_{Dn}/d_n is a constant for a given total voltage and calculate the ratio for a voltage on the N side equal to (i) 1 V, (ii) 0.1 V.

2.7. This question concerns the depletion-layer approximation for the linearly graded junction. Equation (2.44) is a simplified expression for the free carrier concentration versus distance. The simplest representation is a straight line on a log-linear graph, with a carrier concentration equal to n_i at $x = 0$ and equal to N_D at $x = d_n$. The net charge versus distance x is the difference between this $n(x)$ curve and the impurity concentration $N(x) = ax$. Draw these two curves on log-linear plots and the resulting space-charge on a linear-linear plot for two cases (i) $N_D(x = d_n) = 10^{12}$, (ii) $N_D(x = d_n) = 10^{14}$ cm^{-3}. Assume $n_i = 10^{10}$ cm^{-3}. What conclusions can you draw about the applicability of the depletion-layer approximation in the linearly graded junction? Calculate the values of voltage, depletion-layer width, and junction gradient for the above two cases, assuming zero bias conditions.

2.8. Consider an exponential impurity profile of the form $N(x) = N \exp - (x/x_0) - N_{epi}$, giving a PN junction similar to that shown in Fig. 2.7(a). Assume substantial reverse bias such that the depletion layer in the N_{epi} layer terminates at a depth x_n from the surface, where x_n is much greater than the junction depth X_j, and solve Poisson's equation to obtain an approximate expression for x_p, the depletion-layer boundary in the exponential region in terms of x_n, x_0, X_j, N_0, N_{epi}. Obtain also the ratio of doping levels at the two depletion-layer edges. Calculate the results for a profile given by $N_0 = 10^{21}$ cm^{-3}, $N_{epi} = 10^{16}$ cm^{-3}, $X_j = 1$ μm, reverse bias $= 40$ V.

2.9. Using Eqs. (2.47) through (2.50) for a double gaussian impurity profile (similar to that for the emitter-base junction of a transistor, or of some varactor diodes to be discussed in Chap. 5), calculate the values of junction gradient "a" for the following realistic conditions: $N_e = 10^{21}$, $N_b = 10^{19}$, $N_{epi} = 10^{15}$ cm^{-3} and three sets of junction depths:

(i) $X_{jeb} = 0.1$ μm, $X_{jcb} = 0.25$ μm
(ii) $X_{jeb} = 1.0$ μm, $X_{jcb} = 2.5$ μm
(iii) $X_{jeb} = 10$ μm, $X_{jcb} = 25$ μm

2.10. Show, using the solution to Poisson's equation for a one-sided abrupt junction, that the use of the definition of capacitance $C = dQ/dV$ [Eq. (2.51)] yields the result given in Eq. (2.52) (use $N_A \ll N_D$ for convenience).

2.11. For the idealized impurity profile shown in the accompanying figure, calculate and sketch the nonlinear exponent and the capacitance versus voltage.

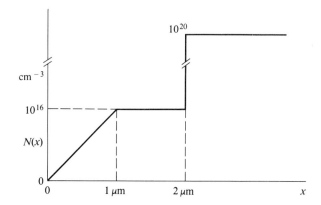

2.12. Consider a high-low junction defined by the impurity profile

$$N(x) = N_0 \exp - \left(\frac{x}{x_0}\right)^2 + N_{epi}$$

Use Eq. (2.81) with $n = N(x)$ to derive an expression for electric field as a function of distance. Hence derive an expression for the space-charge density. Sketch this for the case $N_0 = 10^{18}$, $N_{epi} = 10^{15}$ cm^{-3}, in terms of the normalized distance x/x_0. A good indication of how neutral the transition region is can be found by examining the normalized space charge $(\varepsilon/qN)\, dE/dx$. Evaluate this quantity on either side of the high-low transition region for two cases (i) $x_0 = 1$ μm, (ii) $x_0 = 0.1$ μm and comment on the conclusion stated in the text following Eq. (2.78).

REFERENCES

1. M. S. Adler et al., "Introduction to semiconductor physics," *SEEC*, vol. 1, Wiley (1964).
2. S. Sze, *Physics of Semiconductor Devices*, 2d ed., John Wiley (1981).
3. A. A. Eltoukhy and D. J. Roulston, "An efficient method for the analysis of the space-charge region of diffused junctions," *Solid State Electronics*, vol. 25, pp. 829–831 (1982).
4. S. P. Morgan and F. M. Smits, "Potential distribution and capacitance of a graded *pn* junction," *Bell System Tech. Jnl.*, vol. 39, pp. 1573–1602 (November 1960).
5. S. C. Rustagi and S. K. Chatopadhya, "Characteristics of linearly graded PN junctions," *Solid State Electronics*, vol. 22, pp. 1819–1827 (September 1979).
6. R. M. Sirsi and A. R. Boothroyd, "Characterization of space-charge properties of a linearly graded *pn* junction by an approximate 'regional' analysis method," *IEEE Trans. Electron Devices*, ED-23, pp. 348–353 (March 1976).
7. H. K. Gummel and D. L. Scharfetter, "Depletion layer capacitance of p^+n step junctions," *Journal of Applied Physics*, vol. 38, pp. 2148–2153 (April 1967).
8. C. G. B. Garrett and W. H. Brattain, "Physical theory of semiconductor surfaces," *Phys. Rev.*, vol. 99, p. 376 (1955).
9. I. Getreu, "Modeling the bipolar transistor," Tektronix, Inc., Beaverton, Oregon (1976).
10. R. J. Van Overstraeten, H. J. De Man, and R. P. Mertens, "Transport equations in heavily doped silicon," *IEEE Trans. Electron Devices*, ED-20, pp. 290–298 (March 1973).
11. F. A. Lindholm, "Simple phenomenological modeling of transition-region of forward biased PN junction diodes and transistor diodes," *Jnl. Appl. Phys.*, vol. 53, pp. 7606–7608 (November 1982).
12. K. Negus and D. J. Roulston, "Simplified modeling of delays in the emitter-base junction," *Solid State Electronics*, vol. 31, pp. 1464–1466 (September 1988).
13. R. M. Warner and B. L. Grung, *Transistors—Fundamentals for the Integrated Circuit Engineer*, John Wiley & Sons, New York (1983).
14. D. P. Kennedy and R. R. O'Brien, "On the mathematical theory of the linearly graded *pn* junction," *IBM Jnl. Res. and Dev.*, pp. 252–270 (May 1967).
15. W. L. Engl, O. Manck, and A. W. Wieder, "Modeling of bipolar devices" in Process and Device Modeling for IC design. (*NATO Advanced Study Institute Series E-21*) Leyden, The Netherlands, Noordhoff (1977).
16. C. T. Sah, "Effects of electrons and holes on the transition layer characteristics of linearly graded *pn* junctions," *Proc. IEEE*, vol. 49, pp. 603–618 (March 1961).

CHAPTER
3

PN JUNCTION
DIODES

3.1 INTRODUCTION

In this chapter we shall take the results for PN junction space-charge regions from the previous chapter and extend them to determine the terminal *I-V* characteristics of several real PN junction diodes. First, we consider moderate forward bias for three different structures. Then high current operation—high level injection—is considered. Low forward bias where space-charge recombination dominates is treated next; finally reverse bias and an overview of avalanche multiplication and breakdown is presented.

3.2 THE NARROW BASE DIODE

A narrow base diode is defined as one in which there is negligible recombination throughout the neutral region and a plane of infinite surface recombination velocity at the extremity of each region. Let us consider the P^+N diode shown in Fig. 3.1. Note that neutrality forces the excess hole and electron concentrations to be the same at every value of x outside the (narrow) space-charge region.

Since the continuity equation for holes is

$$\frac{dJ_p}{dx} = -\frac{qp'}{\tau_h} \tag{3.1}$$

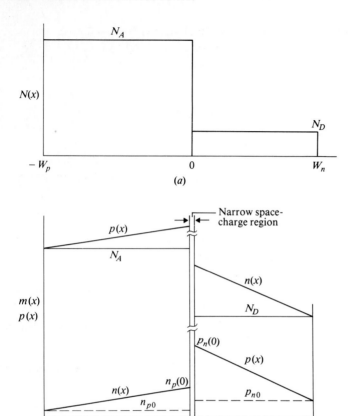

FIGURE 3.1

(a) The ideal doping profile of a P^+N narrow base diode; zero recombination is assumed to exist in both neutral regions, and a plane of infinite recombination velocity is assumed to exist at both extremities of the structure; (b) the carrier concentration distribution.

where p' is the excess hole concentration (above the thermal equilibrium value $p_{n0} = n_i^2/N_D$), it follows that zero recombination, obtained by having infinite lifetime, leads to the result that

$$\frac{dJ_p}{dx} = 0 \tag{3.2}$$

or, on the assumption that drift current can be neglected (see Sec. 3.2.1),

$$J_p = -qD_p \frac{dp}{dx} = \text{const} \tag{3.3}$$

A plane of infinite recombination means that there can be no excess carriers at or beyond the extremities of the two neutral regions. This means that

$$p'(W_n) = 0 \tag{3.4}$$

and

$$n'(-W_p) = 0 \tag{3.5}$$

where, for convenience, the region widths W_n for the N side and W_p for the P side are measured from the depletion-layer edges. In fact, since we are considering forward bias, this is for all practical purposes the same as defining the widths from the metallurgical junction, because the depletion layer will be quite narrow.

In practice, a plane of high recombination velocity can be obtained in one of three ways: a good ohmic contact; a separate semiconductor region with very low lifetime; a depletion-layer boundary of a reverse biased junction. Although the first two are easy to visualize, it is the third method which is of most practical significance. This is because it corresponds to the "diode connected transistor" used in nearly all integrated circuits where the diode function is required. It is also for this reason that our treatment of the narrow base diode is so important; the lightly doped region is that which corresponds to the base region of a transistor and is also the region which determines most of the electrical terminal characteristics.

To return to our determination of terminal current, we note that the magnitude of the gradient of hole concentration on the N side neutral region is now given by

$$\frac{dp}{dx} = \frac{p'(0) - p'(W)}{W_n}$$

$$= \frac{p_n(0) - p_{no}}{W_n} \tag{3.6}$$

where $p_n(0)$ is the injected carrier concentration. The hole concentration distribution is thus

$$p(x) = [p_n(0) - p_{no}]\left[1 - \frac{x}{W_n}\right] + p_{no} \tag{3.7}$$

The magnitude of the current flowing due to the gradient is therefore

$$J_p = qD_p \frac{p_n(0) - p_{no}}{W_n} \tag{3.8}$$

Substituting our result in Eq. (2.71) for $p_n(0)$ as a function of applied voltage and multiplying by the cross-sectional area of the diode gives the magnitude of the hole current

$$I_p = \frac{qAD_p p_{no}}{W_n}\left[\exp\left(\frac{V_a}{V_t}\right) - 1\right] \tag{3.9a}$$

A similar equation can be written by inspection for the P^+ side

$$I_n = \frac{qAD_n n_{p0}}{W_p} \left[\exp \left(\frac{V_a}{V_t} \right) - 1 \right] \qquad (3.9b)$$

If recombination is not negligible, it may be included approximately by considering a recombination current $I_{pr} = Q_p/\tau_p$ where the charge Q_p is the area of the triangular hole distribution on the N side shown in Fig. 3.1. This gives the recombination current

$$I_{pr} = \tfrac{1}{2} qA[p(0) - p_{n0}] \frac{W_n}{\tau_p} \qquad (3.10)$$

The total hole current crossing the junction is now given by the sum of Eqs. (3.9a) and (3.10). The exact solution for hole current injected into the N region, including arbitrary recombination, is

$$I_p = \frac{qAD_p p(0)/L_p}{\tanh (W_n/L_p)} \qquad (3.11)$$

It is clear from the above results that if the P side doping level is much higher than the N side doping level, and provided the region widths do not differ too significantly, then the terminal current is determined for all practical purposes by the lightly doped N side. The converse would apply to an N^+P diode, in which case the terminal current would be determined by the lightly doped P side. It is seldom, if ever, necessary to consider both sides in order to design a diode for specific characteristics—we shall see later on that it is also the lightly doped side which determines the maximum reverse voltage that can be applied before breakdown occurs.

3.2.1   Minority Carrier Drift Current in the P^+N Narrow Base Diode

This section examines the assumption that minority carrier (hole drift) current can be neglected on the N side of the P^+N diode, at least for moderate or low currents.

In the P^+N narrow base diode, the current is predominantly hole current. To a good approximation, therefore, we can say that throughout the moderately doped N region the electron current is zero

$$J_n = 0 = qD_n \frac{dn}{dx} + q\mu_n nE \qquad (3.12)$$

The electric field is therefore given by

$$E = -V_t \frac{1}{n} \frac{dn}{dx} \qquad (3.13)$$

Charge neutrality gives us the electron concentration from our previously derived [Eq. (3.7)] hole concentration distribution $p(x)$:

$$n(x) = N_D + p(x) \qquad (3.14)$$

Therefore the field becomes

$$E(x) = -V_t \frac{1}{N_D + p(x)} \frac{dp}{dx} \qquad (3.15)$$

The hole drift current $J_p = q\mu_p pE$ is therefore

$$J_{p\,\text{drift}} = -qD_p \frac{dp}{dx} \frac{p(x)}{N_D + p(x)} \qquad (3.16)$$

This is much smaller than the hole diffusion current already derived [Eq. (3.8)], at least up to a certain bias level, as can be seen by taking the ratio r_{dd}

$$r_{dd} = \frac{J_{p\,\text{drift}}}{J_{p\,\text{diff}}} = \frac{p(x)}{N_D + p(x)} \qquad (3.17)$$

In fact this ratio enables us to write the condition of validity of the narrow base diode analysis. The result [Eq. (3.8)] is valid only if

$$p(0) \ll N_D \qquad (3.18)$$

This is referred to as *low level injection*. Subsequently, we will examine the behavior of the diode when this condition is violated—the case of *high level injection*.

3.3 THE WIDE BASE DIODE

This structure is of no practical importance, for reasons which will soon be apparent. It does however explain operation in an extreme case and as such is of some pedagogical interest. Consider the situation shown in Fig. 3.2.

Clearly, if the N region width is much greater than a hole diffusion length L_p, the shape of the hole distribution in the neutral region will be determined by

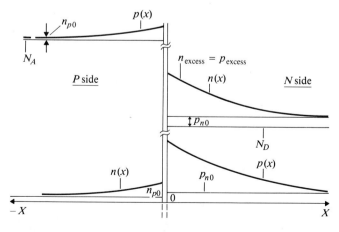

FIGURE 3.2
Free carrier concentration distributions in the P^+N wide base diode structure. The region widths are several times greater than the carrier diffusion lengths.

the recombination process. Applying the hole continuity equation gives

$$\frac{dJ_p}{dx} = -\frac{qp'}{\tau_p} \tag{3.19}$$

Using Eq. (1.18) for hole diffusion current

$$J_p = -qD_p \frac{dp}{dx} \tag{3.20}$$

and combining these two equations gives

$$\frac{d^2p}{dx^2} = \frac{p'(x)}{L_p^2} \tag{3.21}$$

where we have substituted the diffusion length L_p for $\sqrt{D_p \tau_p}$. The solution to this equation has the general form

$$p'(x) = A \exp\left(-\frac{x}{L_p}\right) + B \exp\left(\frac{x}{L_p}\right) \tag{3.22}$$

Assuming the same boundary conditions as for the narrow base diode, there is no physical possibility to allow for a carrier concentration which rises moving away from the junction. It follows that $B = 0$. The hole concentration at the depletion-layer boundary is given as before and this determines the value of the constant A, which is $p'(0) = p(0) - p_{n0}$. The carrier concentration distribution is thus

$$p(x) = p_{n0} + [p(0) - p_{n0}] \exp\left(-\frac{x}{L_p}\right) \tag{3.23}$$

and the current is determined at $x = 0$ by taking the derivative

$$I_{p(x=0)} = qAD_p \frac{dp}{dx}\bigg|_{x=0} \tag{3.24}$$

This minority carrier (hole) current will decay exponentially as x increases toward the contact. The recombination means that it will gradually be replaced by majority carrier (electron current). For the purposes of finding the terminal electrical characteristics, all we need to know is the current at any one point; the easiest position to take is at $x = 0$ (note that this is strictly the depletion-layer boundary and not the actual metallurgical junction). The actual current distribution is as shown on Fig. 3.3. The final expression for hole current at $x = 0$ as a function of voltage, from the preceding results, is

$$I_p = \left(\frac{qAD_p p_{n0}}{L_p}\right)\left[\exp\left(\frac{V_a}{V_t}\right) - 1\right] \tag{3.25}$$

A similar expression may be written by inspection for electron current on the P side. The only difference between this result and that for the narrow base diode [Eq. (3.9a)] lies in the substitution for W_n by L_p. In fact, if we consider the

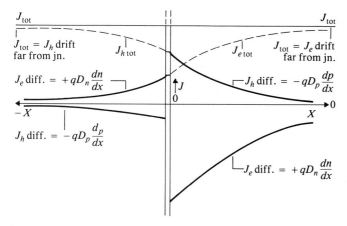

FIGURE 3.3
Current distributions in the P$^+$N wide base diode.

requirement for a long region $W_n \gg L_p$, then at even quite moderate current densities there will be a significant voltage drop due to the majority carrier current flowing in the low doped (and, therefore, moderate-to-high resistivity) N region. The narrow base diode, in contrast, has a narrow N region and thus a much lower neutral region voltage drop. It has excellent *I-V* characteristics and is by far the better choice for any real application; this fact, combined with the ease of fabrication using a diode-connected transistor in integrated circuit technology, means that the wide base (or long base) diode will not be considered any further. We will, however, deal next with another type of diode in which the current is dominated by bulk recombination—the P$^+$NN$^+$ or PIN diode.

3.4 THE P$^+$NN$^+$ DIODE

This is a widely used structure, especially in its high current mode as a PIN diode. The basic device is shown in Fig. 3.4

In order to derive a simple expression for the *I-V* characteristics, we must first find relations for the currents crossing the P$^+$N and NN$^+$ junctions. This will be done in the following two sections. Then we will be in a position to solve for the terminal characteristics.

3.4.1 Current Flow Across the NN$^+$ Low-High Junction

In order to determine the value of the minority carrier current which flows across the low-high junction, we need only consider flow in the heavily doped N$^+$ region. On the assumption that it is a thick substrate (typically two to three hundred microns for mechanical strength of the silicon wafer) and bearing in mind that lifetime (and thus diffusion length) decrease at high doping levels, it is often a very good

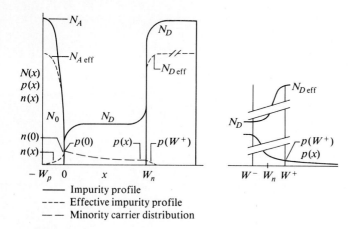

——— Impurity profile
- - - - Effective impurity profile
— — Minority carrier distribution

FIGURE 3.4
P^+NN^+ diode structure doping profile and carrier concentration distributions.

approximation to treat this region as part of a wide base diode. Note that this does not contradict our earlier statement that wide base diodes per se have no practical applications; here we are dealing with only one current component of the P^+NN^+ diode and, as we are about to see, it will be a negligible component in most good devices. The current crossing the low-high junction is (neglecting a small amount of recombination in the low-high transition region) the same as the "wide base diode" current in the N^+ region; this is given by:

$$J_p(W^+) = \frac{qD_{p+}\,p(W^+)}{L_{p+}} = J_p(W^-) \tag{3.26a}$$

where we use + for subscripts for diffusion coefficient and diffusion length in the N^+ region to distinguish from the values in the center N layer to be used shortly. By combining Eq. (3.26a) with the previous result for the low-high junction [Eq. (2.85)] we obtain the more useful relation

$$J_p(W^+) = \frac{qD_{p+}\,p(W^-)}{L_{p+}}\frac{N_{epi}}{N_{D\,eff}^+} = J_p(W^-) \tag{3.26b}$$

It is sometimes convenient to express this in terms of a recombination velocity S_{nn+} at the position W^-

$$J_p(W^-) = qp(W^-)S_{nn+} \tag{3.27a}$$

where

$$S_{nn+} = \frac{D_{p+}}{L_{P+}}\frac{N_{epi}}{N_{D\,eff}^+} \tag{3.27b}$$

The reader may readily check that values lie typically in the range 0.1 to 100 cm/s; note that by constrast the recombination velocity at a metal contact is of order 10^6 cm/s.

3.4.2 The Diffused P$^+$ Region

The surface layer of most diodes is created by diffusion of a shallow region of, in this case, acceptor atoms. Clearly, as shown in Fig. 3.4, the doping profile thus created gives rise once again to a retarding field. In this case, however, the metal contact is quite close to the point of injection and we must consider the boundary conditions in order to find the value of the injected minority carrier current. The results to be derived here are important for future work, and this serves as a convenient point to introduce the concepts.

Bearing in mind that we are now dealing with a P region whose doping level rises in the direction of normal electron current flow, the previously derived result for retarding field [Eq. (2.81)] gives

$$E = V_t \frac{1}{p} \frac{dp}{dx} \tag{3.28}$$

where we define for convenience x to be positive moving away from the junction. A simple but nevertheless useful result can be obtained analytically if we consider the case of an exponential law for the effective doping level (note that we are studying minority carrier current and therefore must include bandgap narrowing effects discussed in Chap. 1 and represented by the effective doping level)

$$N_{A\,\text{eff}}(x) = N_0 \exp\left(\frac{x}{x_0}\right) \tag{3.29}$$

where N_0 is the value of the effective doping at the edge of the space-charge layer at which point the injected electron concentration has a value given by

$$n(0) = n_{p0} \exp\left(\frac{V_a}{V_t}\right) \tag{3.30}$$

and the thermal equilibrium electron concentration at this point is given by

$$n_{p0} = \frac{n_i^2}{N_0} \tag{3.31}$$

The value of retarding electric field may be written in terms of the characteristic length x_0 by combining Eqs. (3.28) and (3.29) as

$$E = \frac{V_t}{x_0} \tag{3.32}$$

This is now used in the normal equation for electron current

$$J_n = qD_n \frac{dn}{dx} + q\mu_n nE \tag{3.33}$$

to give

$$J_n = qD_n \left[\frac{dn}{dx} + \frac{n}{x_0}\right] \tag{3.34}$$

Note once again that since dn/dx is negative (decaying carrier concentration), the field clearly acts against normal current flow. If we neglect recombination, that is, J_n independent of distance, this leads to the second-order equation

$$\frac{d^2n}{dx^2} + \frac{1}{x_0}\frac{dn}{dx} = 0 \qquad (3.35)$$

with the general solution

$$n(x) = C_1 x_0 \exp\left(-\frac{x}{x_0}\right) + C_2 \qquad (3.36)$$

The boundary condition $n(W_p) = 0$ (imposed by the assumption of an ohmic contact at $x = W_p$) enables C_2 to be determined as a function of C_1 and the condition $n(x = 0) = n(0)$ then leads to the determination of C_1. The solution for carrier concentration versus distance is, therefore,

$$n(x) = \frac{n(0)[\exp(-x/x_0) - \exp(-W_p/x_0)] + n_{p0}[1 - \exp(-x/x_0)]}{1 - \exp(-W_p/x_0)} \qquad (3.37)$$

Substituting this result into the expression for electron current [Eq. (3.34)] at $x = 0$ and neglecting the thermal equilibrium term, i.e., assuming $V_a \gg V_t$, gives

$$J_n = \frac{qD_n n(0)}{x_0}\frac{\exp(-W_p/x_0)}{1 - \exp(-W_p/x_0)} \qquad (3.38)$$

On the assumption that the P region width is at least a couple of x_0 away from the junction (which, for an exponential effective profile, implies that the effective doping level has risen to a value at least ten times N_0) this result simplifies to

$$J_n = \frac{qD_n n(0)}{x_0 \exp(W_p/x_0)} \qquad (3.39)$$

An alternate way of looking at this is to write the result as

$$J_n = \frac{qD_n n(0)}{W_p}\frac{W_p/x_0}{\exp(W_p/x_0)} \qquad (3.40)$$

It is easy to observe that this is the same as a "narrow base diode" current reduced by the factor $(W_p/x_0)/\exp(W_p/x_0)$. Since N_{eff} peaks at a level of order 10^{18} cm^{-3} (see Chap. 1 and Fig. 3.4) and N_0 is typically of order 10^{16} cm^{-3} at the depletion-layer boundary under normal forward bias, we see that this factor lies in the range $1/10$ to $1/100$.

Although we have neglected recombination in this analysis, since the minority carrier charge distribution is known, the additional current due to recombination may be written by inspection as

$$J_{nr} = \frac{Q_n}{\tau_n} \qquad (3.41)$$

The total minority carrier charge is found by integration of the excess electron distribution [Eq. (3.37)]. For the approximation already used of W_p at least two times x_0, and for $V_a \gg V_t$ such that the thermal equilibrium concentration may be neglected, we obtain the simple result

$$Q_n = qn(0)x_0 \tag{3.42}$$

and the corresponding recombination current

$$J_{nr} = \frac{qn(0)x_0}{\tau_n} \tag{3.43a}$$

The total current J_{nt} will be the sum of Eqs. (3.40) and (3.43a):

$$J_{nt} = J_n + J_{nr} \tag{3.43b}$$

Note that this result is valid provided the recombination is not so great that the distribution is significantly modified. This is valid for recombination currents at least equal to J_n given by Eq. (3.40). This point will be examined in more detail when we discuss the emitters of bipolar transistors, where the value of the total injected current is more important. Here, in the case of diodes, as we shall soon see, knowledge of the exact value is not usually important since it is normally the center N region current which will dominate. We shall therefore use Eq. (3.40) in the present chapter and omit the effect of recombination expressed by Eq. (3.43a).

3.4.3 The Terminal Current in the P$^+$NN$^+$ Diode

We are going to assume (a) that the current crossing the P$^+$N junction is entirely hole current and that (b) the current crossing the NN$^+$ junction is entirely electron current. These assumptions may be readily justified for a wide range of practical devices if we start our analysis by jumping ahead and using the end result for hole current due to recombination in the central N region. This latter current may be estimated by recognizing that the hole current diffusing into the N region from the P$^+$N junction will (with the assumption of zero hole current at $x = W_n$) be due to recombination in the N region of width W_n. If the injected hole concentration is $p'(0)$, then the current due to the charge in the central region recombining with a carrier lifetime τ_p is approximately (see Fig. 3.4)

$$J_p = qp'(0) \frac{W_n}{\tau_p} \tag{3.44}$$

Now, from the result just obtained in the previous section for the electron current J_n injected into the P diffused region, the ratio of J_n/J_p is

$$\frac{J_n}{J_p} = \frac{n'(0)}{p'(0)} \frac{D_n \tau_p}{W_p W_n} \frac{W_p/x_0}{\exp(W_p/x_0)} \tag{3.45}$$

For a forward-biased junction, as we have seen previously, the depletion-layer region tends to a narrow thickness and the doping levels become approximately equal on both sides; it follows that $n(0)$ becomes comparable to $p(0)$. For a typical P diffused region, W/x_0 is of order 2 to 3. The diffusion coefficient in the P^+ region has an effective value of about 3 to 10 cm²/s. Carrier lifetimes in the lightly doped N region vary typically from a few microseconds to a few tens of microseconds. Substitution of these values into the above ratio shows that it will normally (but not always) be less than unity. For example, if W_e is 2 μm and W_n is 10 μm with D_n equal to 3 cm²/s, the ratio is approximately 1/50. We are therefore justified, for a large range of structures, in stating that the current crossing the P^+N junction is predominantly hole current.

For the second assumption, let us consider the ratio of hole current J_p injected into the left side of the N region, Fig. 3.4, to hole current $J_p(W^-)$ injected into the N^+ substrate using Eqs. (3.26b) and (3.44):

$$\frac{J_p}{J_p(W^-)} = \frac{p(0)}{p(W^-)} \frac{W_n L_p^+}{L_p^2} \frac{D_p}{D_{p+}} \frac{N_{D\,\text{eff}}^+}{N_{\text{epi}}} \tag{3.46}$$

For $W_n < L_p$, the hole concentration will not alter significantly over the distance W_n, and therefore $p(0) \sim p(W^-)$. Consequently, for typical values of diffusion lengths and region widths, it is clear that the ratio [Eq. (3.46)] of hole current at the left side to hole current at the right side is often much greater than unity. Our original assumptions are thus valid for a wide range of diode structures. If the two assumptions concerning the values of currents crossing the two junctions are not valid, this means simply that the terminal current will be somewhat higher than our calculated value.

As an aside, it is worth noting at this point that "exact" values of current are seldom required in designing a diode. This is because: (i) it is seldom possible during manufacture to accurately control all the process parameters and variables such as recombination lifetime, and diffused layer doping profiles can vary significantly from batch to batch; (ii) in applications, the circuit designer will seldom use a diode with a forced voltage source—current will normally be determined by the circuit via resistors. Good *estimates* of current for a given voltage are all that are normally sought at the design stage.

Let us now proceed to an "exact" solution for current flow in the center N region, using the above conditions. The hole continuity equation is

$$\frac{1}{q}\frac{dJ_p}{dx} = -\frac{p'}{\tau_p} \tag{3.47}$$

Assuming that only diffusion current exists (there being no evident electric field to create hole drift current), and substituting $J_p = -qD_p\,dp'/dx$ gives

$$\frac{d^2 p'}{dx^2} = \frac{p'}{D_p \tau_p} \tag{3.48}$$

With the conditions existing at $x = 0$ and $x = W_n$

$$J_F = J_{px=0} = -qD_p\left.\frac{dp}{dx}\right|_{x=0} \tag{3.49}$$

and

$$\left. \frac{dp}{dx} \right|_{x=W_n} = 0 \tag{3.50}$$

the solution for p as a function of x is [2]

$$p'(x) = \frac{L_p J_F}{q D_p} \frac{\cosh\ [(W_n - x)/L_p]}{\sinh\ (W_n/L_p)} \tag{3.51}$$

From the equation for hole current at $x = 0$,

$$J_F = q D_p \frac{dp}{dx}$$

we obtain the expression for hole current at $x = 0$

$$J_F = \frac{q D_p p'(0)}{L_p} \tanh \frac{W_n}{L_p} \tag{3.52}$$

For

$$W_n \ll L_p \tag{3.53}$$

this simplifies to

$$J_F = q D_p p(0) \frac{W_n}{L_p^2} = q p(0) \frac{W_n}{\tau_p} \tag{3.54}$$

This is, of course, the original expression which we assumed would be a good estimate for the hole current injected into the center N region of the P^+NN^+ diode. It follows that the I-V characteristic is similar to Eq. (3.9) with

$$I_p = \frac{q A p_{n0} W_n}{\tau_p} \left[\exp\left(\frac{V_a}{V_t}\right) - 1 \right] \tag{3.55}$$

3.5 HIGH FORWARD BIAS

3.5.1 High Current Operation of the P^+N Narrow Base Diode

Let us now consider what happens when the forward bias voltage V_a is high enough that the low level injection condition derived in Eq. (3.18) is violated; that is, $p(x) \gg N_D$. Figure 3.5 shows the situation now existing. The space-charge neutrality condition now becomes:

$$n(x) = N_D + p(x) \sim p(x) \tag{3.56}$$

Since the electron concentration $n(x)$ is now much greater than the background doping level over most of the N region (in order to maintain charge neutrality) the conductivity of this region is increased. It is said to be

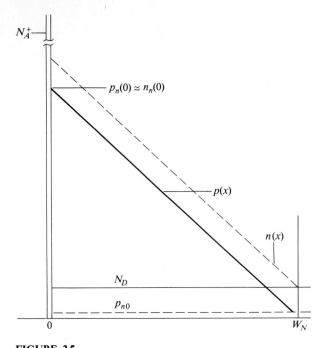

FIGURE 3.5
The P^+N narrow base diode under high level injection: doping profile and carrier concentration distributions.

conductivity-modulated and the terms *conductivity modulation* and *high level injection* are often used interchangeably to denote this situation in different devices. Assuming that the highly doped P^+ region remains in low level injection, we can write

$$p_p(0) = N_A \tag{3.57}$$

or, using the generalized law of the junction [Eq. (2.66)]

$$\frac{N_A}{p_n(0)} = \exp\left(\frac{V_{jt}}{V_t}\right) \tag{3.58}$$

Note that, on the assumption of a narrow depletion-layer width, as usual we use $x = 0$ as the reference point for both neutral regions. The total junction voltage V_{jt} may be split into two components

$$V_{jt} = V_{bi} - V_{aj} \tag{3.59}$$

where V_{aj} is the fraction of the applied bias voltage V_a which appears across the depletion layer (part of V_a is now "lost" as a potential drop in the neutral N region as we shall see). Equation (3.58) can thus be rewritten by combining Eqs.

(2.24) and (3.58) as

$$p_n(0) = p_{no} \exp\left(\frac{V_{aj}}{V_t}\right) \tag{3.60}$$

The total voltage drop in the neutral N region is no longer negligible and can be evaluated by remembering that in a P^+N diode, the current is predominantly hole current. Setting the electron current to zero as in Sec. 3.2.1 gives a Boltzmann-type relation:

$$E(x) = -V_t \frac{1}{n} \frac{dn}{dx} \tag{3.61}$$

Integrating from 0 to W_n gives the neutral region voltage drop V_{an}:

$$V_{an} = V_t \ln\left(\frac{p_n(0)}{N_D}\right) \tag{3.62}$$

Combining Eqs. (3.59), (3.60), and (3.62) we obtain

$$V_a = V_{an} + V_{aj} = V_t\left[\ln\left\{\frac{p_n(0)}{N_D}\right\} + \ln\left\{\frac{p_n(0)}{p_{no}}\right\}\right] \tag{3.63}$$

Combining terms and substituting $p_{no} = n_i^2/N_D$, we thus have the result

$$V_a = 2V_t \ln\left[\frac{p_n(0)}{n_i}\right] \tag{3.64}$$

In other words, the injected carrier concentration $p_n(0) = n_n(0)$ is given by

$$p_n(0) = n_n(0) = n_i \exp\left(\frac{V_a}{2V_t}\right) \tag{3.65}$$

We note the very important property of this diode (and others) operating under high level injection, that the injected carrier concentration is a function only of the intrinsic carrier concentration n_i and not of the doping level. This is perhaps not too surprising a result, since the background doping is "swamped" by the now much higher injected free carrier concentration. Furthermore, we note the change in the exponential law (or slope on a semilog plot) from a $1/V_t$ to a $1/2V_t$ behavior (see Fig. 3.6). This is symptomatic of high level injection in nearly all PN junction devices.

In order to obtain the total current under high level injection conditions, we combine the hole diffusion and drift terms, using Eq. (3.61) for the electric field to obtain

$$J_p = -qD_p \frac{dp}{dx} + q\mu_p pE \tag{3.66}$$

$$= -qD_p\left[\frac{dp}{dx} + \frac{p}{n}\frac{dn}{dx}\right] \tag{3.67}$$

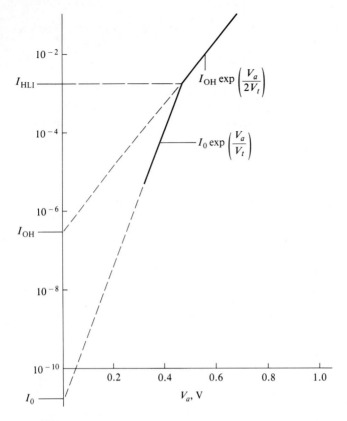

FIGURE 3.6
I-V characteristics of P^+N diode showing low level and high level injection regions.

From Fig. 3.5 it is clear that since the hole and electron concentrations are nearly equal near the junction on the N side, this may be simplified to

$$J_p = -\frac{q(2D_p)p_n(0)}{W_n} \qquad (3.68)$$

Note that this resembles the normal diffusion current, with a multiplying factor of 2; in fact, the current is now made up of equal magnitudes of diffusion and drift current.

The high level injection part of the *I-V* characteristic is obtained by combining Eqs. (3.65) and (3.68) to give:

$$I = I_{OH} \exp\left(\frac{V_a}{2V_t}\right) \qquad (3.69)$$

where

$$I_{OH} = \frac{2qAD_p n_i}{W_n} \qquad (3.70)$$

Figure 3.6 shows the forward biased *I-V* characteristic, with the two regions—low level and high level injection. It is instructive to note that the current corresponding to the onset of high level injection I_{HLI} may be written as

$$I_{\text{HLI}} = \frac{4qAD_p N_D}{W_n} \tag{3.71}$$

and that the following simple relation applies:

$$I_0 I_{\text{HLI}} = I_{\text{OH}}^2 \tag{3.72}$$

This assumes, of course, that the high level value of D_p is the same as the low level value. This is not strictly true, since carrier-carrier scattering will reduce the mobility (and hence the diffusion coefficient) at high injection levels.

3.5.2 The PIN Diode

This structure is perhaps the most widely used of all the diodes in discrete component form. It is actually a P^+NN^+ structure operating under high level injection conditions; but because this is the way it is normally used, and also because the analysis is quite straightforward, we treat it here as a separate device. It finds applications varying from high voltage rectifiers to microwave limiters.

Figure 3.7 shows the relevant diagram of carrier concentration distributions. We now have a forward biased P-I junction and a forward biased I-N junction. Injection will now occur equally on both sides of the center I region. In

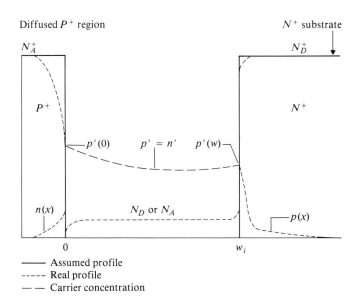

— Assumed profile
---- Real profile
— — Carrier concentration

FIGURE 3.7
Carrier distributions in the PIN diode.

the following treatment, we will use $n = p$ for excess carrier concentrations, since the analysis is only valid for high level injection in any case. The continuity equation for electrons is

$$\frac{dJ_n}{dx} = \frac{qn}{\tau} \tag{3.73}$$

and for holes

$$\frac{dJ_p}{dx} = -\frac{qp}{\tau} \tag{3.74}$$

where, from the lifetime model discussed in Chap. 1, for high level injection

$$\tau = \tau_{n0} + \tau_{p0} \tag{3.75}$$

After substituting the expressions for drift and diffusion currents into the above, we obtain:

$$-\frac{n}{\tau} + \mu_n E \frac{dn}{dx} + \mu_n n \frac{dE}{dx} + D_n \frac{d^2 n}{dx^2} = 0 \tag{3.76}$$

$$-\frac{p}{\tau} - \mu_p E \frac{dp}{dx} - \mu_p p \frac{dE}{dx} + D_p \frac{d^2 p}{dx^2} = 0 \tag{3.77}$$

For high level injection, as in the P^+N diode, the carrier concentrations are much greater than the background doping level so we have $p(x) = n(x)$ for charge neutrality. The approximation is even better for this diode as can be seen from the diagram, because there is no "sink" at the right-hand side. The above two equations may thus be combined, eliminating the electric field term, to give

$$\frac{p}{\tau} = D \frac{d^2 p}{dx^2} \tag{3.78}$$

where D is known as the ambipolar diffusion coefficient:

$$D = \frac{2 D_p D_n}{D_p + D_n} \tag{3.79}$$

Using arguments similar to those used in Sec. 3.4 for the P^+N junction of the P^+NN^+ diode, it is easy to show that the ratio of currents crossing the PI junction is

$$\frac{J_n}{J_p} = \frac{D_n \tau}{W_p W_i} \frac{W_p / x_0}{\exp(W_p / x_0)} \frac{n(0)}{p(0)} \tag{3.80}$$

This ratio is typically (but not always) less than unity; for example, if $W_i = 50 \ \mu m$, $W_p = 5 \ \mu m$, $\tau = 1 \ \mu s$, and $D_n = 3 \ cm^2/s$, and if we assume $n(0) \sim p(0)$, then we have a ratio of order $1/10$. Note that the assumption of negligible electron current crossing the P^+N junction is only used to simplify the analysis. A greater ratio will mean simply that the terminal current will be underestimated by our analysis.

For the IN^+ junction, the ratio of hole to electron current, assuming once again a "wide base diode" substrate as in Sec. 3.4, is

$$\frac{J_p}{J_n} = \frac{qD_{p+}\,p(W^+)/L_{p+}}{qn(W_i)W_i/\tau} \tag{3.81}$$

$$= \frac{D_{p+}\,\tau}{L_{p+}\,W}\frac{p(W^+)}{n(W_i)} \tag{3.82}$$

Applying the relationship for total voltage across the IN^+ junction [Eqs. (2.75) and (2.76)], we have

$$V_{lh} = V_t \ln\left[\frac{N_{eff}^+}{n(W_i)}\right] = V_t \ln\left[\frac{p(W_i)}{p(W^+)}\right] \tag{3.83}$$

Notice that now this low-high voltage is no longer constant (as it was in the P^+NN^+ treatment of Sec. 3.4) but depends now on the carrier concentration in the neutral I region. Furthermore, since $n(W_i) = p(W_i)$ the ratio of hole-to-electron current at the IN^+ junction is

$$\frac{J_p}{J_n} = \frac{D_{p+}\,\tau}{L_{p+}\,W_i}\frac{p(W_i)}{N^+} \tag{3.84}$$

It is left to the reader to show once again that this ratio is typically quite a bit less than unity. We conclude from the above treatment that it is fair to continue the analysis assuming (i) zero electron current at the P^+I junction and (ii) zero hole current at the IN^+ junction, although noting that these conditions may quite easily be violated and recognizing that the effect of such a violation would be to increase the current to a value above that which we will derive.

Using J_F to denote the terminal current, equating the electron current to zero at $x = 0$ gives

$$\left.\frac{dp}{dx}\right|_{x=0} = -\frac{J_F}{2qD_p} \tag{3.85}$$

and equating the hole current to zero at $x = W_i$ gives

$$\left.\frac{dp}{dx}\right|_{x=W_i} = \frac{J_F}{2qD_n} \tag{3.86}$$

With these boundary conditions, the solution of Eq. (3.78) can be shown [2] to be

$$p(x) = \frac{LJ_F}{2q}\left[\frac{\cosh\,(x/L)}{D_n\,\sinh\,(W_i/L)} + \frac{\cosh\,[(W_i - x)/L]}{D_p\,\sinh\,(W_i/L)}\right] \tag{3.87}$$

where we have used $L = \sqrt{D\tau}$. For $W_i \ll L$ this simplifies to

$$p(x) = \frac{L^2 J_F}{qDW_i} \tag{3.88}$$

The hole distribution thus becomes independent of position for the case of center I layers whose width is small compared to a diffusion length. This is similar to the P^+NN^+ diode, where the charge distribution also became "rectangular" for narrow center layers. In this case, the current density may be expressed as

$$J_F = \frac{Q}{\tau} \tag{3.89}$$

where Q is the free carrier charge in the I region $= qp(0)W_i$.

As mentioned above, the applied voltage V_a splits equally between the two junctions P^+I and IN^+. From the above expression for the low-high junction voltage, we have

$$V_{lh} = V_t \ln \left[\frac{N_{eff}^+}{n(W_i)} \right] \tag{3.90}$$

This is made up of the difference between the fixed built-in barrier voltage $V_t \ln [N_{eff}^+/n_i]$ and one half of the applied voltage $V_a/2$. In other words, we have

$$V_t \ln \left[\frac{N_{eff}^+}{n(W_i)} \right] = V_t \ln \left[\frac{N_{eff}^+}{n_i} \right] - \frac{V_a}{2} \tag{3.91}$$

from whence we obtain the equation for injected carrier concentration as a function of applied bias voltage

$$n(W_i) = n_i \exp \left(\frac{V_a}{2V_t} \right) \tag{3.92}$$

Since $n(x) = p(x)$ for all x in the center I layer, it follows that this is also the value for $p(W_i)$, $n(0)$, $p(0)$.

The final I-V law for this PIN diode is thus given by

$$I = I_0 \left[\exp \left(\frac{V_a}{2V_t} \right) - 1 \right] \tag{3.93}$$

where

$$I_0 = \frac{qAn_i W_i}{\tau} \tag{3.94}$$

and where we have introduced the -1 term to satisfy the necessary condition that zero current flows when the applied voltage is zero (a necessary addition to the equation because in writing all the equations in this treatment, we assumed high level injection and implicitly equated total concentrations to excess concentrations).

In fact, in a real device, zero center layer doping does not exist in silicon. The PIN device will always revert to the P^+NN^+ characteristics at low forward bias and the background doping level N_{epi} will then determine the I-V characteristic.

A word is in order here concerning the voltage drop across the center I region at high currents. This may be estimated quite simply by calculating the

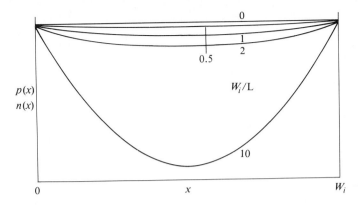

FIGURE 3.8
Carrier concentration distributions in the center layer for a PIN diode for different values of W_i/L.

voltage drop V_i that would occur due to a drift current I flowing in the region whose conductivity σ is given (on the assumption of a "rectangular" charge distribution) by

$$\sigma_i = q\mu n \qquad (3.95)$$

The value of V_i due to the I-R drop is ($V_i = AI/\sigma_i W_i$):

$$V_i \sim \frac{qAn(0)W_i}{\tau} \frac{W_i}{qA\mu n(0)} \qquad (3.96)$$

$$= V_t\left(\frac{W_i}{L}\right)^2 \qquad (3.97)$$

It is clear that for center layer widths less than a diffusion length, this drop can for all practical purposes be neglected (the carrier lifetime in a moderately doped layer of silicon is typically between 1 and 30 μs, giving diffusion lengths between about 30 and 200 μm). In fact, this result brings to light one important design requirement of such diodes. Poor forward bias characteristics will be obtained if the center layer thickness W_i is much greater than the high level injection diffusion length. In the limit where W_i is more than about five diffusion lengths, the structure will begin to behave as two wide-base diodes connected in series, with the associated high resistance and the center layer voltage drop higher than that given by Eq. (3.97) because the carrier concentration is on average now considerably less than $n(0)$. Figure 3.8 shows carrier concentration distributions for different values of W_i/L.

3.6 LOW FORWARD BIAS IN DIODES (SPACE-CHARGE RECOMBINATION)

Under low forward bias conditions, one source of current dominates in all silicon diodes, namely, space-charge recombination in the PN junction. Since the PN

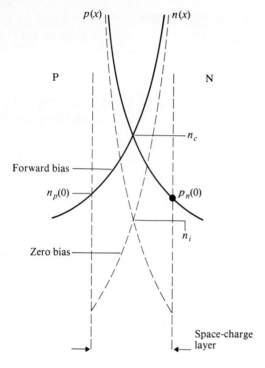

FIGURE 3.9
The space-charge region of a low forward biased linear graded PN junction and the free carrier concentration distribution.

junction space-charge region is almost identical in all the diodes considered so far, one treatment suffices to determine this low forward bias behavior. Let us consider the charge distribution shown in Fig. 3.9 for a forward biased linear graded junction.

As we have already seen in Sec. 2.4, the free carrier concentration distribution in the space-charge region of a linearly graded PN junction can be approximated by an exponential decay on either side of the metallurgical junction given, for holes on the N side, from Eqs. (2.28), (2.44), and (2.45) by

$$p(x) = p_c \exp\left(-\frac{x}{x_c}\right) \tag{3.98}$$

where

$$x_c = \frac{4}{3} \frac{V_t d_n}{V_{bi} - V_a} \tag{3.99}$$

where we have extended the previous result to arbitrary forward bias so that the total voltage across the junction is

$$V_{tot} = V_{bi} - V_a \tag{3.100}$$

In Sec. 2.6.1, we derived the generalized law of the junction and showed that the *pn* product at any position within the space-charge layer is given by Eq. (2.70)

$$n(x)p(x) = n_i^2 \exp \left(\frac{V_a}{V_t} \right) \tag{3.101}$$

In other words, at the metallurgical junction $x = 0$, the equal hole and electron concentrations are given by

$$n_c = p_c = n_i \exp \left(\frac{V_a}{2V_t} \right) \tag{3.102}$$

The minority carrier hole charge on the N side within the space-charge region is found by integration from 0 to d_n. Since we have already shown that d_n is considerably greater than the characteristic length x_c, it follows that the result of this integration will give:

$$Q_p = qA \int_{x=0}^{d_n} p_c \exp \left(-\frac{x}{x_c} \right) dx \sim qA \int_0^\infty p_c \exp \left(-\frac{x}{x_c} \right) dx \tag{3.103}$$

$$= qAp_c x_c \tag{3.104}$$

If we assign some "effective" carrier lifetime τ_{DE} to these minority carriers, we can write the current due to hole recombination from the charge continuity equation as

$$I_{pr} = \frac{Q_p}{\tau_{DE}} \tag{3.105}$$

Adding the corresponding current due to electrons recombining on the P side and combining Eqs. (3.102), (3.103), and (3.104) gives the total space-charge recombination current:

$$I_{scr} = I_{pr} + I_{er} = \frac{2qAn_i}{\tau_{DE}} \frac{4}{3} d_n \frac{V_t}{V_{jt}} \exp \left(\frac{V_a}{2V_t} \right) \tag{3.106}$$

$$= I_{OR} \exp \left(\frac{V_a}{2V_t} \right)$$

where $V_{jt} = V_{bi} - V_a$. The above picture is somewhat simplified. However, if we compare the result with the rigorous treatment of Sah [3] we find that for the case of recombination centers at the intrinsic level (see the discussion on the Shockley-Read-Hall recombination model in Chap 1), the above result is correct if we define τ_{DE} as $2\sqrt{\tau_{p0} \tau_{n0}}$. Since the individual lifetime parameters are rarely, if ever, known for a given process in the vicinity of the junction, our result is quite adequate for nearly all engineering purposes.

Let us examine now the relative magnitudes of this new current compared to the previously derived terminal current due to diffusion in the neutral region. We shall use the P^+NN^+ diode as an example. Let r_{sc} be the ratio of normal

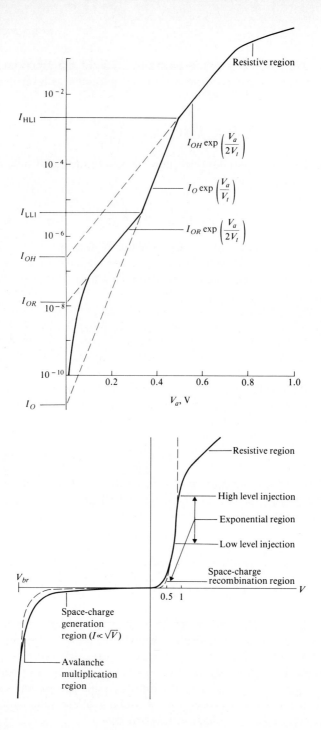

FIGURE 3.10
I-V characteristics showing the three regions: low current space charge recombination, medium current (ideal) range, high level injection: (*a*) log-linear scale; (*b*) linear-linear scale (ideal case shown by broken curve).

78

neutral region diffusion current to space-charge recombination current. From Eq. (3.55) and the above result, we obtain

$$r_{sc} = \frac{J_{p\,neut}}{J_{sc}} \tag{3.107}$$

$$= \left(\frac{3}{8}\right)\left(\frac{\tau_{DE}}{\tau_{p0}}\right)\left(\frac{n_i}{N_D}\right)\left(\frac{W_n}{d_n}\right)\left(\frac{V_{jt}}{V_t}\right)\exp\left(\frac{V_a}{2V_t}\right) \tag{3.108}$$

At zero bias, using typical values of $\tau_{p0} = 20$ μs, $\tau_{DE} = 0.1$ μs, $W_n = 20$ μm, $d_n = 0.1$ μm, $n_i = 10^{10}$ cm^{-3}, $N_D = 10^{16}$ cm^{-3}, $V_{jt} = V_{bi} = 0.7$ V, we obtain

$$r_{sc} \sim 10^{-4} \tag{3.109}$$

This means that the neutral region diffusion current will be less than the space-charge recombination current until $\exp(V_a/2V_t) \sim 10^4$ or, in other words, until V_a reaches a value of about 0.4 V. This is typical of many silicon diodes.

It is interesting to note that the space-charge recombination current obeys the same "law" as high level injection current. Figure 3.10 shows typical I-V characteristics where it can be seen that in some cases the "ideal" $\exp(V_a/V_t)$ will never be observed, i.e., when $I_{OR} > I_{OH}$. It should also be noted that according to Sah's theory, space-charge recombination current can have a value proportional to $\exp(V_a/mV_t)$ where m can have values other than 2. This can occur, for example, over a finite range of bias if the recombination centers are not at the midgap level. Fortunately, from an applications point of view, the exact behavior at very low forward bias is not normally important in diodes, and we shall not reconsider space-charge recombination current until dealing with low current gain falloff in bipolar transistors.

We may note also in passing that because of the n_i term in the numerator of Eq. (3.108), for materials like germanium where n_i is of order 10^{13} at room temperature, the value of r_{sc} is not likely to fall below unity; the same applies to silicon at high temperatures. Under these conditions, the low current space-charge recombination current characteristic will not usually be observed.

3.7 REVERSE BIAS

In a reverse biased diode, the theory already developed indicates a constant current I_0 [(Eqs. (3.9) and (3.55)]. This theory was based on diffusion current in the neutral regions (including neutral region recombination in the wide base, P$^+$NN$^+$, and PIN diodes). In fact, a process analogous to space-charge recombination just considered occurs under reverse bias and usually dominates in silicon devices—this is called *space-charge generation*. Referring to Fig. 3.11, and recognizing that the law of the junction tells us that the total carrier concentration is depressed to zero at each space-charge layer boundary for voltages more negative than a few kT/q [see Eq. (3.101)], it is clear that the carrier concentration is less

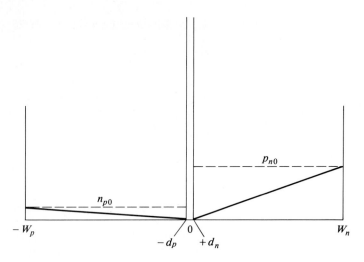

FIGURE 3.11
Carrier distributions in a reverse biased narrow base diode.

than n_i inside the depletion layer. The dopant atoms have had their free carrier contributions removed (the reason for this being a "depletion layer") by the field, and therefore the thermally generated carriers n_i create a generation current at a rate

$$G = -U = -\frac{n_i}{\tau_g} \tag{3.110}$$

Using the complete Shockley-Read-Hall model from Chap. 1 gives the same result and enables us to define the generation lifetime as

$$\tau_g = \frac{n_1 \tau_{p0} + p_1 \tau_{n0}}{n_i} \tag{3.111}$$

This is approximately $2\tau_{p0}$ if we assume $\tau_{p0} \sim \tau_{n0}$ and $n_1 \sim p_1 \sim n_i$. Once again, these parameters are seldom known for a fabrication process and it is customary to obtain the equivalent generation lifetime directly from a reverse current measurement after fabrication and then to assume that it will characterize that particular process.

The reverse current is now obtained simply by integrating the generation over the width of the depletion layer

$$I_g = qA \int_{d_p}^{d_n} U \, dx \tag{3.112}$$

$$= -\frac{qAn_i(d_n + d_p)}{\tau_g} \tag{3.113}$$

This may be compared to the normal minority carrier diffusion current. Taking the P^+NN^+ diode as an example, we can again define a ratio of r_{gd} as the reverse bias neutral region diffusion current to the generation current

$$r_{gd} = \frac{J_{p\,diff}}{J_g} = \frac{qp_{n0}\,W_n/\tau_p}{qn_i\,d_n/\tau_g} \tag{3.114}$$

which simplifies to

$$r_{gd} = \frac{n_i}{N_D}\,\frac{W_n}{d_n}\,\frac{\tau_g}{\tau_{p0}} \tag{3.115}$$

We have neglected the depletion-layer width d_p on the heavily doped P^+ region, since under reverse bias this will normally be much smaller than the lightly doped N side depletion-layer width d_n. For a typical diode, with the values above in Sec. 3.6, and assuming a zero bias depletion-layer width d_n of 1 μm, with $\tau_g \sim \tau_{p0}$, we see that the ratio is of order $1/10^5$.

In fact, it is clear that basically the same terms are involved as in the determination of the ratio r_{sc} in the previous section. This is not too surprising, since the generation process is characterized by the same lifetime model as the recombination process.

This generation current will vary slowly with bias because of the dependence on depletion-layer width, which increases with the square root of the total voltage $V_{bi} - V_a$ where V_a is now negative. This slow increase will continue until avalanche multiplication starts to have an effect.

3.8 AVALANCHE MULTIPLICATION AND BREAKDOWN IN DIODES

In a space-charge region where the field is fairly high for substantial reverse bias, free carriers can be accelerated by the electric field to a point where they acquire sufficient energy to create additional carriers when collisions occur. In order to acquire this energy there are two requirements: firstly, the electric field must be high enough; secondly, there must be a sufficient distance for the carriers to be accelerated to a high enough velocity. The process of generation of extra carriers during a collision is referred to as impact ionization, leading to avalanche multiplication, and the overall increase in current is described by an avalanche multiplication factor. For example, referring to Fig. 3.12, a hole current entering a space-charge region with a value I_{h0} at $x = 0$ travels a distance W (the depletion-layer width) and acquires a new value $M_h I_{h0}$; similarly for electrons, entering at W with a current I_{e0} and exiting at $x = 0$ with a new value $M_e I_{e0}$. To calculate the values of M_h and M_e in a given situation, the electron and hole ionization rates α_h and α_e must be known. The α_e and α_h values are defined as the number of electron-hole pairs generated by an electron (hole) per unit distance traveled.

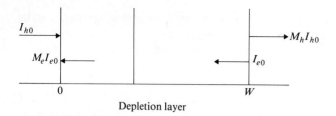

Depletion layer

FIGURE 3.12
The avalanche multiplication process.

The relation between incremental current and distance for electrons is

$$dI_e = I_e \alpha_e \, dx + I_h \alpha_h \, dx \tag{3.116}$$

This can be rearranged in the form

$$\frac{dI_e}{dx} - (\alpha_e - \alpha_h)I_e = \alpha_h I \tag{3.117}$$

where I is the total current (sum of $I_e + I_h$).

Considering the case where the electron current enters the depletion layer at $x = 0$ with a value $I_e(0)$ and exits at $x = W$ with a value $M_e I_e(0) = I_e(W)$, we can obtain the relation [4]

$$\int_0^W \alpha_e \exp\left[-\int_0^x (\alpha_e - \alpha_h) \, dx'\right] dx = 1 - \frac{1}{M_e} \tag{3.118}$$

A similar equation can be written for hole current entering at $x = W$.

Once the electric field versus distance within the depletion layer is known (for the uniformly doped case, we have obtained the analytic solution), the ionization integral given by Eq. (3.118) can be evaluated numerically. The condition for which M_e becomes infinite gives the breakdown voltage of the junction $[I_e(W) = \infty \times I_e(0)]$.

Several sets of ionization coefficient data have been published which enable the multiplication factor M to be computed from a solution of Poisson's equation giving the field E versus distance x. An empirical law has been established and the coefficients measured by Van Overstraeten and De Man [5] using Chynowith's empirical formula [6]:

$$\alpha_e = a_e \exp\left(-\frac{b_e}{E}\right) \tag{3.119}$$

$$\alpha_h = a_{h1} \exp\left(-\frac{b_{h1}}{E}\right) \qquad \text{for} \qquad E \leq 4 \times 10^5 \text{ V/cm} \tag{3.120}$$

$$\alpha_h = a_{h2} \exp\left(-\frac{b_{h2}}{E}\right) \qquad \text{for} \qquad E > 4 \times 10^5 \text{ V/cm} \tag{3.121}$$

where the constants are given in the following table (units of 10^5); a_e, a_{h1}, a_{h2}, are in units of cm^{-1}; b_e, b_{h1}, b_{h2}, are in units of V/cm.

a_e	b_e	a_{h1}	b_{h1}	a_{h2}	b_{h2}
7.03	12.31	15.82	20.36	6.71	16.93

The ionization coefficients may also be calculated using Crowell and Sze's empirical approximation to Baraff's theory [7], fitting the data of Lee et al. [8] with slightly modified values for mean free path for electrons α_e and holes α_h given by Conradi [9].

The expressions thus obtained at room temperature (300 K) for silicon are

$$\alpha_e = \exp\left(\frac{\chi_e}{\lambda_e}\right) \qquad \alpha_h = \exp\left(\frac{\chi_h}{\lambda_h}\right)$$

where

$$\lambda_e = 0.637 \times 10^{-6}\ \text{cm} \qquad \lambda_h = 0.411 \times 10^{-6}\ \text{cm}$$

and

$$\chi_e = -0.0257\left(\frac{0.251 \times 10^7}{E}\right)^2 - 0.325\left(\frac{0.251 \times 10^7}{E}\right) - 0.252\ \text{cm}$$

$$\chi_h = -0.0257\left(\frac{0.389 \times 10^7}{E}\right)^2 - 0.325\left(\frac{0.389 \times 10^7}{E}\right) - 0.252\ \text{cm}$$

Figure 3.13(a) gives computed values of M for various doping levels for a one-sided abrupt junction; both the above sets of ionization coefficients are used and compared with the simple empirical law

$$I = MI_0 \tag{3.122}$$

where

$$M = \frac{1}{1 - (V/V_{br})^n} \tag{3.123}$$

It is seen that a value of n close to 2 or 3 gives a similar shape to the computed curves.

For many applications, it is not necessary to evaluate the ionization integral. It turns out that breakdown occurs for a field at the junction (where, as we have already seen, it has its peak value) E_{br} whose value varies only slowly with doping levels as shown in Fig. 3.13(b). For a moderately doped silicon junction (10^{14} to 10^{16} cm^{-3}), this breakdown value of peak field is $E_{br} \sim 3 \times 10^5$ V/cm. An empirical relation for a wider range of doping levels for an abrupt one-sided junction is

$$E_{br} = 2.5 \times 10^5\left\{\frac{1 + [(\log_{10} N_{epi} - 14)^3]}{14}\right\} \tag{3.124}$$

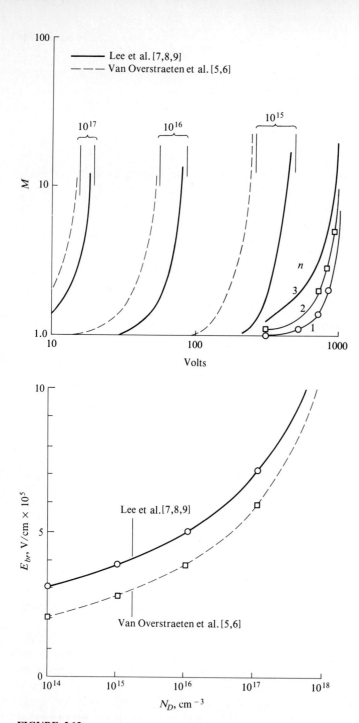

FIGURE 3.13
(a) Values of multiplication factor M versus voltage computed using two sets of published ionization coefficients; also shown are the results obtained using the simple empirical law for $n = 1, 2, 3$ for 1000 V breakdown; (b) breakdown value of electric field at the junction for a one-sided abrupt junction for the two sets of ionization data.

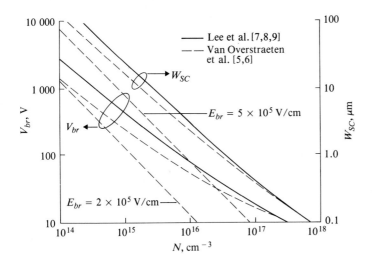

FIGURE 3.14
Computed breakdown voltage V_{br} and depletion layer width W_{sc} for a one-sided abrupt junction using two sets of ionization coefficients and using the simple empirical law [Eq. (3.125)].

The analytic solution to Poisson's equation for the abrupt junction [Eqs. (2.16) and (2.17)] thus enables one to determine approximately the breakdown voltage V_{br} by setting the peak value of electric field E_{max} equal to the breakdown value E_{br}

$$V_{br} = \frac{\varepsilon}{2q} \frac{E_{br}^2}{N_{epi}}$$
(3.125)

where N_{epi} is the (epitaxial) doping level on the lightly doped side.

Figure 3.14 shows computed breakdown voltages versus doping levels for the one-sided abrupt junction using the two sets of ionization coefficients discussed above; for comparison, the results obtained with the simple empirical formula are also shown.

If the diode is heavily doped on both sides of the junction, the depletion layer is too narrow for enough energy to be acquired due to acceleration, and the impact ionization and avalanche multiplication process no longer determines breakdown. For values of peak field at the junction above 10^6 V/cm, breakdown is determined by direct disruption of the covalent bonds. This is the Zener mechanism and is dominant for breakdown voltages less than 6 V; this corresponds to doping levels on both sides of the junction above about 10^{18} cm^{-3}.

For a linearly graded junction, we can combine Eqs. (2.39) and (2.40) to relate the peak value of electric field to the total junction voltage; this gives

$$V_{br} = \frac{4}{3}\left(\frac{2\varepsilon}{q}\right)^{1/2} E_{br}^{3/2} a^{-1/2}$$
(3.126)

For typical gradients in the range 10^{21} to 10^{22} cm^{-4}, the value of E_{br} is close to 4×10^5 V/cm, increasing for larger gradients.

It should be noted that breakdown voltage calculations, even solving numerically the ionization integral for a given impurity profile, can seldom be used for exact design purposes. This is because the ionization coefficients themselves are not "absolute" constants, but depend to some extent on the material. Some workers report that De Man's data gives better agreement for industrial grade silicon, whereas the data based on the work of Lee, Sze et al. is valid for high purity silicon. Furthermore, the above treatment assumes one-dimensional space charge; most real diodes have a two-dimensional region at the edge of the diode, which leads to breakdown voltages lower than the theoretical maximum for the vertical impurity profile. This point will be discussed further when we study real diode devices.

The above discussion relates to avalanche multiplication. If *both* sides of a PN junction are heavily doped, a second breakdown mechanism comes into play—the Zener or tunnel effect. For the case of heavy doping on both sides of a junction, the depletion layer is narrow and the carriers do not acquire sufficient energy to create significant avalanche multiplication. Zener breakdown then dominates. This can be thought of as direct disruption of the covalent bonds and has a temperature coefficient which is negative (opposite to that for avalanche breakdown). Zener breakdown comes into play when the electric field exceeds about 10^6 V/cm. Using this constant value for high doping levels enables breakdown voltages to be calculated using the empirical formulas given above. Figure 3.15 shows the *I-V* characteristics as the doping level on both sides is progressively increased. A case of particular interest occurs when the doping level is of order 10^{19} cm^{-3} on both sides of the junction. A simple calculation shows that the breakdown voltage and built-in barrier voltage are each close to one volt. This means that if a reverse bias of even a tenth of a volt is applied, breakdown will occur. For forward bias, on the other hand, the electric field is decreased and the normal *I-V* characteristic is obtained. Since, for small applied bias, this structure appears to conduct more readily in the reverse direction than in the forward direction, it is referred to as a *backward diode*.

As the doping level increases still further, the electric field due to the built-in barrier potential exceeds the breakdown value. Even small forward bias is not enough to bring the diode out of the breakdown region, and a large current will flow for both forward and reverse bias as shown in curve (iii) of Figure 3.15. If the forward bias is further increased, the value of electric field (proportional to $V_{bi} - V_a$) decreases below the breakdown value and the current drops, before increasing due to injection with an approximately exponential characteristic. This is the tunnel diode characteristic and may be understood in terms of the energy band diagram shown in Fig. 3.16. For the high doping levels under consideration, both sides of the junction are degenerate and the Fermi level lies in the conduction and valence bands for the N and P material, respectively. We recall that the space-charge region is now extremely narrow because of the high doping, as given by Eq. (2.18). Thus there exists a large (of order E_g) but very narrow barrier.

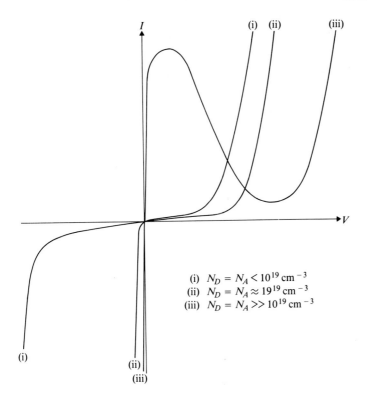

FIGURE 3.15
I-V characteristics for a symmetrically doped abrupt junction (i) doping level below 10^{19} cm^{-3}, Zener breakdown; (ii) doping level close to 10^{19} cm^{-3}; (iii) doping level of order 10^{20} cm^{-3}, tunnel diode characteristics.

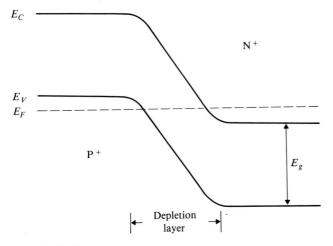

FIGURE 3.16
Band diagram for tunnel diode.

Band-to-band tunneling can take place where an electron near the Fermi level in the conduction band is opposite (at a similar energy level) empty states in the valence band, and vice versa. Application of bias in either direction can therefore cause current to flow. This situation exists until the forward bias reduces the tunneling current toward zero and normal injection eventually occurs as shown by the double-valued I-V characteristic of Fig. 3.15 [curve (iii)].

3.9 CONCLUSIONS

In this chapter we have examined the dc characteristics of a range of diode structures, ranging from the ideal narrow base diode to the PIN diode. We have noted the theoretical behavior in different ranges of bias, from reverse breakdown to reverse space-charge generation; low forward bias where space-charge recombination current dominates, to the "ideal" medium forward current region, to finally high current where high level injection dominates. We have shown that the current voltage law is always of the form $I = I_0 \exp(V_a/mV_t)$ but the physical significance of the I_0 and m values depends on the region of operation.

In the next chapter we examine high frequency and transient behavior of diodes; this is followed in Chap. 5 by a discussion of practical diodes for various applications.

PROBLEMS

3.1. Consider the current in a narrow base diode given by Eqs. (3.8) and (3.10). Calculate the total current by adding these two contributions and compare with Eq. (3.11) for W_n/L_p values equal to 0.3, 1.0, and 3.0. The current can be expressed in normalized form with respect to $qD_p p(0)/W_n$. Sketch $p(x)$ for the three cases.

3.2. Consider a P^+N diode with a wide N region. Derive an expression for the ohmic voltage drop in the N region assuming $W_n \ll L_p$, in terms of V_t, $p(0)/N_D$, and (W_n/L_p). Calculate the value of voltage applied across the junction V_a and the additional voltage drop in the W_n region for a bias corresponding to the onset of high level injection for a diode with $N_D = 10^{16}$ cm^{-3}, $W_n = 10L_p$.

3.3. Equation (3.15) gives the electric field versus distance for the neutral region of a narrow base diode. Use this equation to derive an expression for the normalized space-charge density $(\varepsilon/qN_D)\,dE/dx$. Show that for bias up to the high level injection limit, this quantity is much less than unity and that hence the assumption of a "neutral" region is excellent.

3.4. Equation (3.27b) was derived for the effective recombination velocity at a low-high junction. Consider $N_{epi} = 10^{15}$ cm^{-3} and four values of doping in the N$^+$ region: 10^{17}, 10^{18}, 10^{19}, and 10^{20} cm^{-3}. Calculate values for the recombination velocity using Eqs. (1.36), (1.37), and (1.38) for recombination, with $\tau_0 = 1$ μs, $N_{ref} = 10^{17}$ cm^{-3}, $C_N = 10^{-31}$ cm^6 s^{-1} and using Eq. (1.22) for mobility. Perform the calculations first neglecting the effect of bandgap narrowing $(N_{eff} = N_D)$ and then using Slotboom's formula for ΔE_g given by Eq. (1.34).

3.5. Use Eqs. (3.40) and (3.43) to calculate the ratio of J_{nr}/J_{total} for a diffused P region with the following data: $N_0 = 10^{16}$ cm^{-3}, $N_{\text{eff}} = 3 \times 10^{18}$ cm^{-3}, $D_n = 5$ cm^2/s, $\tau_n = 100$ ns, x_0 values of 0.2 μm, 1.0 μm, and 5.0 μm.

3.6. (a) Examine the range of validity of $J_n/J_p < 1$ in Eq. (3.45) for the current injected into a diffused layer compared to the current injected into the center layer. Reasonable ranges for the quantities involved are: 30 μm $< L_p <$ 140 μm, 0.3 μm $< W_p <$ 10 μm, 1 μm $< W_n <$ 100 μm, $n(0)/p(0) = 1$.

  (b) Examine the range of validity of the condition given by Eq. (3.46) for current at either end of the center layer. In the heavily doped substrate, assume a carrier lifetime of 0.1 μs and a diffusion coefficient of 5 cm^2/s. Let W_n vary from 1 to 100 μm.

3.7. (a) Calculate and sketch the I-V characteristic on a semilog graph for a P$^+$NN$^+$ diode of area 10^{-4} cm^2, $W_n = 10$ μm, $N_{\text{epi}} = 10^{15}$ cm^{-3}, $\tau_p = \tau_n = 1$ μs. Include both low level and high level injection regions.

  (b) For the same diode, complete the plot for low current space-charge recombination using three values of τ_{DE}: 0.01 μs, 0.3 μs, and 10 μs.

3.8. Derive the ambipolar diffusion equation (3.78) from the preceding two equations.

3.9. Examine the range of conditions for which the hole to electron current ratio given by Eq. (3.84) is less than unity. Consider W_i values equal to 1, 10, and 100 μm and use N$^+$ equal to the effective doping value of 3×10^{18} cm^{-3}. Assume $\tau = 10$ μs, $D_{p^+} = 5$ cm^2/s.

3.10. Equations (3.125) and (3.126) tell us that for junctions which are highly doped on both sides, or have a large doping gradient, the breakdown voltage decreases. Calculate for both cases the conditions (doping level, gradient) for which the breakdown voltage is close to one volt (use $E_{br} = 10^6$ V/cm). Since the built-in barrier voltage increases (albeit slowly) with doping level in this situation, it is clear that "breakdown" can occur before any reverse bias is applied. Sketch the I-V characteristic for this case. If the doping level (or gradient) is further increased, it is apparent that the field can be reduced below the "breakdown" value only by the application of forward bias. Although the current-voltage behavior under these conditions depends on quantum mechanical tunneling, some idea of the I-V law can be obtained by calculating the forward bias which brings the diode out of the "breakdown" condition. Sketch this law for the case of symmetric doping of 10^{20} cm^{-3}.

REFERENCES

1. J. Lindmayer and C. Y. Wrigley, *Fundamentals of Semiconductor Devices*, Van Nostrand-Reinhold, New York (1965).
2. R. C. Varshney and D. J. Roulston, "Transient behavior of a range of P$^+$NN$^+$ diodes with narrow center regions," *Solid State Electronics*, vol. 13, pp. 1081–1095 (1970).
3. C. T. Sah, R. N. Noyce, and W. Shockley, "Carrier generation and recombination in PN junctions and PN junction characteristics," *Proc. IRE*, vol. 45, pp. 1228–1243 (September 1957).
4. S. Sze, *Physics of Semiconductor Devices*, 2d ed., John Wiley (1981).
5. R. van Overstraeten and H. de Man, "Measurement of the ionization rates in diffused silicon PN junctions," *Solid State Electronics*, vol. 13, pp. 583–608 (May 1970).
6. A. G. Chynowith, "Ionization rates for electrons and holes in silicon," *Phys. Rev.*, vol. 109, pp. 1537–1540 (1958).

7. C. R. Crowell and S. M. Sze, "Temperature-dependence of avalanche multiplication in semiconductors," *Appl. Phys. Lett.*, vol. 9, pp. 242–244 (1966).
8. C. A. Lee et al., "Ionization rates of holes and electrons in silicon," *Phys. Rev.*, vol. 134, pp. A761–A773 (1964).
9. J. Conradi, "Temperature effects in silicon avalanche diodes," *Solid State Electronics*, vol. 17, pp. 99–106 (1974).

CHAPTER
4

TRANSIENT AND HIGH FREQUENCY BEHAVIOR OF DIODES

4.1 INTRODUCTION

This chapter explains the operation of various types of diode when driven by voltage or current steps. Approximate methods are used so that the reader gains a clear insight into the transient operation. More exact solutions are presented where it is felt necessary for engineering purposes. The chapter starts with an analysis of the narrow and wide base diodes under classical "current-switching" conditions. This is followed by a study of the P^+NN^+ and PIN diodes under normal switching conditions. The chapter then deals with the PIN step recovery or snap-off phenomenon, which occurs under high forced current conditions. The effect may be used to advantage in microwave applications (frequency multiplication and division); it can be a serious disadvantage in high voltage rectifier circuits. We continue with a brief outline of the open circuit voltage decay method for lifetime measurement because of its interest for lifetime extraction. The chapter concludes with a treatment of the diode under sinusoidal drive conditions: firstly for small signals (with a discussion of the small signal equivalent circuit), secondly for the large signal situation.

4.2 NARROW BASE DIODE

4.2.1 General Considerations

Let us consider the behavior of an ideal narrow base P^+N diode when a forward bias voltage is applied to its terminals at some time $t = 0$ as shown in Fig. 4.1(a). The current waveform and carrier concentration at some instant t_1 after applying the voltage is shown in Fig. 4.1(b),(c). If we neglect voltage drops in the neutral region and external circuit, the applied voltage will appear instantaneously across the space-charge layer giving rise to a step increase in the injected carrier concentration to a value $p(0)$. If t_1 is much less than t_B, the diffusion transit time of the

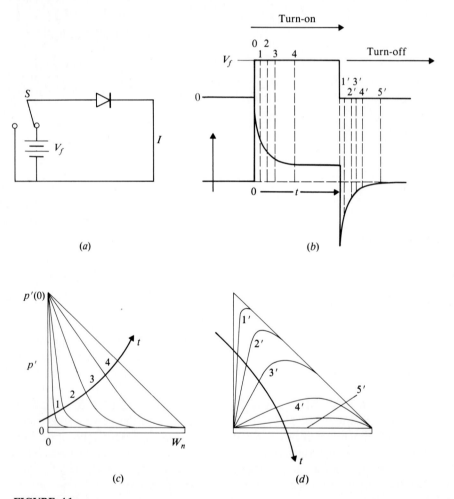

(a) (b)

(c) (d)

FIGURE 4.1
(a) Circuit for voltage switching; (b) current waveform in the P^+N narrow base diode; (c) carrier concentrations versus distance during turn-on; (d) carrier concentrations versus distance during turn-off.

carriers from 0 to W_N, the carrier concentration must decay to zero at a very short distance from the junction as depicted in curve 1 of Fig. 4.1(c). A short time later the excess carriers will have diffused toward the ohmic contact. For a time t much greater than the transit time t_B, steady state conditions will be reached. We thus have a succession of curves as shown in Fig. 4.1(c).

Since the terminal current is equal to the minority carrier diffusion current at the edge of the space-charge layer on the N side, it is quite obvious that the current will start at some very high positive value (it will be infinite in the ideal, but nonrealizable case, of zero series resistance) and decay to the steady state value after a time $t \gg t_B$. This behavior is illustrated in Fig. 4.1(b). Forcing the voltage abruptly back to zero gives a high negative current transient due to the reversed concentration gradient at the space-charge layer boundary as shown in Fig. 4.1(d).

In practice of course, there will always be finite series resistance due to both generator, circuit, and diode neutral regions. In fact, it is perfectly feasible to "force" a current step through a diode. This case is illustrated in Fig. 4.2(a) where, provided the voltage V is large compared to the final steady state forward

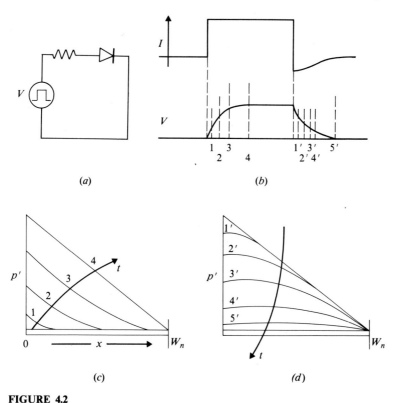

(a)   (b)   (c)   (d)

FIGURE 4.2
(a) Circuit for "current" drive (voltage source in series with resistance); (b) current and voltage waveforms; (c) carrier concentration versus distance during turn-on; (d) carrier concentration versus distance during turn-off (voltage source reduced to zero).

bias across the diode terminal, the step current waveform (b) is obtainable. In this case the constant current condition implies a constant slope for the minority carrier concentration at the space-charge layer boundary. We therefore obtain the situation shown in Fig. 4.2(c) where, again, the steady state triangular distribution is reached at some time $t > t_B$. Note that if the source voltage is reduced to zero, the current reverses due to the finite forward voltage already across the diode (of order 0.7 V), as shown in Fig. 4.2(b),(d).

The most important turn-off condition from a practical viewpoint is that depicted in Fig. 4.3. Here, we force a reverse current through the diode by reversing the source voltage to some large (compared to the forward voltage drop

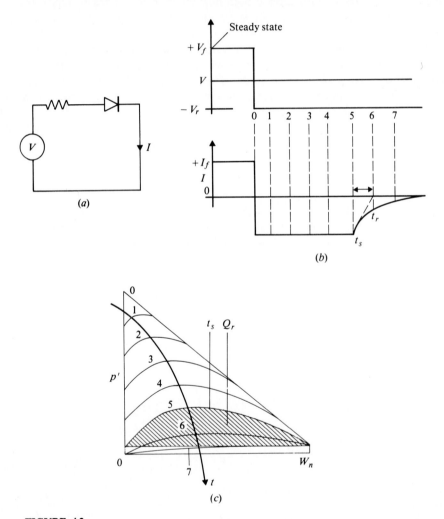

FIGURE 4.3
(a) Circuit for forced current during turn-off; (b) voltage and current waveforms; (d) carrier concentration versus time during constant current turn-off.

across the diode) negative value, and extract the excess stored charge. The slope of the carrier concentration at the space-charge layer boundary will now be reversed in direction (due to the current reversal) but since it takes a finite time t_s for $p'(0)$ to become zero, the current will remain approximately constant for a finite time (comparable to t_B if $I_f \approx -I_r$). When $p'(0) = 0$ the total terminal voltage across the diode is zero (from the results of Sec. 2.7).

As the voltage becomes negative, the carrier concentration at $x = 0$ tends very rapidly to a value much less than the thermal equilibrium value p_{n0}. However, the current cannot reach its reverse saturation value I_0 instantaneously since the residual charge at time t_s must first be evacuated. Also the junction capacitance C and circuit resistance R form a time constant which affects this part of the transient. In the narrow base diode, as we shall see below, the "recovery time" t_r (cf Fig. 4.3) is also comparable to t_B if $I_f \approx -I_r$. In most cases of interest, the currents will be many orders of magnitude greater than the reverse saturation current, and we will henceforth neglect the difference, on a linear scale, between p_{n0} and zero.

4.2.2 Approximate Theoretical Derivations

Consider the simplified triangular representation of instantaneous charge distribution shown in Fig. 4.4.

We wish to determine the reverse constant current storage time t_s and the reverse recovery time t_r. Let us assume a steady state current I_f to exist at time $t = 0$. The current is switched instantaneously to a reverse value $-I_r$. We know that the slope (dp/dx) at $x = 0$ is the same for all times $0 < t < t_s$ and is determined by the relation

$$I_r = -qAD_p \frac{dp}{dx} \tag{4.1}$$

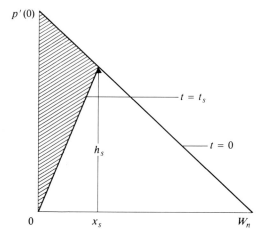

FIGURE 4.4

Approximate carrier concentration versus distance in narrow base region at time $t = t_s$.

At time t_s, defined as the time at which the excess carrier concentration becomes zero at $x = 0$, the following expression for the current is evident from Fig. 4.4:

$$I_r = -qAD_p \frac{h_s}{x_s} \tag{4.2}$$

By considering the original steady state distribution we may write

$$\frac{h_s}{W_n - x_s} = \frac{p'(0)}{W_n} \tag{4.3}$$

Eliminating h_s from Eqs. (4.2) and (4.3) and substituting

$$I_f = \frac{qAD_p p'(0)}{W_n} \tag{4.4}$$

gives

$$x_s = \frac{W_n}{1 + I_r/I_f} \tag{4.5}$$

The charge removed during this time, shown as the hatched area on Fig. 4.4, is given approximately by

$$Q_1 \approx I_r t_s \tag{4.6}$$

and by

$$Q_1 = \tfrac{1}{2} qAp'(0)x_s \tag{4.7}$$

Combining Eqs. (4.6), (4.7), and (4.5) gives the storage time t_s

$$t_s = t_B \left[\frac{I_f/I_r}{1 + I_r/I_f} \right] \tag{4.8}$$

where

$$t_B = \frac{W_n^2}{2D_p} \tag{4.9}$$

This is the diffusion transit time for holes across the N region. Since current is charge flowing per unit time, it is trivial to show that the same expression for t_B is also given by Q_F/I_F where Q_F is the total stored charge under forward bias. This expression will be increasingly accurate for small values of t_s, that is, large values of I_r, since recombination will be negligible under these conditions (let us recall that by definition, in the narrow base diode, $t_B \ll \tau_p$ where τ_p is the minority carrier lifetime) and because in the above analysis we neglect the redistribution of the carriers to the right of the peak value in Fig. 4.4 during evacuation of the charge.

For the reverse recovery, or decay, phase $(t > t_s)$ we have to extract a residual charge Q_r given by

$$Q_r = Q_s - Q_1 \tag{4.10}$$

where Q_s is the original steady state stored charge. This gives:

$$Q_r = t_B \left[\frac{I_r}{1 + I_r/I_f} \right]$$ (4.11)

If we approximate the decay by an exponential

$$i = I_r \exp \left(-\frac{t}{t_r} \right)$$ (4.12)

with t measured from the time t_s, we can write

$$Q_r \approx \int_0^\infty I_r e^{-t/t_r} \, dt$$ (4.13)

$$= I_r t_r$$ (4.14)

This enables us to express the recovery time as

$$t_r = \frac{Q_r}{I_r}$$ (4.15a)

The reverse recovery time in terms of diffusion transit time is therefore

$$t_r = t_B \frac{1}{1 + I_r/I_f}$$ (4.15b)

As for the storage phase, Eq. (4.15b) becomes increasingly valid for large values of reverse current I_r. Kingston [1] gives an exact solution for the narrow base diode, in the form of an infinite series. For the storage time t_s the above results are within 20 percent of the exact results for $I_r/I_f \geq 1.0$. For the recovery time, Kingston's results are defined in terms of the time to decay by 90 percent of the constant reverse current value. The reported values using this definition tend to $0.5t_B$ for $I_r/I_f \leq 1.0$, to $0.14t_B$ for $I_r/I_f = 10$, and to $0.006t_B$ for $I_r/I_f = 100$. The very simple theory used here is thus remarkably useful for estimating approximate storage and recovery times using Eq. (4.8) for $I_r/I_f \geq 1.0$ and Eq. (4.15b) for a very wide range of current ratios.

The depletion-layer capacitance will also affect the recovery phase through the RC time constant formed by the circuit resistance and junction capacitance. This will not normally be important if the diode is driven at high currents, but can dominate the recovery phase if the current associated with the neutral region is small. The charge on the junction capacitance C_j at a reverse voltage V_R may be derived by integrating the product CV using Eq. (2.53)

$$Q_c = 2C_j(V_R)V_R$$ (4.16)

Provided $Q_c \ll Q_r$, this effect can be neglected. Otherwise it may be included as an additional time constant. The effect on measurement of the forward time constant is discussed in [11].

4.3 WIDE BASE DIODE

Although the wide base diode does not correspond to any "useful" device, we include it here because it represents an asymptotic case which enables us to have a very clear insight into the operation of lifetime-dependent diodes in general. We shall restrict the discussion to (a) a forward current step, (b) the reverse storage and recovery phases.

Figure 4.5 represents the same case as that considered for the narrow base diode, Figs. 4.2(a) and 4.3(a), where the diode is fed from a voltage generator of magnitude much greater than the steady state forward voltage drop across the diode, with a resistance R in series.

For a forward current step, the excess carrier concentration rises as shown, with a constant slope

$$\left(\frac{dp}{dx}\right)\bigg|_{x=0} = -\frac{I_F}{qAD_p} \tag{4.17}$$

at the space-charge layer boundary. We may use the charge continuity equation (see App. 6) directly

$$I = \frac{dQ}{dt} + \frac{Q}{\tau_p} \tag{4.18}$$

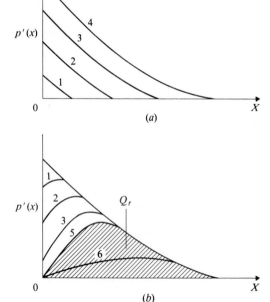

(a)

(b)

FIGURE 4.5
P$^+$N wide base diode: current driven carrier concentration curves (a) during turn-on; (b) during turn-off.

The general solution of Eq. (4.18) for a constant current is

$$Q(t) = I\tau_p + c \exp\left(-\frac{t}{\tau_p}\right) \tag{4.19}$$

Substituting the conditions

$$I = I_f \tag{4.20}$$

$$Q = 0 \quad \text{at} \quad t = 0 \tag{4.21}$$

we find

$$c = -I_f\tau_p \tag{4.22}$$

The charge, as a function of time, is thus given by

$$Q(t) = I_f\tau_p\left[1 - \exp\left(-\frac{t}{\tau_p}\right)\right] \tag{4.23}$$

Notice that after a time $t \ll \tau_p$, steady state conditions are reached and we have

$$Q_f \to I_f\tau_p \tag{4.24}$$

We can obtain approximate information about the voltage across the diode by assuming

$$Q(t) \approx qAp'(0, t)L_p \tag{4.25}$$

This is a very approximate relation, however, and must be used with caution. From Eq. (4.25) and the junction law, we can write the voltage versus time as

$$V_a(t) = V_t \ln\left[1 + \frac{p'(0, t)}{p_{n0}}\right] \tag{4.26}$$

If we now consider the case where the generator voltage (and hence the diode current) abruptly changes sign, we can study the reverse turn-off phase.

The following conditions now apply:

$$I = -I_r \tag{4.27}$$

$$Q = Q(t_1) \tag{4.28}$$

at some time $t = t_1$. This gives

$$Q = I_r\tau_p + Q(t_1) \tag{4.29}$$

or, in other words, the solution of the charge continuity equation (4.19) becomes

$$Q(t) = -I_r\tau_p + [Q(t_1) + I_r\tau_p] \exp\left(-\frac{t}{\tau_p}\right) \tag{4.30}$$

In the special case where steady state forward bias conditions have been attained at $t = t_1$ (i.e., when $t_1 \gg \tau_p$), we have

$$Q(t) = -I_r\tau_p + (I_f + I_r)\tau_p \exp\left(-\frac{t}{\tau_p}\right) \tag{4.31}$$

Note that if we further *assume that all the charge is evacuated at some time* t_s' after switching from a current I_f to a current $-I_r$, and that the *current remains constant* up to this time t_s', we can evaluate t_s' by setting $Q(t)$ in Eq. (4.31) equal to zero. This gives

$$0 = -I_r \tau_p + (I_f + I_r)\tau_p \exp\left(-\frac{t_s'}{\tau_p}\right) \tag{4.32}$$

or

$$\frac{t_s'}{\tau_p} = \ln\left(1 + \frac{I_f}{I_r}\right) \tag{4.33}$$

This gives an approximate solution for the reverse constant current storage time t_s. We know, however, that the voltage V_a will reach zero before all the charge is extracted and that, from this time onward, the current will decay. The assumptions involved in writing Eq. (4.33) are therefore invalid. Kingston [1] solved the continuity equation for this case and his exact solution for t_s may be used:

$$\text{erf}\sqrt{\frac{t_s}{\tau_p}} = \left(1 + \frac{I_r}{I_f}\right)^{-1} \tag{4.34}$$

It may be noted by calculating values of t_s from Eqs. (4.33) and (4.34) that the approximate solution [Eq. (4.33)] becomes increasingly valid for small values of I_r/I_f (cf Fig. 4.5 where, for small values of I_r/I_f, the slope at $x = 0$ will be small and the residual charge becomes negligible).

If we use the exact solution for the constant current storage time given by Eq. (4.34) we may again use the charge continuity equation to solve for the recovery (decay) time by approximating it to an exponential as for the narrow base diode [Eq. (4.12)]. The residual charge Q_r is obtained from Eq. (4.31)

$$Q_r = I_r \tau_p\left[\left(1 + \frac{I_f}{I_r}\right)\exp\left(-\frac{t_s}{\tau_p}\right) - 1\right] \tag{4.35}$$

Using Eqs. (4.13) and (4.14) for the charge Q_r evacuated during an exponential decay enables us to write the reverse recovery time constant as

$$t_r = \tau_p\left[\left(1 + \frac{I_f}{I_r}\right)\exp\left(-\frac{t_s}{\tau_p}\right) - 1\right] \tag{4.36}$$

In addition, there will be a time constant associated with the charge Q_c on the junction capacitance given by Eq. (4.16) [11].

4.3.1   Comparison of Results for Narrow Base and Wide Base Diodes

It is apparent that for the case of $I_f/I_r \approx 1$, that is, reverse currents comparable to forward currents, the narrow base diode has a reverse storage and a decay time each of the order of t_B, the diffusion transit time. In the wide base diode, each of these two times is comparable to the minority carrier *lifetime*, τ_p. Since the narrow base diode has $t_B \ll \tau_p$ (by definition) it is quite obvious that it will give a much shorter total "switching time" than the wide base diode, other parameters being comparable.

In the next section we shall see that for P^+NN^+ structures, the ratio of decay to storage time (t_r/t_s) is controlled by the center layer width of the device but that the sum of the two, i.e., the total switching time, is always comparable to the minority carrier lifetime τ_p. It is therefore the narrow base diode structure (defined as a diode in which the lightly doped region is much shorter than a diffusion length and terminated by a plane of high surface recombination velocity) which is best suited for high speed switching applications. In practice, a high speed switching junction diode will have t_B comparable to τ_p, that is, the narrow width W_N will be accompanied by an "artificial" reduction in the minority carrier lifetime τ_p (for example, by the controlled introduction of gold atoms).

4.4 P^+NN^+ AND PIN DIODES

4.4.1 P^+NN^+ Diodes

As before, we shall assume: (i) the current crossing the P^+N junction is entirely hole current (this implies a heavily doped P^+ region); (ii) the current crossing the NN^+ plane (substrate boundary) is entirely electron current (implying a heavily doped substrate); (iii) the voltage drop across the center N region is negligible. The analysis will be concerned with the application of a forward step of current density I_F followed, after steady state conditions have been reached, by a reverse step of current density $-I_R$ for the duration of the reverse storage phase t_s; after this time, an exponential decay [2] of time constant t_r will be assumed.

Consider the P^+NN^+ device with abrupt P^+N and NN^+ transitions under low level injection conditions. Figure 4.6 represents the excess hole concentration in the homogeneous N region. Under the above stated assumptions, the terminal current during the forward phase is

$$I_F = I_p = -qAD_p\left(\frac{dp}{dx}\right)\bigg|_{x=0} \tag{4.37}$$

For very short times $t < t_c < \tau_p$ [see Eq. (4.39) below where τ_p is the hole lifetime] after application of the step I_F, the hole concentration can be approximated by the triangular distribution shown in Fig. 4.6(a). The area under this distribution represents the instantaneous stored minority carrier charge per unit cross section Q which is also, for short times, given by the product $I_F t$. We can thus write

$$x = \frac{2I_F t}{qAp(0)} \tag{4.38}$$

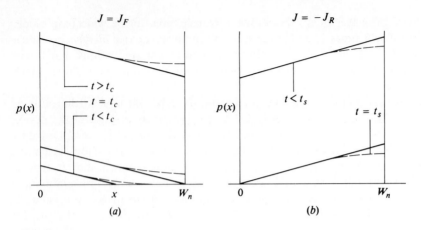

FIGURE 4.6
P^+NN^+ diode approximate carrier concentration curves (a) during turn-on; (b) during turn-off.

It should be noted that the triangular distribution is a fair approximation (i.e., to within a factor of order two) relating Q, t, and I_p at $x = 0$. It is obviously incorrect as far as hole current at the NN^+ boundary is concerned, since $I_p = 0$ at $x = W_n$ for condition (ii) above. This fact does not, however, enter into the analysis. When the hole concentration at $x = W_n$ starts to rise above zero, we can define a time t_c (which in this case is simply the diffusion transit time from $x = 0$ to $x = W_n$) by combining Eqs. (4.37) and (4.38) with $x = W_n$

$$t_c = \frac{W_n^2}{2D_p} = \left(\frac{W_n}{L_p}\right)^2 \frac{\tau_p}{2} \tag{4.39}$$

For $t < t_c$, combining Eqs. (4.37) and (4.38), we have

$$p(0) = t^{1/2} \frac{I_F}{A} q^{-1} \left(\frac{D_p}{2}\right)^{-1/2} \propto t^{1/2} \tag{4.40}$$

$$p(0) = Q^{1/2} \left(\frac{2I_F}{AD_p}\right)^{1/2} q^{-1} \propto Q^{1/2} \tag{4.41}$$

Assuming the usual law to hold across the P^+N depletion-layer boundaries, we have

$$p(0) = p_{n0} \exp\left(\frac{qV}{kT}\right) \tag{4.42}$$

where p_{n0} is the thermal equilibrium hole concentration and V the instantaneous terminal voltage. For times greater than t_c it is apparent that we can write

$$Q \approx qAW_n p(0, t) \tag{4.43}$$

Using the previously stated assumption of zero hole current at $x = W_n$, the charge continuity equation applies

$$I = \frac{dQ}{dt} + \frac{Q}{\tau_p} \tag{4.44}$$

For times greater than t_c we can thus use Eqs. (4.42), (4.43), and (4.44) as a valid set of equations describing the current/voltage/time relations of the device. Under steady state conditions, the current is simply $I_F = qAW_n p(0)/\tau_p$.

For the reverse current step, Fig. 4.6(b), applied after steady state conditions have been attained, we replace I_p by $-I_R$ in Eq. (4.37).

After the time t_s it is apparent from Fig. 4.6(b) and Eq. (4.37) that the residual charge per unit cross section, using the triangular approximation, is

$$Q_r = \tfrac{1}{2}qAp(W_n)W_n = I_R \frac{W_n^2}{2D_p} \tag{4.45}$$

Defining a recovery (decay) phase by $I = -I_R \exp\left(-t/t_r\right)$ we obtain

$$t_r = \frac{W_n^2}{2D_p} \tag{4.46}$$

Note that in most P^+NN^+ diodes $t_r < \tau_p$ (since $t_r/\tau_p = \tfrac{1}{2}(W_n/L_p)^2$). The storage time t_s in Fig. 4.3 is thus given approximately by Eq. (4.33).

$$\frac{t_s'}{\tau_p} = \ln\left(1 + \frac{I_f}{I_r}\right)$$

The approximation is excellent for low voltage microwave structures and starts to fail where the lightly doped layer thickness W_n becomes comparable to the carrier diffusion length L_p as in high voltage rectifiers. In addition, the junction-depletion layer capacitance will influence the total recovery time and for low currents, i.e., small values of charge Q_r, the charge given by Eq. (4.16) must be taken into account.

4.4.2   PIN Diode

For the PIN structure, Fig. 4.7(a) and (b) (which of course corresponds also to the P^+NN^+ device under high level injection conditions), under the same assumptions as before, with $p = n$ (space-charge neutrality condition), we obtain:

$$I_F = I_P = -2qD_p\left(\frac{dp}{dx}\right)\bigg|_{x=0} = 2qD_n\left(\frac{dn}{dx}\right)\bigg|_{x=W_n} \tag{4.47}$$

Considering, for simplicity, the case where $D_n = D_p = D$, we obtain the triangular distributions shown in Fig. 4.7(a). Defining t_c in a similar way to before gives

$$t_c = \frac{W_n^2}{8D} \tag{4.48}$$

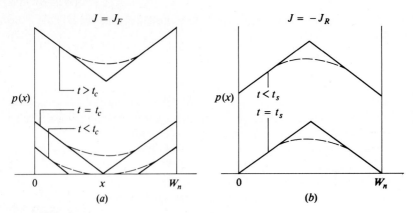

FIGURE 4.7
PIN diode approximate carrier concentrations (a) during turn-on; (b) during turn-off.

It should be noted that in intrinsic material a large transient voltage drop and space-charge exist in the I region immediately after application of the pulse. Restricting our argument, for simplicity, to the case where sufficient time has elapsed for this voltage to become negligible (see below) we obtain for $t < t_c$

$$p(0) = t^{1/2} I_F q^{-1} (2D)^{-1/2} \propto t^{1/2} \tag{4.49}$$

$$p(0) = Q^{1/2} I_F^{1/2} q^{-1} (2D)^{-1/2} \propto Q^{1/2} \tag{4.50}$$

The total terminal voltage is

$$V = \frac{2kT}{q} \ln \left(1 + \frac{p(0)}{n_i} \right) \tag{4.51}$$

For $t > t_c$ it is apparent that Eqs. (4.43) and (4.44) apply with τ_p replaced by the high level lifetime τ. The carrier distribution becomes essentially rectangular for $t \gg t_c$ provided that $t_c \ll \tau$, that is, for $(W_n/L)^2 \ll 8$ where $L = \sqrt{D\tau}$ is the carrier diffusion length.

The reverse storage time is given again by Eq. (4.33) with the appropriate high level injection value of lifetime

$$\frac{t_s'}{\tau} = \ln \left(1 + \frac{I_f}{J_r} \right)$$

For the reverse recovery phase we replace I_F in Eq. (4.47) by $-I_R$. Using the triangular approximation of Fig. 4.7(b) and applying the relation $Q_r = I_r t_r$ where Q_r is determined by the area of the triangle, we find

$$t_r = \frac{W_n^2}{8D} \tag{4.52}$$

This is valid provided the reverse current is not large; see Sec. 4.5. It is clear from the above that the PIN diode is essentially the same as the P^+NN^+ diode with one-half the center layer width. This is because in the PIN structure, carriers are injected from both sides simultaneously.

4.4.2.1 INITIAL VOLTAGE SPIKE AT START OF TURN-ON. For a given current, there will be an initial voltage drop across the center I region of doping level N_D. The value of this voltage is obtained as the product of the current and the resistance of the center layer width W_n. Writing the final steady state current as

$$I = \frac{qAp(0)W_n}{\tau}$$

and the resistance of the center layer at $t = 0$ (no conductivity modulation) as

$$R_i = \frac{W_n}{q\mu_n N_D}$$

the voltage drop V_i at $t = 0$ is

$$V_i = V_t \left[\frac{p(0)}{N_D} \right] \left[\frac{W_n}{L} \right]^2 \tag{4.53}$$

Typically, N_D may be of order 10^{14} cm^{-3}, while the steady state value $p(0)$ may be one hundred times higher. For a diode with $W_n/L \sim 1/3$ we see that the initial value of this center layer voltage drop can be of order 0.25 V. For a final carrier concentration equal to 10^{17} cm^{-3} the voltage drop for this diode will be 2.5 V.

As the center layer charge builds up according to Fig. 4.7(a), the center layer resistance decreases due to conductivity modulation. For a time equal to one-tenth of the carrier lifetime τ, the value of $p(0, t)$ will be approximately one-tenth of the final value $p(0)$; in the above example with final concentrations of 10^{16} and 10^{17} cm^{-3}, the center layer voltage drop will therefore be reduced to 0.025 V and 0.25 V, respectively, at this time. For the greater part of the turn-on transient, it is therefore safe to neglect this voltage V_i. Figure 4.8 shows some calculated curves of the voltage V_i and the forward voltage versus time.

4.4.3 General Remarks

In summary, it is apparent that the two idealized structures considered, each having the common property of a narrow effective center layer, give surprisingly similar results in their transient behavior. The above results have been verified by computer-aided solutions of the semiconductor equations assuming only quasi-space-charge neutrality and give remarkably good agreement with the simple treatment outlined above [2]. In particular, the square law [Eqs. (4.40), (4.41), and (4.49), (4.50)] and linear [Eq. (4.43)] regions agree with computed results and the values of t_c and t_r agree within a factor of about three (typical diodes used in experiments had $t_c \approx \tau_p/100$). It should be noted that since t_c and t_r are small [Eq. (4.39)] for small ratios of W_n/L_p, Eqs. (4.42) and (4.43) may be combined to form a very simple mathematical model for this general class of device

$$Q(t) = I_0 \tau \left[\exp\left(\frac{qV(t)}{mkT} \right) - 1 \right] \tag{4.54}$$

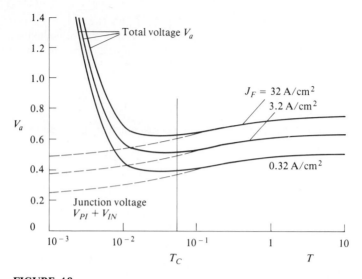

FIGURE 4.8

Center layer voltage drop versus normalized time compared to total voltage in a PIN diode under current turn-on conditions for $W_n = 7 \ \mu$m, $L = 20 \ \mu$m. (*After* [2] *Reprinted with permission from Solid State Electronics, Pergamon Press PLC.*)

This, together with Eq. (4.44), is sufficient to solve for all forward transients involving frequencies less than $(2\pi t_c)^{-1}$. Note that in Eq. (4.54), I_0 and m effectively define the relation between $p(0)$ and V. They may be obtained directly from the dc characteristics of the device for the range of currents for which the model is required (corresponding to high or low level injection). The lifetime τ should also be measured under the correct conditions. This technique, using Eqs. (4.44) and (4.54), has been successfully applied to predict the behavior of various circuits containing step-recovery and similar silicon epitaxial diodes.

4.5 TRANSIENT TURN-OFF BEHAVIOR OF PIN STEP-RECOVERY DIODES

The PIN diode covers a very wide range of applications, as we have already mentioned. In nearly all of these applications the transient turn-off is of some importance. The PIN structure exhibits under certain conditions a rather complicated turn-off behavior toward the end of the "normal" decay time discussed previously (and equal to the diffusion transit time). In order to understand this phenomenon let us consider again the normal "resistive" switching circuit and waveforms shown in Fig. 4.9.

Diagram (c) splits the voltage waveform into several regions. The step 1-2 is due to the reversal of current in the finite diode series resistance. Region 2-3 corresponds to the constant current phase discussed above (note that point 2 and point 3 may be positive or negative depending on the magnitude of the reverse

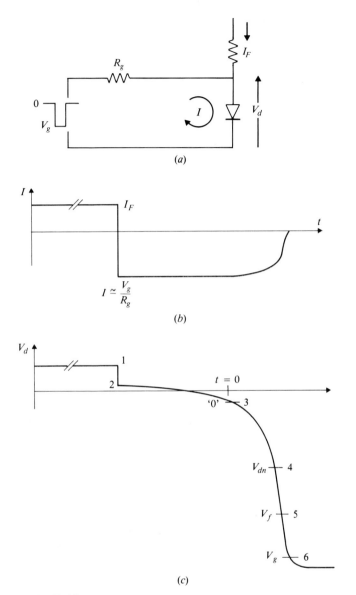

FIGURE 4.9
(a) Circuit for fast step recovery transient; (b) current waveform; (c) voltage waveform. (*After Varsh-ney and Roulston* [3]. *Reprinted with permission from Solid State Electronics, Pergamon Press PLC.*)

current and diode resistance). Region 3 to 4 corresponds to the diffusion transit time, i.e., the evacuation of the residual charge, for the case where the current is held approximately constant, i.e., for a large enough reverse source voltage. Let us examine this phase more closely than we did in Sec. 4.4.

Figure 4.10(a) shows the excess carrier concentration diagram for the case

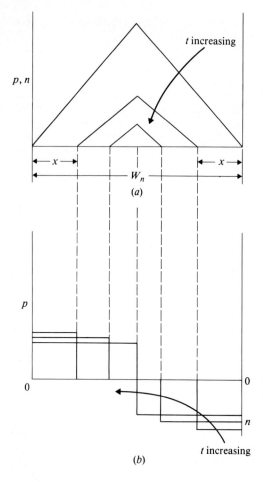

FIGURE 4.10
(a) Neutral ($n = p$) carrier concentration curves during slow evacuation; (b) uncompensated carrier concentrations during slow evacuation. (*After Varshney and Roulston* [3]. *Reprinted with permission from Solid State Electronics, Pergamon Press PLC.*)

of approximately constant current during this part of the decay phase. Whether or not the current can be maintained approximately constant will depend on the circuit and diode parameters (see below). Figure 4.10(b) shows the way in which the charge must build up in the space-charge regions, which advance from each junction toward the center, with time. Assuming that the field is sufficiently high for saturated drift velocity (v) conditions to exist, the space-charge hole concentration must be given by

$$p = \frac{I}{qAv} \tag{4.55}$$

where v is the velocity of the carriers. At the end of the phase corresponding to the evacuation of the residual charge, i.e., after a time t_c (cf Sec. 4.4) where

$$t_c = \frac{W_n^2}{8D} \tag{4.56}$$

we have an approximately rectangular distribution of free carriers in the space-charge region as shown in Fig. 4.10(b), point 4 of Fig. 4.9(c).

During this "slow" transient, Poisson's law in the space-charge region in the left half is simply

$$\frac{dE}{dx} = \frac{qp}{\varepsilon} \tag{4.57}$$

Integrating this twice gives the voltage across each space-charge layer.

Multiplying by two and substituting from Eq. (4.55) gives the total diode voltage

$$V_d = \frac{x^2 I}{\varepsilon v A} \tag{4.58}$$

Notice that x and hence V_d are time-dependent in Eq. (4.58). At the end of this phase (condition 4 in Fig. 4.9), i.e., at a time $t = t_c$, we have

$$x = \frac{W_n}{2} \tag{4.59}$$

and hence the voltage V_{dn} at point 4 in Fig. 4.9(c) is given by

$$V_{dn} = \frac{W_n^2 I}{4\varepsilon v A} \tag{4.60a}$$

This may be written in the form

$$V_{dn} = \frac{W_n I}{4Cv} \tag{4.60b}$$

where C is the capacitance of the diode for large reverse bias, i.e.,

$$C = \frac{\varepsilon A}{W_n} \tag{4.61}$$

Notice that the current through the diode will be approximately constant during this "slow" transient only if the generator voltage V_g is much greater than V_{dn}.

The situation existing between points 4 and 5 of Fig. 4.9 is shown in Fig. 4.11. The free carriers in the space-charge regions are extracted and the field builds up to a uniform value. The total current I during this phase at plane B is a combination of particle current, $I_{particle}$

$$I_{particle} = qApv \tag{4.62}$$

and displacement current I_{disp}

$$I_{disp} = A\varepsilon \frac{dE}{dt} \tag{4.63}$$

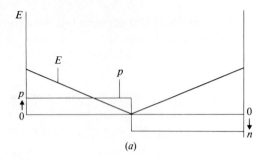

(a)

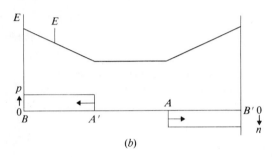

(b)

(c)

FIGURE 4.11
Fast turn-off phase. (a) Space-charge and electric field at start of fast recovery; (b) space-charge and electric field during fast recovery phase; (c) electric field during last part of transient (RC phase). (*After Varshney and Roulston* [3]. *Reprinted with permission from Solid State Electronics, Pergamon Press PLC.*)

so that

$$I = I_{\text{disp}} + I_{\text{part}} \tag{4.64}$$

The algebraic sum of these two currents must equal the terminal current. At the end of a time t_v between points 4 and 5 (beyond the end of the slow transient t_c) the situation depicted in diagram (c) will exist. All the free carriers will have been evacuated. Since these carriers are (we have assumed high field conditions) moving with their saturated drift velocity, the time t_v required to evacuate them from the space-charge layers will be simply the drift transit time, i.e.,

$$t_v = \frac{W_n}{2v} \tag{4.65}$$

From this time onward (i.e., beyond point 5 in Fig. 4.9), the current is entirely displacement current

$$E = \frac{V}{W_n} \qquad I_{\text{disp}} = \frac{A\varepsilon}{W_n}\frac{dV}{dt} = C\frac{dV}{dt}$$

The rate of decay will be determined by the time constant

$$t_d = R_g C \qquad (4.66)$$

where R_g is the external circuit resistance and C [Eq. (4.61)], the diode capacitance (we assume that the depletion layer "reaches through" the I region).

The exact shape of the current decay and voltage rise beyond point 4 of Fig. 4.9 is relatively easy to determine by combining the preceding equations with Kirchoff's law relating the total (time-dependent) diode voltage and current to the fixed generator voltage and circuit resistance.

It is intuitively obvious (and may be rigorously shown) that for the special case

$$R_g C = \frac{W_n}{4v} \qquad (4.67)$$

the fastest overall current decay, point 4 to point 6, Fig. 4.9, is obtained.

Figure 4.12 shows calculated curves of normalized diode voltage V_d/V_g versus normalized time $t_1 = W_n/2v$ for various values of a parameter K defined by

$$K = \frac{4vCR_g}{W_n} \qquad (4.68)$$

It is clear from this figure that the overall fastest transient occurs for $K = 1$, that is, for the condition specified above by Eq. (4.67). We will discuss this step-recovery behavior further when considering practical diodes in the next chapter.

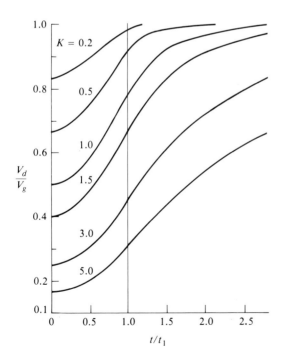

FIGURE 4.12
Normalized diode voltage as a function of normalized time for various values of K [Eq. (4.68)]. (*After Varshney* [3]. *Reprinted with permission from Solid State Electronics, Pergamon Press PLC.*)

As we mentioned, it is an important design problem, to be eliminated in high voltage diode circuits, where it was first observed [4].

4.6   OPEN CIRCUIT VOLTAGE DECAY (OCVD) TRANSIENT

In devices such as high voltage PIN diodes and solar cells, the lifetime in the wide I region is an important parameter. It is frequently obtained by measurement of the voltage decay under open circuit conditions [5]. Figure 4.13 shows the way in which the carrier concentration decays with time in a wide base diode. Note that because of the open circuit condition, the concentration gradient must be zero at the depletion-layer boundaries. From the charge continuity equation, the charge is obtained as a function of time:

$$Q(t) = Q(0) \exp\left(-\frac{t}{\tau_p}\right) \tag{4.69}$$

where the charge is the integral at any time over the whole region. If we make the approximation that this charge is instantaneously proportional to the injected carrier concentration, as in Eq. (4.25), we obtain

$$\frac{p(0, t)'}{p(0, 0)'} = \exp\left(-\frac{t}{\tau_p}\right) \tag{4.70}$$

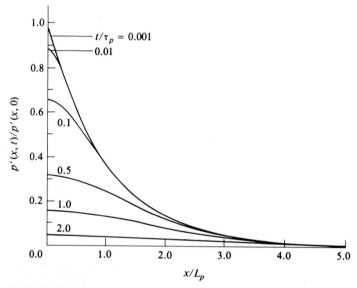

FIGURE 4.13
Variation of carrier concentration with time under open circuit conditions for a one-sided wide base diode. (*After Jain, Heasell, Roulston* [6]. *Reprinted with permission from Progress in Quantum Electronics, Pergamon Press PLC.*)

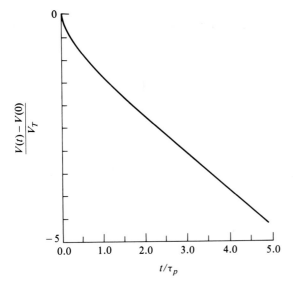

FIGURE 4.14
Open circuit voltage decay for lifetime measurement. (*After Jain, Heasell, Roulston* [6]. *Reprinted with permission from Progress in Quantum Electronics, Pergamon Press PLC.*)

Applying the law of the junction as in Eq. (4.26) gives

$$\frac{V(t) - V(0)}{V_t} = -\frac{t}{\tau_p} \tag{4.71}$$

Thus a plot of voltage versus time enables the carrier lifetime to be determined, as shown in Fig. 4.14.

For this case of the wide base diode, for which the technique was originally used, there are two approximations involved: (i) the charge $Q(t)$ is not exactly proportional to the value of $p(0, t)$; the exact solution has been published and the approximation is quite good provided $\exp [V(t)/V_t] \gg 1$ [5]; (ii) the minority carrier (electron) concentration in the P^+ region is coupled via the Boltzmann relations to the above hole concentration $p(0, t)$; this can give rise to a very different situation, and the "time constant" measured may be quite different from the lifetime in the wide lightly doped region.

Coupling between Regions in Transient Turn-Off

Figure 4.15 shows the situation during open-circuit voltage decay in a P^+N diode with uniform doping on both sides. We will restrict the discussion to low level injection and neglect space-charge recombination. Since, according to Eqs. (2.66) and (2.67), the voltage across the junction must be the same whether derived from hole or electron concentrations, it follows that

$$\frac{n(0^-, t)}{p(0^+, t)} = \frac{N_D}{N_A^+} = \text{const} \tag{4.72}$$

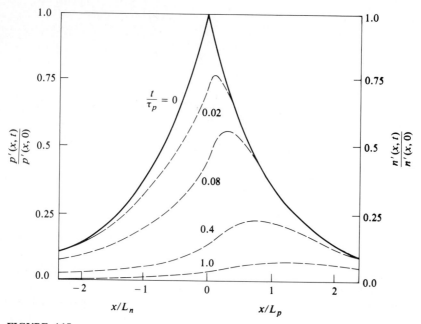

FIGURE 4.15
Coupling effect between carriers in P and N regions during transient turn-off. (*After Jain and Murlidharan* [7]. *Reprinted with permission from Solid State Electronics, Pergamon Press PLC.*)

Furthermore, the zero total current condition corresponding to open-circuit decay implies

$$J_n(0^-) = -J_p(0^+) \tag{4.73}$$

For very short times, these currents will be individually equal to zero. As time elapses, these currents will assume finite values, with the two components just balancing to satisfy Eq. (4.73). This explains the finite gradient on either side of the junction shown in Fig. 4.15. It is clear that the time-dependent decay will be a strong function of the p^+ region if the dc contribution of this region is important.

This coupling effect is of course present in all diodes, and for current switching as well as open circuit voltage decay; measurements of lifetime using these techniques must therefore be treated with extreme caution.

The technique may also be applied to the PIN and P^+NN^+ diodes. In this case, the charge varies with time as shown in Fig. 4.16 for a diffused P^+NN^+ diode. Note the nonzero carrier concentration gradient in the diffused layer. This is due to the retarding field discussed in Sec. 3.4.2 and will persist throughout the transient. The approximation of proportionality between injected minority carrier concentration $p(0, t)$ and total charge $Q(t)$ is even better than for the wide base diode, as was discussed already, so Eq. (4.71) could be expected to apply with good accuracy. The measured time constant will only be the carrier lifetime,

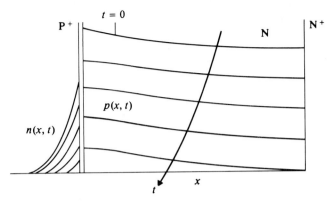

FIGURE 4.16
Carrier concentrations during open circuit decay in a diffused P^+NN^+ diode.

however, if two conditions are satisfied: (i) the coupling effect of the charge in the diffused P^+ region is negligible; (ii) the recombination velocity at the NN^+ interface is small. If these conditions are not satisfied, an equivalent lifetime will be measured and must be interpreted accordingly.

4.7 LARGE SIGNAL SINUSOIDAL HIGH FREQUENCY BEHAVIOR

4.7.1 Diodes under Sinusoidal Drive Conditions

Figure 4.17 shows a diode in series with a resistance and sinusoidal voltage drive, $V_g \sin \omega t$ with a dc bias V_B. For the case where the ac voltage is small enough, the small signal equivalent circuit may be simply derived.

The small signal resistance r is the ratio of ac voltage across the diode to ac current through the diode. This is the same as the inverse conductance g which is the reciprocal of the slope of the dc I-V characteristic. Hence

$$g = \frac{1}{r} = \frac{dI}{dV}$$

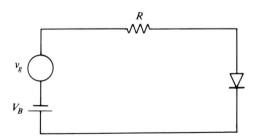

FIGURE 4.17
Circuit for sinusoidal drive.

From the general I-V characteristic discussed in Chap. 3, this may be readily evaluated to give, for $V \gg V_t$,

$$g = \frac{1}{r} = \frac{I}{mV_t} \qquad (4.74)$$

The small signal capacitance is made up of two components, the diffusion and the depletion-layer capacitance. The former is due to the stored minority carrier charge, the latter has already been discussed in Sec. 2.6. The diffusion capacitance is a result of the change of free carrier charge when the applied voltage is altered. For the P^+NN^+ and PIN diodes it is therefore given by

$$C_{\text{diff}} = \frac{dQ}{dV_a}$$

Since the charge is proportional to the dc current I [see Eqs. (3.54) and (3.89)], this may be rearranged in the form

$$C_{\text{diff}} = \frac{dI}{dV_a} \frac{dQ}{dI}$$

$$= g\tau \qquad (4.75)$$

where τ is the appropriate value of lifetime and the small signal conductance is, from Eq. (3.9), etc., equal to I/mV_t. The complete small equivalent circuit of the diode is shown in Fig. 4.18. The series resistance is due to a combination of center layer resistance [in reverse bias, to be discussed see Eq. (5.3)], substrate, and contact resistance.

Although this equivalent circuit may be used for any PN junction diode, the expression for the diffusion capacitance is different for the narrow base and wide base diodes. It may be shown [8] that for the narrow base diode

$$C_{\text{diff}} = \tfrac{2}{3} g t_B$$

where t_B is the diffusion transit time [Eq. (4.9)]. The reason for the factor 2/3 lies in the fact that some of the charge due to the free carriers is evacuated through the ohmic contact when the applied voltage is changed. This is not the case in the P^+NN^+ and PIN diodes where the total charge is "recovered" at each voltage change.

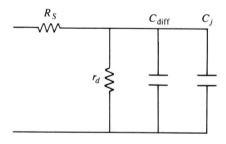

FIGURE 4.18
Small signal equivalent circuit for a junction diode.

For the wide base diode a similar situation exists, and in this case it may be shown [8] that the diffusion capacitance is given by

$$C_{\text{diff}} = \tfrac{1}{2}g\tau$$

where τ is the minority carrier lifetime.

4.7.2 Large Signal Sinusoidal Drive

Figure 4.19 shows the current waveforms for a PIN diode in the circuit of Fig. 4.17 at three different frequencies under large signal drive conditions. At low frequencies, the diode acts as a perfect rectifier (albeit with a small offset voltage determined by the I-V characteristic). As the frequency is increased, the storage time discussed in Sec. 4.4 makes the current flow in the reverse direction for a short time before reducing to zero. At frequencies higher than $1/2\pi\tau$, conduction in the reverse direction lasts almost a complete half period.

Clearly, the value of rectified current decreases at high frequencies—in fact, the diode acts as a capacitance at high enough frequencies. The dependence of rectified current on frequency may in fact be used to determine the carrier lifetime [9].

If a dc reverse bias is added in series with the ac drive as in Fig. 4.17, the situation shown in Fig. 4.20 exists at frequencies much greater than $1/2\pi\tau$ when all of the charge injected during the forward part of the cycle is recovered. This is

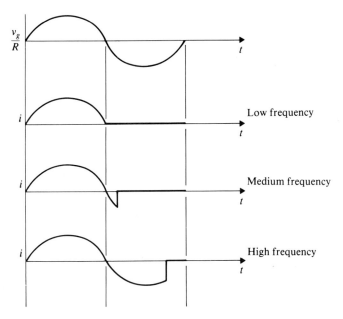

FIGURE 4.19
Current waveforms for large signal sinusoidal drive conditions.

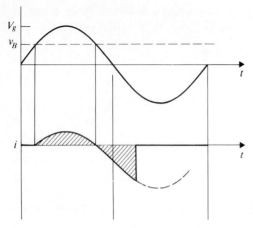

FIGURE 4.20
Current waveform at high frequency in the presence of a reverse bias (cf Fig. 4.17) with $V_g > V_B$.

the basis of the charge storage or step-recovery diode frequency multiplier, although for high efficiency operation a more complicated inductive circuit arrangement is used [10].

4.8 CONCLUSIONS

In this chapter we have examined the transient behavior of diodes. We saw that the narrow and wide base diode behavior during current switching can be explained by fairly simple theory to a useful degree of approximation for engineering purposes. The dominant time constant is the base diffusion transit time $W_p^2/2D_n$, for the N^+P narrow base diode and the carrier lifetime τ for the wide base diode.

We then studied the very important P^+NN^+ and PIN diode switching behavior and saw that the two structures are basically similar, with two exceptions: firstly, the PIN diode can have an initial voltage spike during current turn-on; secondly, the PIN diode exhibits in certain cases a very rapid "step recovery" or "snap-off" turn-off at the end of the reverse current switching phase under "high current" forced drive conditions. Otherwise, both structures are characterized by the carrier lifetime for the "storage phase" and by the diffusion transit time across the center layer for the dominant "decay phase" time constant.

The treatment of transients concluded with a brief overview of the open circuit voltage decay method for obtaining lifetime from measurements. A discussion of the small signal equivalent circuit followed by large signal sinusoidal drive behavior concluded this chapter.

PROBLEMS

4.1. Consider a P^+N diode with an ohmic contact at a distance W_n from the junction on the N side. Assume the following values: $W_n = 50\ \mu m$, $A = 1\ cm^2$, $N_D = 10^{14}\ cm^{-3}$,

$\tau_p = 20$ μs. The diode is fed from a source voltage V_g in series with a resistance $R_g = 1000$ Ω. At time $t = 0$, V_g rises instantaneously from a value V_R to a value $V_F = 20$ V, remains at the value V_F for a time T_{on} and then falls abruptly to a value V_R. Calculate the storage and delay times, using the formulas developed for the narrow base diode, and sketch the voltage waveform across the diode (include junction capacitance effects making suitable approximations) for the following cases: (i) $V_R = -20$ V, $T_{on} = 100$ μs, (ii) $V_R = -20$ V, $T_{on} = 1$ μs, (iii) $V_R = -200$ V, $T_{on} = 100$ μs.

4.2. For a P^+NN^+ diode with the same details as the diode in Prob. 4.1 except for the NN^+ junction replacing the ohmic contact, and for the same circuit arrangement, but with $R_g = 10$ Ω, calculate the significant transient parameters and sketch the current and voltage waveforms for the three cases used in Prob. 4.1.

4.3. A P^+NN^+ diode has $W_n = 10$ μm, $N_D = 10^{14}$ cm^{-3}, $A = 10^{-4}$ cm^2, $\tau_p = 1$ μs. Make any necessary calculations and sketch carefully the diode current and voltage versus time when driven from a step source voltage starting at V_g and changing abruptly to $-V_g$ in series with a resistance R_g for two cases: (i) $V_g = 20$ V, $R_g = 2$ kΩ, (ii) $V_g = 8$ V, $R_g = 800$ Ω, (iii) $V_g = 2$ V, $R_g = 200$ Ω. Calculate also the value of R_g for the fastest step recovery transient.

4.4. Equation (4.9) was stated to be the diffusion transit time of carriers across the region W_n. By considering the diffusion velocity $dx/dt = I/qAp(x)$ in steady state, where $p(x)$ is the triangular distribution discussed in Chap. 3, show by integration that $W_n^2/2D_p$ is rigorously the diffusion transit time.

4.5. For the diode used in Prob. 4.3, consider the application of a current step I_f applied at time $t = 0$. Use Eqs. (4.39) and (4.48) as appropriate with the associated derivations and sketch carefully the carrier distributions at times $t_c/10$, t_c, $10t_c$, $100t_c$ for (i) $I_f = 1$ μA, (ii) $I_f = 1$ mA.

4.6. Consider a P^+NN^+ diode of center layer width $W_n = 10$ μm and carrier lifetime $\tau_p = 1$ μs. Equation (3.27) defined the recombination velocity S_{nn+} at the NN^+ interface. By considering the steady state solutions for carrier distributions used in Chap. 3, estimate approximately the storage time t_s as S_{nn+} varies from 100 to 100 000 cm/s. (Note that this range of recombination velocities could arise either from structural parameter changes or from variations in recombination centers at the low-high interface due to changes in processing conditions, such as substrate surface condition before epitaxial layer deposition.)

4.7. Consider a PIN diode used in a half-wave rectifier circuit. The carrier lifetime in the I region is 1 μs. Sketch the current waveforms and estimate using approximate "ballpark" methods the value of rectified current for a peak voltage of 10 volts in series with 10 ohms, at frequencies of 0.1, 1.0, and 10.0 MHz.

REFERENCES

1. R. H. Kingston, "Switching time in junction diodes and junction transistors," *Proc. IRE*, vol. 42, pp. 829–834 (May 1954).
2. R. C. Varshney and D. J. Roulston, "Transient behaviour of a range of P^+NN^+ diodes with narrow center regions," *Solid State Electronics*, vol. 13, pp. 1081–1095 (1970).
3. R. C. Varshney and D. J. Roulston, "Turn-off transient behaviour of p-i-n diodes," *Solid State Electronics*, vol. 14, pp. 735–745 (1971).
4. H. Benda and E. Spenke, "Reverse recovery processes in silicon power rectifiers," *Proc. IEEE*, vol. 55, p. 1331 (August 1967).

5. S. C. Choo and R. G. Mazur, "Open circuit voltage decay behavior of junction diodes," *Solid State Electronics*, vol. 13, pp. 553–564 (May 1970).

6. S. C. Jain, E. L. Heasell, and D. J. Roulston, "Recent advances in the physics of silicon *p-n* junction solar cells including their transient response," *Progress in Quantum Electronics*, vol. 11, no. 2, pp. 105–204 (1987).

7. S. C. Jain and R. Murlidharan, "Effect of emitter recombinations on the open circuit voltage decay of a junction diode," *Solid State Electronics*, vol. 24, pp. 1147–1154 (December 1981).

8. J. Lindmayer and C. Y. Wrigley, *Fundamentals of Semiconductor Devices*, Van Nostrand-Reinhold, New York (1965).

9. D. J. Roulston and J. V. Hanson, "Lifetime measurements of step recovery diodes using sinusoidal input," *Proc. IEEE*, vol. 57, pp. 1201–1202 (June 1969).

10. D. J. Roulston, "Frequency multiplier using a charge storage diode in an inductive circuit," *Proc. IEEE*, vol. 55, pp. 1220–1221 (July 1967).

11. K. Venkateswaran and D. J. Roulston, "Effect of depletion-layer capacitance on lifetime measurements of P^+NN^+ diodes," *Electronics Letters*, vol. 6, pp. 681–683 (October 1970).

CHAPTER
5

STRUCTURE
AND THEORY
OF PRACTICAL
DIODES

5.1 INTRODUCTION

In this chapter, we consider a range of practical diodes. At one extreme, there are microwave diodes for use in various roles at frequencies of up to tens of GHz and several volts breakdown; the applications include limiters (where the diode acts as a low impedance when forward biased and as a high (capacitive) impedance when zero or reverse biased); varactor frequency multiplication and parametric amplification, both of which utilize the nonlinear capacitance voltage characteristic; the hyperabrupt diode is also considered at this point (although not usually a microwave device, its properties also involve the capacitance voltage characteristic); the step-recovery (snap-off) diode for frequency multiplication or division (using the effect discussed in Sec. 4.5). At the other extreme is the very wide range of low frequency (including 50 or 60 Hz) rectifier or protection structures, often requiring very high breakdown voltages (up to over one thousand volts).

Each of the applications utilizes different structures, either in the choice of vertical impurity profile or horizontal geometry, focusing on several different theoretical properties of the device. In some cases the theory developed in earlier chapters suffices to define an optimal design, while in other cases a modified theoretical approach is necessary in order to obtain the design information in a suitable form. In some cases only qualitative results are available and the design engineer must resort to empirical methods and iterative fabrication steps in order to obtain a structure which satisfies specified performance criteria.

We commence the chapter with two of the simpler structures used in microwave applications—the varactor and parametric diode—and include the hyperabrupt varactor as a logical extension; we then proceed with a discussion of the other microwave structures (including a treatment of the PIN diode limiter insertion loss and the need for an exact solution of Poisson's equation). Attention is then focused on the low frequency (up to several tens of kHz) high voltage rectifier diode including such important design aspects as gold doping to enhance the switching speed, and the use of guard rings for obtaining high breakdown voltages.

5.2 DIODE STRUCTURES

Although planar technology is sometimes used for low performance diodes, invariably a design constraint is the peak inverse voltage, which will be essentially the avalanche breakdown voltage. In order to extract the best possible performance, the sidewall region can be removed. This is usually done using a mesa etch, with subsequent passivation using oxide, plus nitride or phosphorus glass [15]. The final structure is as shown in Fig. 5.1. This mesa technique is widely used for microwave devices; it is also used for some high voltage (power) diodes, but we will discuss the guard ring concept in relation to high voltage diodes later in this chapter. In addition to giving a breakdown voltage which approaches the theoretical limit for bulk one-dimensional breakdown, the mesa structure eliminates the sidewall junction capacitance, an advantage in all cases, since this is essentially an unwanted parasitic. Finally, for cases in which the diode is driven into forward bias, minority carrier charge in the region outside the active device area is eliminated by use of the mesa etch fabrication method.

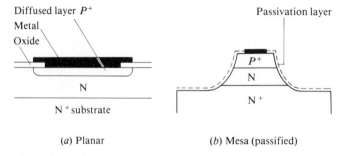

(a) Planar   (b) Mesa (passified)

FIGURE 5.1
(a) The planar diode structure; (b) the mesa diode structure.

5.3 MICROWAVE AND RF DIODES

5.3.1 The Varactor Capacitance-Voltage Law and Series Resistance

Figure 5.2 shows the impurity profile of a diffused P^+N diode on an N^+ substrate. The moderately doped N layer is almost universally formed by epitaxial growth and in this case is typically from 1 to 10 μm in thickness, denoted in the following text by W_{epi}. In Chap. 2, Sec. 2.6, we discussed the way in which the depletion-layer capacitance varies with applied bias; the treatment was given for both the symmetrical abrupt and one-sided abrupt junction, and also for the linearly graded junction. In the case presently under discussion, where the diode is used under reverse or near-zero bias conditions, the one-sided abrupt junction is often (but not always) a good approximation. Figure 5.3 is a diagram of the capacitance-voltage characteristic on both a linear and a logarithmic scale. The junction capacitance is simply

$$C_j = \frac{\varepsilon A}{w_{scl}} \tag{5.1}$$

where A is the cross-sectional area of the diode and w_{scl} is the depletion-layer thickness for a given bias voltage. We will neglect the thin depletion layer in the P^+ region.

The capacitance-voltage law is given by combining Eqs. (2.53) and (2.56):

$$C_j(V) = \varepsilon A \sqrt{\frac{qN_D}{2\varepsilon}} \, [V'_{bi} - V_a]^{-\gamma} \tag{5.2}$$

where, according to Eq. (2.57), $\gamma = 1/2$. Both the varactor frequency multiplier and the parametric amplifier require a high value of capacitive nonlinearity, and hence a large value of γ; the one-sided abrupt junction is therefore preferred to the linearly graded junction [1, 2].

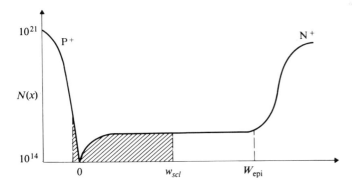

FIGURE 5.2
Impurity profile of a typical diffused diode made using an epitaxial layer on an N^+ substrate. The depletion layer is shown hatched.

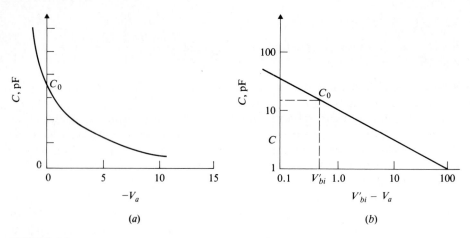

FIGURE 5.3
Capacitance voltage diagram for a varactor or parametric diode (a) linear scale; (b) logarithmic scale.

Both devices also require a low value of series loss resistance. From Fig. 5.2 it is clear that the series resistance at a given reverse bias is given by

$$R_s = \rho_{epi} \frac{W_{epi} - w_{scl}}{A} \tag{5.3}$$

where ρ_{epi} is the resistivity of the epitaxial layer $= 1/q\mu_n N_{epi}$. Both the power conversion efficiency for the varactor multiplier and the noise figure for the parametric amplifier can be shown to depend on the diode cut-off frequency f_c, defined by $1/2\pi R_s C_j$ and hence given by combining the above equations for capacitance and resistance:

$$f_c = \frac{1}{2\pi R_s C_j} = \frac{1}{\varepsilon \rho_{epi}[(W_{epi}/w_{scl}) - 1]}$$

Clearly, this cut-off frequency is bias-dependent, tending to infinity when the epitaxial region is fully depleted. In a real structure, finite contact resistance and the finite resistivity of the bulk N^+ substrate will yield nonzero values of resistance at maximum reverse bias.

Both the varactor diode frequency multiplier and the parametric amplifier are used in circuits in which the diode is driven by a large (sinusoidal) source voltage; the capacitance-voltage variation is thus swept from some reverse bias (limited by the onset of breakdown) to some small forward bias (limited by injection of minority carriers and increased loss).

Normally, the device will be designed so that the avalanche breakdown voltage occurs at approximately the "reach-through" condition. In order for the voltage swing not to drive the diode into or near breakdown, a reasonable "average" value of bias voltage (the operating point) will be less than one-half this breakdown value. A convenient reference condition for defining the cut-off frequency is the bias at which $w_{scl} = W_{epi}/2$. In this case, the cut-off frequency is

given by

$$f_c = \frac{1}{2\pi\rho_{\text{epi}}\,\varepsilon} = \frac{1}{2\pi\tau_{\text{rel}}} \tag{5.4}$$

where the dielectric relaxation time τ_{rel} was previously defined [Eq. (2.11)] with reference to the extrinsic Debye length.

It is clear from this result that the only way of increasing the value of the cut-off frequency for a given material is to use higher epitaxial layer doping levels. However, we recall [Eq. (3.125)] that the avalanche breakdown voltage varies approximately inversely with doping level. The cut-off frequency may thus be written in the form:

$$f_c = \frac{\mu_n E_{br}^2}{4\pi V_{br}} \tag{5.5}$$

From the data for silicon, this reduces to the approximate form

$$f_c = \frac{10 \times 10^{12}}{V_{br}} \tag{5.6}$$

For a given operating voltage (a characteristic of concern to the circuit designer) the only remaining way of increasing the cut-off frequency is the use of material with higher mobility. Several high performance diodes of this type are available using gallium arsenide instead of silicon and, for comparable breakdown voltages, give at least five times improvement in performance because of the increased electron mobility of the N layer. Table 5.1 shows computed results for both silicon and gallium arsenide diodes. It should be noted that these results assume one-dimensional bulk breakdown—a condition only approached using mesa technology.

5.3.2 The Hyperabrupt Varactor Diode

In some applications involving the nonlinear variation of capacitance with junction voltage, the goal is to maximize the nonlinear coefficient γ independently of

TABLE 5.1
Breakdown voltage versus varactor diode cutoff frequency for silicon and gallium arsenide diodes. The breakdown voltages are given using first the ionization coefficient data of Van Overstraeten and De Man, followed by the values using the data of Lee, Crowell, and Sze (see Chap. 3).

N_{epi}, cm^{-3}	V_{br}, V	res (Si), $\Omega\cdot$cm	f_c(Si), GHz	res (GaAs), $\Omega\cdot$cm	f_c(GaAs), GHz
10^{17}	16/21	0.09	1700	0.013	11 800
10^{16}	55/84	0.54	284	0.085	1 800
10^{15}	270/500	4.60	33	0.757	202

any effect on the series loss resistance. Applications include RF tuning circuits where the diode capacitance is used to control the oscillation frequency. Values of γ higher than 0.5 can be achieved with an impurity profile of the type shown in Fig. 5.4. This structure is referred to as a "hyperabrupt junction." Its operation may be understood by considering a reverse bias voltage such that the depletion layer ends in the region where the doping level decreases with increasing depth (corresponding to increasing total reverse junction voltage). An increase in reverse voltage makes the depletion layer terminate at a yet lower doping level; extrapolating from the linearly graded junction, through the constant doped case to this hyperabrupt case, it is easy to visualize from Fig. 5.4 that the value of γ will change from 1/3 (bias point 1) through 1/2 (bias point 2) to some value greater than 1/2 (bias point 3).

This behavior of γ as a function of doping gradient may be seen mathematically if we consider a particularly simple representation of doping level versus distance [3]:

$$N(x) = Fx^m \tag{5.7}$$

Substituting this into Poisson's equation

$$\frac{dE}{dx} = \frac{qN(x)}{\varepsilon}$$

and integrating twice with the appropriate boundary conditions as in Chap. 2, gives the depletion-layer thickness (width) d, as a function of total junction voltage

$$d = \left[\frac{\varepsilon(m + 2)}{qF} \right]^{1/(m+2)} V^{1/(m+2)} \tag{5.8}$$

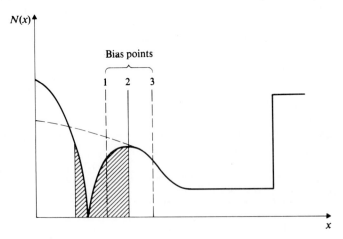

FIGURE 5.4
Impurity profile of a hyperabrupt PN junction varactor diode.

Since the capacitance is given by $C = \varepsilon A/d$, it follows that

$$C(V) \propto V^{-1/(m+2)} \tag{5.9}$$

In terms of the previously used exponent $\gamma = 1/(m+2)$, the following results summarize the behavior for different values of m

$m = 1$	$C \propto V^{-1/3}$	$\gamma = 1/3$	linear
$m = 0$	$C \propto V^{-1/2}$	$\gamma = 1/2$	abrupt
$m = -1$	$C \propto V$	$\gamma = 1$	hyperabrupt
$m = -3/2$	$C \propto V^2$	$\gamma = 2$	hyperabrupt

This is clearly a very simplistic representation of what could be obtained in practice but nevertheless correctly predicts the achievable trends.

The real hyperabrupt profile may be readily obtained by a double diffusion, although in order to better control the shape, ion implantation is often used [4]. The design goal is to obtain as wide a variation of capacitance as possible over a small voltage range. For some bias point corresponding to the depletion layer terminating as shown in Fig. 5.4, point 3, it is apparent that because the doping level decreases beyond the depletion-layer edge, the series resistance will in general be higher than for the case of constant doping level at the same capacitance (same depletion-layer thickness). The hyperabrupt varactor is not therefore normally used for high efficiency frequency multiplication or parametric amplification purposes.

The maximum change in capacitance for a given voltage change (i.e., the value of γ) is severely constrained as can be seen from the following discussion. To obtain a large value of γ the doping level must decrease rapidly (over a short distance); this implies a large peak value of doping. If the peak value of doping level is made too high, however, breakdown will occur before the depletion layer reaches the region where the impurity profile has a negative gradient [we recall that for a junction with a linear gradient a, that is, the region with positive gradient on the N side in Fig. 5.4, the breakdown voltage varies inversely as $\sqrt{a}$ as given by Eq. (3.126)].

It is relatively simple to determine the parameters of the positive gradient region so that operation is below breakdown for a voltage which brings the depletion layer to the peak of the doping profile, i.e., to the value at which the value of γ just starts to exceed the value 1/2. Consider the simplified impurity profile of Fig. 5.5, where a one-sided linear graded region is followed by the hyperabrupt (negative gradient) region. The value of x_0 corresponding to the peak doping level N_0 for a specified peak field E_0 and junction voltage V_0 is given by

$$x_0 = \frac{3}{2} \frac{V_0}{E_0} \tag{5.10a}$$

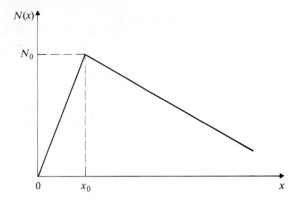

FIGURE 5.5
Simplified impurity profile diagram for a hyperabrupt junction.

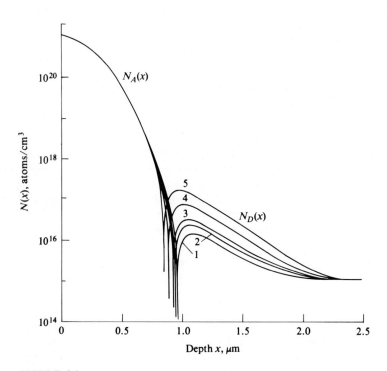

FIGURE 5.6
Impurity profiles for typical hyperabrupt junctions. Only the surface concentration N_{DS} of the N diffused layer is altered. Cases 1 through 5 N_{DS} values are: 10^{17}, 1.5×10^{17}, 2×10^{17}, 4×10^{17}, and 8×10^{17} cm^{-3}. Corresponding computed breakdown voltages (De Man and Van Overstraeten data) are: 49, 48, 47, 34, and 14 volts. The doping level is assumed to rise to a high value beyond $x = 2.5$ μm, corresponding to the N$^+$ substrate.

The corresponding values of gradient and peak doping level are

$$a = \frac{8\varepsilon E_0^3}{9qV_0^2} \qquad (5.10b)$$

$$N_0 = \frac{4\varepsilon E_0^2}{3qV_0} \qquad (5.10c)$$

Note that in order to obtain a useful hyperabrupt region, the epitaxial layer doping level would have to be substantially less than 10^{16} cm^{-3}.

There does not appear to be any simple design procedure for the hyperabrupt junction. The compromises between breakdown voltage, maximum desired value of γ, and range over which the large γ value is required are complex. Figure 5.6 shows (computed) real hyperabrupt impurity profiles formed by using two gaussian diffusions. Only the surface concentration of the diffusion is altered, from 10^{17} to 8×10^{17} cm^{-3}. The corresponding plots of capacitance and γ versus voltage are shown in Fig. 5.7. Notice that, as expected, when increasing the peak doping level, curves 1 through 4, the maximum value of γ increases. Values in the range 1 to 2 are typical for commercial hyperabrupt varactor diodes. However, in passing from case 4 to case 5, avalanche breakdown occurs before a value of γ much above 0.5 is attained. This demonstrates the rather critical nature of the impurity profile for this type of diode. In attempting to increase the value of γ by increasing the peak doping level, the breakdown voltage is lowered, eventually creating the impossibility of actually reaching the hyperabrupt part of the profile before the breakdown voltage is encountered.

Notice also from Fig. 5.7(a) that for higher voltages in cases 1 through 4, the capacitance saturates because the depletion layer has penetrated through to the N$^+$ substrate—the reach-through condition. A well-designed hyperabrupt profile will have little, if any, region corresponding to the original epitaxial-layer doping level, since this would yield a region with γ equal to 0.5 and add unnecessary series resistance. As in the parametric and varactor diodes, the series resistance (which determines the Q factor of the junction capacitance) can be reduced by using gallium arsenide instead of silicon.

5.3.3 The Microwave Step-Recovery Diode

In Chap. 4 we saw that under certain conditions, a very fast "step recovery" transient could be obtained when turning off a PIN diode. One major application of this phenomenon is harmonic generation to provide significant power at frequencies well above the maximum oscillation frequency of an available active device (bipolar transistor, JFET, MESFET) [5, 6]. Since the fast recovery time is given by Eq. (4.65)

$$t_v = \frac{W_i}{2v}$$

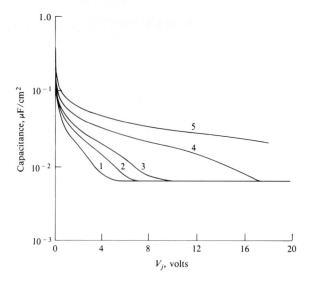

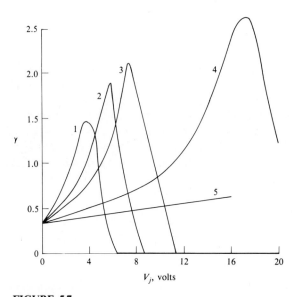

FIGURE 5.7
(a) Capacitance versus voltage for the hyperabrupt profiles of the previous figure; (b) capacitance voltage coefficient versus voltage.

where W_i is the center I layer width and v the limit velocity of the carriers, and since v is close to 10^7 cm/s, it is clear that the only design parameter available is the center layer width W_i. The breakdown voltage is directly related to W_i; assuming a lightly doped center layer such that the region is in reach-through well before the breakdown condition is attained, as shown in Fig. 5.8, the break-

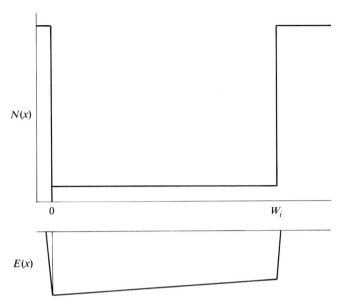

FIGURE 5.8
Step-recovery diode profile and field distribution for moderate reverse bias.

down voltage is given by

$$V_{br} = E_{br} W_i \tag{5.11}$$

The value of E_{br} is not rigorously constant, as we have already seen. A good average value for lightly doped center layer regions with W_i less than 2 μm is 4×10^5 V/cm. This gives the relation between step-recovery time and breakdown voltage as

$$t_v = \frac{V_{br}}{8} \quad \text{ps} \tag{5.12}$$

In other words, for 1 ps fast recovery time, the breakdown voltage will be 8 volts. Table 5.2 gives computed values for a selected range of lightly doped center

TABLE 5.2
Computed values of breakdown voltage for step-recovery diodes
(1) using De Man and Van Overstraeten ionization coefficients;
(2) using Lee, Sze, etc. data—see Chap. 3

w, μm	t_r, ps	$V_{br}(1)$, V	$V_{br}(2)$, V	$E_{br}(1)$, (V/cm) × 10⁵	$E_{br}(2)$, (V/cm) × 10⁵
2.0	10	60	80	3	3.9
1.0	5	34	45	3.4	4.5
0.5	2.5	21	26	4.0	5.0

layer widths. It should be noted that for center layer widths less than 1 μm the effect of the impurity profile gradient of both the diffused surface layer and of the NN$^+$ low-high junction can seriously affect the field distribution and (in forward bias) the free carrier distribution. For this reason, the surface layer is often formed by epitaxial growth (before the mesa etch and passivation step). Ion implantation is an alternate solution. This is also true for the next category of microwave diode to be discussed.

5.3.4 The Microwave Diode Limiter

Figure 5.9 shows the carrier concentration distribution for a PIN diode under forward bias and the effect of a high frequency signal modulating this distribution. If the frequency of the RF modulation f is much greater than the reciprocal of the diffusion transit time [Eq. (4.48)], that is, if

$$f \gg \frac{1}{2\pi t_d} = \frac{4D}{\pi W_i^2}$$

then the applied RF voltage will have varied through many cycles before the carriers diffuse across the center layer W_i. The envelope of the carrier distribution will therefore be as shown in Fig. 5.9. Since most of the center layer carrier distribution does not get affected by the RF signal, it follows that the resistance presented to this signal is essentially that due to the carrier concentration of the distance W_i

$$R_{hf} = \frac{W_i}{Aq\mu n} \tag{5.13}$$

where we assume high level injection $n = p$ and where μ is the ambipolar mobility. This is the same as the dc resistance associated with the center layer voltage

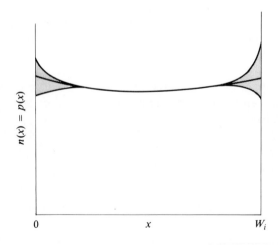

FIGURE 5.9
Carrier distribution in a PIN diode under forward bias. The shaded area represents the modulation envelope due to a sinusoidal modulation at a frequency high compared to the diffusion transit time.

drop discussed in Sec. 3.5.2, given by Eq. (3.97). The value of the resistance is therefore approximately

$$R_{hf} \approx \frac{V_t}{I} \left(\frac{W_i}{L} \right)^2 \tag{5.14}$$

where L is the ambipolar diffusion length. Since this resistance is inversely proportional to the dc bias, it provides a current controlled resistance; this finds applications in microwave variable attenuators [7].

A particular and very important case is that of the microwave diode limiter. This is a diode which, suitably coupled to a 50 ohm transmission line, acts ideally as a short circuit when forward biased and as an open circuit when zero or reverse biased. While the design of such a diode may appear very straightforward, there are some complicating aspects. Firstly, in some applications, it is convenient to obtain the "open circuit" condition at zero bias. The diode area must therefore be small enough to ensure that the capacitive impedance is large (compared to, for example, a 50-ohm line impedance). The operating breakdown voltage will, as always, determine the center layer width as in Eq. (5.11) and Table 5.2. This voltage is also related to switching speed [Eq. (4.33)], on the assumption that the carrier lifetime can be reduced within the constraint $W_i/L \sim 1$ by using gold doping or other techniques.

Insertion loss in thin center layer structures is a particular problem in high performance diode limiters. Since this is of some theoretical interest too, we shall examine the problem in more detail. Consider the typical PIN diode limiter whose impurity profile is shown in Fig. 5.10.

The carrier distribution obtained from an exact solution of Poisson's equation (including the free carriers) is included on the figure. This demonstrates clearly the spill-over effect mentioned in Chap. 2. It is only for reverse bias that the depletion layer actually extends into the heavily doped region. This clearly complicates determination of the exact capacitance value. Numerically the

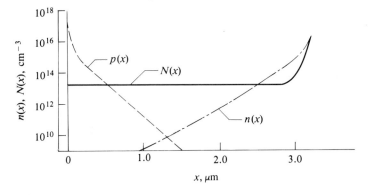

FIGURE 5.10
Impurity profile of typical X band limiter P^+NN^+ diode. The broken curve is the computed carrier distribution at zero bias.

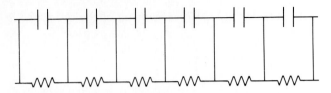

FIGURE 5.11
RC representation of the limiter diode.

capacitance $= \Delta Q / \Delta V$, may always be obtained from two successive charge distributions for slightly different voltages. It may also be obtained from a small signal analysis of the equivalent RC distributed system shown in Fig. 5.11. Each section has a capacitance determined by the dx value. The resistance of each section is determined by the carrier concentration at that particular x value from the solution to Poisson's equation. The complete high frequency impedance may then be transformed into a parallel or series RC equivalent and combined with a 50 ohm impedance to determine insertion loss. It turns out that this loss is not negligible in many cases, particularly at zero bias. Also, although the doping level and center layer thickness are such that the device is in reach-through at zero bias, the capacitance still exhibits a voltage-dependence, because of the effect of the free carriers. Figure 5.12 shows values of capacitance computed for a 10 GHz signal, versus voltage for various cases. Figure 5.13 shows the insertion loss and series resistance versus bias voltage. It is important to note that these results can only be obtained by an exact solution of Poisson's equation including the free carriers (although the zero current approximation still enables a fairly efficient computer solution).

Clearly the choice of fabrication parameters for high performance microwave diode limiters is complex. The above results give some idea of the trade-offs available to the designer, but it is important to note that high performance neces-

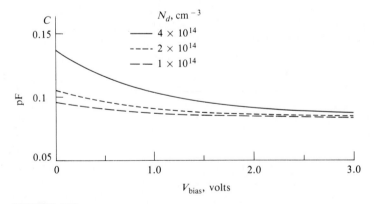

FIGURE 5.12
Computed capacitance at 10 GHz versus voltage for several values of center layer doping level. The diode has a center layer width of 3 μm, and a diameter of 55 μm.

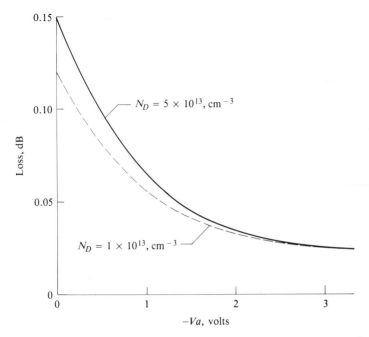

FIGURE 5.13
Computed insertion loss versus bias for a diode with $W_i = 3$ μm, dia $= 55$ μm.

sitates the use of high purity material, with extremely low doping levels in the center region of the PIN diode.

5.4 HIGH VOLTAGE DIODES

The best known high voltage diode application is the 50 or 60 Hz rectifier. There are, however, many more applications, at frequencies up to several tens of kHz or even a few MHz, where high voltage diodes are used, particularly various "protection" circuits involving high power switching circuits. The basic structure is, once again, the PIN device. The higher the breakdown voltage required, the lighter the necessary doping level and the wider (thicker) must be the central I region. Figure 3.14 shows that for 1000 volts breakdown, a doping level of approximately 10^{14} cm^{-3} is required with a center layer width of approximately 100 microns. Since the forward voltage must be as low as possible, and since the center layer voltage drop V_i will increase rapidly for W_i/L much greater than unity [as can be deduced from Fig. 3.8 and Eq. (3.97)], a large center layer width W_i implies the need for a high carrier lifetime τ. Lifetimes in excess of 20 μs are attainable for low doped silicon, corresponding to diffusion lengths of several hundred microns. However, the time taken to turn the diode from the "on" to the "off" state is governed by this carrier lifetime, according to Eq. (4.33). Consequently, it is desirable where possible to use a thinner than nominal center layer

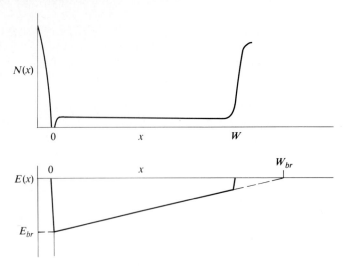

FIGURE 5.14
Doping profile and field distribution for a high voltage diode in reach-through before breakdown occurs. The broken line in the electric field diagram corresponds to a structure with no rise in doping level at $x = W$.

with a corresponding reduced lifetime (usually achieved by carefully controlled diffusion of gold, or by high energy irradiation).

The reduction in breakdown voltage below the maximum possible value can be determined for a device operating under reverse bias reach-through conditions. Figure 5.14 shows the impurity profile and field distribution. It should be noted that for very thick diodes, the center layer is frequently the background wafer material, with the N^+ layer being diffused from the back. The peak value of field is limited by the breakdown value E_{br}. Since the gradient of electric field is given by $dE/dx = qN/\varepsilon$, it is a simple matter to show that the expression for breakdown voltage can be written for $W_i \leq W_{br}$ as

$$V_{br} = \frac{E_{br} W_{br}}{2} \frac{W_i}{W_{br}} \left[2 - \frac{W_i}{W_{br}} \right] \tag{5.15}$$

where W_{br} is the depletion-layer thickness at breakdown under non-reach-through conditions (i.e., the value given in Fig. 3.14); note that this value is a function of the doping level. Figure 5.15 shows a plot of this result. Notice that for a center layer width equal to one-half of the W_{br} value, the breakdown voltage is reduced by only 25 percent, i.e., to 0.75 of its maximum value for that particular doping level. A value of E_{br} equal to 2×10^5 V/cm is reasonable for the range of doping levels encountered in these high voltage structures (see Fig. 3.14).

5.4.1 Fabrication Details

High breakdown voltages are obtained by elimination of the sidewall regions of diffused junctions. A mesa, moat etch, or similar method will give reasonably

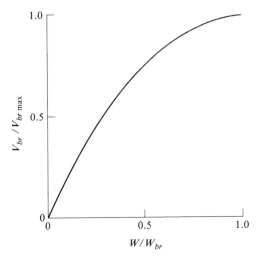

FIGURE 5.15
Reduction in breakdown voltage for reach-through structures.

good results. Using carefully controlled beveling as shown in Fig. 5.16 will enable breakdown voltages approaching the theoretical limit to be obtained. Both positive and negative bevels are used [12], and there exist published curves showing the relationship between bevel angle and breakdown conditions [14, 10].

It is often more economic, however, to use planar technology; sidewall capacitance and some increase in free carrier injection outside the active diode area are not serious problems at low frequencies. The breakdown voltage must, however, be maintained at near maximum value. This can be accomplished while retaining the less expensive planar technology by the use of additional diffused P^+ regions around the P^+ anode as shown in Fig. 5.17 [8]; these regions are known as *floating guard rings* and have the effect of eliminating the undesirable field concentration found in the "cylindrical" sidewall regions by extending the

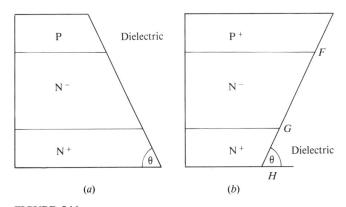

FIGURE 5.16
(a) Negative beveled high voltage diode; (b) positive beveled high voltage diode. (*After Kumar et al.* [13]. *Reproduced with permission, IEEE © 1981.*)

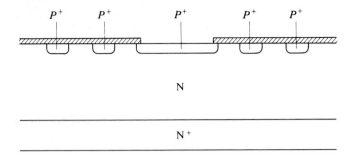

FIGURE 5.17
Use of guard rings to maintain high breakdown voltages in planar diodes.

equipotential contours. Only two-dimensional Poisson solutions with the ionization integral can provide detailed information about the optimum location and lateral dimensions of these guard rings, but the technique is now well established commercially, and breakdown voltages within 90 percent of those obtained using mesa technology are attainable.

Other techniques include the use of field plates [9], and Fig. 5.18 shows a simple technique of increasing the breakdown voltage using planar technology. In this case a deep diffusion is used around the periphery. This effectively produces a deep junction sidewall region, with a larger radius of curvature, and improves the breakdown voltage. Note that the heavily doped shallow diffusion cannot be replaced by a deep moderately doped diffusion since the injection characteristics of this structure would be poor, particularly at high current densities.

Finally, it should be noted that for high currents, the diameter of high voltage diodes often exceeds 1 cm, quite unlike the typical 100 μm diameter of microwave diodes. For further reading on high voltage structures, the reader is referred to the excellent text by B. J. Baliga [10].

5.4.2   The Step-Recovery Phenomenon

As in microwave PIN diodes, it is possible for the step-recovery or snap-off phenomenon to occur when switching off a high voltage PIN rectifier [11]. This is very undesirable because the sudden change in current (often many tens of amps) can create large $L\,di/dt$ spikes in the parasitic inductance which exists in any

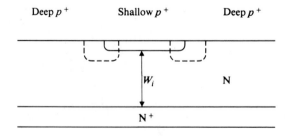

FIGURE 5.18
Use of deep P^+ diffusion to give improved breakdown voltage.

circuit. This can create serious problems, including turning the diode back on again in an oscillatory mode. It is clearly a problem that should be avoided. Let us examine the conditions under which the snap-off effect can occur. Figure 5.19 shows the basic circuit drive conditions and the carrier distribution during steady state "on" conditions and at the end of the reverse constant current storage phase t_s. In Chap. 4 we studied the step-recovery transient for an ideal PIN diode with zero center layer doping. In practice, the center layer will be doped to some level N_{epi}. The step-recovery transient will occur if two conditions are satisfied:

1. The diode is still in high level injection at the end of the constant reverse current (storage) phase.
2. The reverse generator voltage has a value higher in magnitude than the voltage V_{dn} existing across the diode at the end of the following "slow" evacuation phase given by Eq. (4.60).

From Fig. 5.19(b), we can specify the first condition by a "critical" current density using the triangular approximation for free carriers shown by the broken curve in Fig. 5.19(b)

$$J_r \gg J_{rc} = \frac{2qDN_{epi}}{W_i} \tag{5.16}$$

The second condition is given by Eq. (4.60):

$$V_r \gg V_{dn} = \frac{W_i^2 J_r}{4\varepsilon v} \tag{5.17}$$

where J_r is the reverse current density during the slow turn-off phase.

In order to evaluate these conditions, let us pose the value of injected carrier concentration in the "on" condition at some limit n_{max} which will be of the order of a few 10^{18} cm^{-3}. Values in excess of this will give rise to unacceptably high voltage drops in the forward direction because of two effects: firstly, carrier-carrier scattering giving reduced mobility and hence reduced diffusion

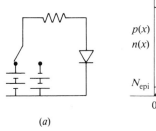
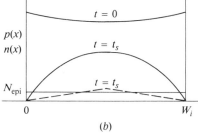

(a)   (b)

FIGURE 5.19
(a) Switching circuit for step-recovery diode; (b) carrier concentrations in steady state (solid line) $t = 0$, and at the end of the constant current evacuation phase $t = t_s$. The broken line is the limiting case for high level injection at $t = t_s$.

constant and reduced diffusion length [from Eq. (3.97) it is seen that this will increase the voltage]; secondly, Auger recombination at high carrier concentrations reduces the carrier lifetime, hence reducing the diffusion length with the same consequences. Using Eq. (3.89) for the forward current density J_f, the ratio of the two current densities is given by

$$\frac{J_{rc}}{J_f} = 2\left(\frac{L}{W_i}\right)^2 \frac{N_{\text{epi}}}{n_{\text{max}}} \tag{5.18}$$

Since the ratio L/W_i is of order unity for a high voltage diode, this current density ratio is of order 10^{-4}. Since the diode is normally, under test conditions, turned off with a reverse current equal to or (typically) ten times greater than the forward current, it is clear that this first condition of high level injection will almost always be satisfied in a high voltage PIN diode. It is interesting to note that such may not necessarily be the case for a microwave step recovery diode with L/W_i typically equal to 100.

The voltage given by Eq. (5.17) may be rearranged by substituting for $J_F = qn_{\text{max}} W_i/\tau$

$$V_{dn} = R\left(\frac{W_i}{L}\right)^2 \frac{qD}{4\varepsilon v} n_{\text{max}} W_i \tag{5.19}$$

where R is the ratio of reverse to forward current. Substituting for the physical constants gives

$$V_{dn} = 4 \times 10^{-14}\left(\frac{W_i}{L}\right)^2 RW_i n_{\text{max}} \tag{5.20}$$

For a 100 micron wide center layer, with $W_i/L = 1.0$, and assuming $n_{\text{max}} \sim 10^{18}$, this gives a voltage of 400R. It is clear that the drive conditions are now critical and that depending on the value of generator voltage, and on the ratio of forward to reverse current drive (plus the forward bias, which determines the value of n_{max}), the very fast step-recovery transient may or may not occur. For example if, using the above numbers, the reverse generator voltage is 400 V, and $R = 1$, no step-recovery will exist. In fact, the current will decay during the early part of the transient ($V_{dn} = 400$ V, $V_g = 400$ V). If, however, the reverse to forward current ratio is 1/10, the step-recovery transient will occur, because $V_{dn} = 40$ V. Or, if the diode is driven less into forward bias, for example, $n_{\text{max}} = 10^{17}$, then $V_{dn} = 40R$, for $R = 1$, giving $V_{dn} = 40$ V.

It may be noted in passing that in the case of the microwave PIN diode, with typically $W_i/L = 0.1$, the value of V_{dn} is typically less than one volt so that the reverse generator voltage can easily be made large enough to produce the fast transient.

For the high voltage diode, conditions favorable to the elimination of the fast recovery and its undesirable voltage spike, are: increase W_i/L ratio (but not enough to degrade seriously the forward "on" voltage), reduce the magnitude of the reverse current, use a lower value of reverse generator voltage.

5.5 CONCLUSIONS

In this chapter we examined various properties of real diode structures and extended the earlier theory where necessary. We saw that while the general theory applies to low voltage microwave devices and to high voltage power rectifiers, the orders of magnitude of the parameters (such as center layer thickness) are quite different.

PROBLEMS

5.1. (a) Why is a varactor or parametric diode usually designed such that breakdown occurs close to reach-through conditions. Discuss other design choices on either side of this choice.

(b) Suggest why a mesa structure is preferable to a planar structure for both varactor and parametric diodes.

5.2. Determine the design parameters for a varactor diode to have a capacitance of 1 pF at 25 V reverse bias (assume this to be one-half of the breakdown voltage). Your answer should give the area, doping level, center layer thickness, and cutoff frequency.

5.3. Determine the design parameters and cutoff frequency of a parametric diode designed to operate at a bias of 2.5 V with a capacitance of 0.1 pF. Repeat the calculation if GaAs is used instead of silicon.

5.4. Derive Eqs. (5.10) for a hyperabrupt varactor diode of Fig. 5.5. Calculate values of x_0, a, and N_0 for a diode in which it is desired that γ exceeds $1/3$ when the applied voltage passes 2 V and the maximum field at the junction is $E_{br}/3$. Use $E_{br} = 3 \times 10^5$ V/cm.

5.5. Consider the idealized impurity profile in the sketch. By extending the solution of Poisson's equation to the case where the field at x_0 is greater than zero, derive an expression for the field at the junction E_j. By setting $E_j = E_{br}$ calculate the breakdown voltage of the device shown in the sketch for the values of x_0 and 'a' found in Prob. 5.4.

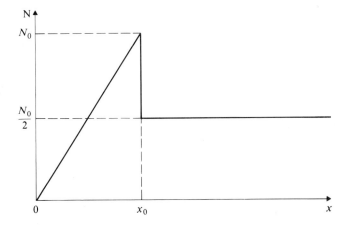

FIGURE P5.5

5.6. This question concerns a microwave diode limiter. Given a required breakdown voltage of 50 V, a center layer lifetime of 10 μs, and a minimum achievable doping level of 10^{13} cm^{-3}, calculate the dc bias current for an ac microwave resistance of 50 Ω; find the value of the zero bias capacitance.

5.7. Design a high voltage rectifier PIN diode for 500 V breakdown assuming mesa or similar construction to maintain breakdown at near the ideal value. Calculate a reasonable value of carrier lifetime for the center layer in order to reduce switching speed to a minimum compatible with low center layer voltage drop. Two choices of design may be considered: (i) breakdown occurs at the same voltage as reach-through, (ii) reach-through occurs at a much lower voltage than breakdown.

5.8. Consider a PIN rectifier diode with a center layer 100 μm thick, $\tau = 20$ μs, doping level $= 10^{13}$ cm^{-3}, and $A = 1$ cm^2. Use the Auger recombination rate given by Eq. (1.36) with $C_N = 10^{-31}$ cm^6 s^{-1} to estimate the increase in center layer voltage drop [Eq. (3.97)] versus injected carrier concentration and current, up to $n = 10^{19}$ cm^{-3}.

5.9. Derive the breakdown reduction formula due to reach-through [Eq. (5.15)].

5.10. A PIN rectifier has $W_i = 100$ μm, $N_d = 10^{14}$ cm^{-3} in center layer, $\tau = 2$ μs, and $A = 1$ cm^2. It is used in a test circuit with a 250 V forward pulse followed by a -250 V turn-off step, in series with a resistance R_g. Calculate the range of values that are acceptable for R_g so that there is no risk of a fast step-recovery or snap-off transient [use Eq. (5.17), etc.]. Find the corresponding values of maximum carrier concentration under forward drive conditions.

REFERENCES

1. P. Penfield, Jr., and R. P. Rafuse, *Varactor Applications*, MIT Press, Cambridge, Mass. (1962).
2. D. J. Roulston, "Very low noise photodetector system using PIN diode and base-band parametric up-converter," *Electronics Letters*, vol. 16, pp. 595–596 (July 1980).
3. M. H. Norwood and E. Shatz, "Voltage variable capacitor tuning—a review," *Proc. IEEE*, vol. 56, pp. 788–798 (May 1968).
4. R. A. Moline and G. F. Foxhall, "Ion-implanted hyperabrupt junction voltage variable capacitors," *IEEE Trans. Electron Devices*, vol. ED-19, pp. 267–273 (February 1972).
5. D. Leenov and A. Uhlir, Jr., "Generation of harmonics and subharmonics at microwave frequencies with PN junction diodes," *Proc. IRE*, vol. 47, pp. 1724–1729 (October 1959).
6. D. J. Roulston, "Frequency multiplication using a charge storage diode in an inductive circuit," *Proc. IEEE*, vol. 55, pp. 1220–1221 (July 1967).
7. K. C. Gupta and A. Singh (eds.), *Microwave Integrated Circuits*, Wiley Eastern (1974).
8. M. S. Adler, V. A. K. Temple, A. P. Ferro, and R. C. Rustay, "Theory and breakdown voltage for planar devices with a single field limiting ring," *IEEE Trans. Electron Devices*, vol. ED-24, pp. 107–113 (February 1977).
9. F. Conti and M. Conti, "Surface breakdown in silicon planar diodes equipped with field plate," *Solid State Electronics*, vol. 15, pp. 93–105 (January 1972).
10. B. J. Baliga, *Modern Power Devices*, John Wiley & Sons, New York (1987).
11. H. Benda and E. Spenke, "Reverse recovery processes in silicon power rectifiers," *Proc. IEEE*, vol. 55, pp. 1331–1354 (August 1967).
12. T. R. Anthony, J. K. Boah, M. F. Chang, and H. E. Cline, "Thermomigration processing of isolation grids in power structures," *IEEE Trans. Electron Devices*, vol. ED-23, pp. 818–823 (August 1978).

13. R. Kumar, D. J. Roulston, and S. G. Chamberlain, "Two-dimensional simulation of a high-voltage *p-i-n* diode with overhanging metallization," *IEEE Trans. Electron Devices*, vol. ED-28, pp. 534–540 (May 1981).
14. M. S. Adler and V. A. K. Temple, "A general method for predicting the avalanche breakdown voltage of negative beveled devices," *IEEE Trans. Electron Devices*, vol. ED-23, pp. 956–960 (August 1976).
15. M. E. Goodge, *Semiconductor Device Technology*, Macmillan Press Ltd., London (1983).

CHAPTER
6

PHOTODETECTOR DIODES AND SOLAR CELLS

6.1 INTRODUCTION

In this chapter we give an overview of various PN junction devices used for converting absorbed optical power into electrical current or power. There are two distinct categories to be considered: firstly, the photodetector diode for low noise detection of optical signals; secondly, the solar cell, for generation of electrical power from sunlight. In the first category, the wavelength is fixed and determined by the signal source; the output is in the form of a modulated current which is then fed into a low noise amplifier; the design goal is high speed and high quantum detection efficiency. In the second category, the input optical power is spread over the entire spectrum of the sun and the design goal is to achieve the maximum power conversion efficiency.

The chapter covers two important types of photodetectors: the PIN photodiode and the avalanche photodiode. Both are widely used in fiber-optic communications systems. We then consider the solar cell, as used in both satellites and on the ground, as a power source. Throughout the chapter, emphasis is placed on the relation between the device technology parameters (impurity profile parameters and background doping levels and thicknesses) and the desired electrical characteristics (quantum efficiency, signal-to-noise ratio, power conversion efficiency). We shall attempt to preserve a balance between the two areas, but it should be noted that the area of solar cells is vast and the reader interested in pursuing the subject in more depth should consult some of the extensive literature available, e.g., [1, 2].

6.2   PHOTODETECTOR STRUCTURES

6.2.1   General Properties—Photon Absorption

If photons with an energy hv (where h is Planck's constant and v the optical frequency) greater than the bandgap energy E_g are incident on a semiconductor, absorption of the light will occur with consequent generation of hole-electron pairs. The law describing the absorption versus depth is exponential and characterized by the absorption coefficient α (dimensions cm^{-1}). Figure 6.1 shows a typical curve of photon absorption versus depth in a PN junction diode. Each absorbed photon generates an electron-hole pair according to the generation rate $G(x)$

$$G(x) = \Phi_0 \, \alpha \, \exp\,(-\alpha x) \tag{6.1}$$

where Φ_0 is the incident photon flux density. Carriers (electrons or holes) reaching the junction depletion region (by diffusion), or generated within the depletion region will be converted to useful current by the electric field in the depletion layer. Carriers generated far away from the junction (further than a diffusion length) may not reach the depletion layer and will hence be "lost" for current generation. In fact, as we shall see below in the photodetector application, it is

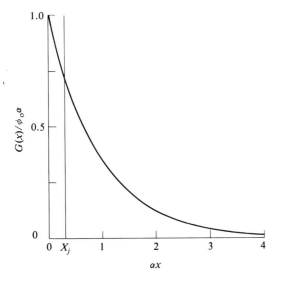

FIGURE 6.1
Generation rate versus depth, due to absorbed photons in a PN junction.

really only the carriers generated within the depletion region which are usable for the high speed photodetector application.

The value of absorption coefficient is strongly dependent on the wavelength of the incoming light for a particular semiconductor. Figure 6.2 shows how α varies versus the frequency ν of the light for the three most common semiconductors, germanium, silicon, and gallium arsenide. The rapid falloff at low photon energies corresponds to the bandgap energy E_g, and the semiconductor is virtually unusable for optical detection for wavelengths above the critical wavelength

$$\lambda_c = \frac{ch}{E_g} = \frac{1.24}{E_g(eV)} \tag{6.2}$$

where c is the speed of light.

At the other extreme of short wavelengths, the absorption coefficient becomes very large and, according to Eq. (6.1), most of the electron-hole pairs will be generated very close to the surface on which the light is incident. Most of them will recombine before reaching the junction depletion layer by diffusion: this is because of both surface recombination and low bulk lifetimes at the high doping levels usually encountered near the surface in a diffused diode. Thus, at both extremes of optical wavelength, the device will be a poor converter of photon flux to electric current. The choice of material is critical for a given detector application (given optical wavelength), since from a technology viewpoint, junction depths will typically be at a depth of order 1/10th of a micron to several microns. Silicon is ideally suited as a detector in the visible or near infrared part of the spectrum. For $E_g = 1.1$ eV, the corresponding cut-off wavelength is $\lambda_c \sim$ 1.1 μm. In particular, it is a near perfect detector for light generated by a GaAs

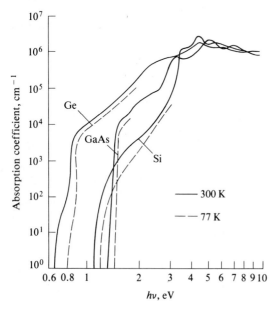

FIGURE 6.2
Absorption coefficients versus photon energy for Ge, Si, and GaAs, using data from Dash and Newman [14], Philipp and Taft [15, 16], Hill [17], Casey et al. [18]. (*After Sze* [3] © *1981, reproduced with permission John Wiley.*)

source, for which the photon energy is the bandgap energy 1.42 eV; this corresponds to a wavelength of about 0.85 μm and is in the near infrared part of the spectrum (and cannot be seen with the naked eye). Germanium, with its bandgap of about 0.7 eV, finds applications in longer wavelength detectors up to about 1.7 μm.

The choice of optical wavelength depends on the application. In fiber-optic systems, a GaAs light source and silicon detector are suitable for Local Area Networks where the attenuation of the fiber is not too important. For longer distances, the optical fiber attenuation minima at 1.3 and 1.5 μm wavelengths are advantageous; in this case silicon photodetectors cannot be used.

6.2.2   Detection Efficiency

Because the main applications of photodetectors are in (relatively) high speed or wide bandwidth applications, photons absorbed outside the PN junction depletion layer are considered "lost" (the diffusion transit time to the depletion-layer edge is proportional to $x^2/2D$ and is long compared to the short high field drift transit time inside the depletion layer). The drift current due to the electron-hole pairs generated within the depletion layer may be calculated simply from Eq. (6.1) if the assumption is made that the surface diffused layer is thin compared to $1/\alpha$

$$I_L = qA \int_0^{w_{scl}} G(x)\, dx$$

$$= qA\Phi_0[1 - \exp(-\alpha w_{scl})] \tag{6.3}$$

where A is the cross-sectional area exposed to the photons and w_{scl} is the depletion-layer thickness. The goal in a photodetector diode is, therefore, quite simply to ensure that w_{scl} is large enough to absorb a high percentage of the incident photons, at the same time having a junction depth small compared to w_{scl}. The internal quantum efficiency is defined as

$$\eta_q = \frac{I_L}{\Phi_0 qA} \tag{6.4}$$

Since the incident optical power is determined by the photon flux density, the area, and the energy of the photons

$$W_{opt} = \Phi_0 Ah\nu, \tag{6.5}$$

the generated current may be expressed as

$$I_L = \frac{\eta_q W_{opt} q}{h\nu} \tag{6.6}$$

As an example, at $\lambda = 0.85$ μm, α for silicon is approximately 30 μm^{-1}. A depletion-layer width of 30 μm would therefore enable 63 percent of the absorbed photons to contribute to useful current; for a width of 60 μm, 86 percent of the photons would contribute to current.

It is apparent that the best structure for this purpose is a PIN type diode. A lightly doped wide base diode would appear to fulfill the same requirements. However, it will necessarily have a large series resistance to the far contact for the above depletion-layer widths (less than 200 μm thick wafers are too thin to handle mechanically). The PIN (or, more accurately, P$^+$NN$^+$) structure can be designed for optimum lightly doped center layer width for a given application, and the device biased to reach through so that the center layer is fully depleted. If (in order to obtain high quantum efficiency) a very wide depletion layer is used, the frequency response becomes limited by the transit time $t_d = w_{scl}/v_d$, where v_d is the saturated drift velocity. It has been shown [3, 4] that at a frequency $f_c = 2.4/2\pi t_d$, the magnitude of the ac photocurrent is reduced by 3 dB. In some cases, this is a real design constraint (e.g., a 60 μm depletion layer will limit the useful bandwidth to about 700 MHz). In many cases, however, other factors dictate the vertical impurity profile. For example, a monolithic integrated photodetector is constrained by compatibility requirements with other parts of the chip; quantum efficiency must sometimes be sacrificed. Furthermore, the transit time is not the only bandwidth limitation. The diode has a junction capacitance

$$C_j = \frac{\varepsilon A}{w_{scl}}$$

and this will combine with the input capacitance of the amplifier and its input resistance to limit the frequency response. This is frequently the dominant mechanism.

Figure 6.3 shows some practical photodetector diode structures. It must be noted that the overall detection efficiency will be less than the above quantum efficiency due to reflection from the surface. This can typically account for a further 15 percent reduction and antireflection coatings are essential for high performance devices. Note the contact arrangement. The diameter of the diode must be small to reduce the junction capacitance, but large enough to recover the incoming signal, for example, from an optical fiber. This latter constraint normally requires the diameter to be 100 μm or more.

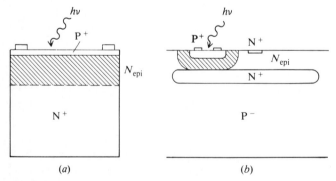

FIGURE 6.3
Practical PIN photodetectors. (*a*) Discrete PIN structure; (*b*) integrated circuit structure.

6.2.3 Signal-to-Noise Ratio

Figure 6.4 shows the equivalent circuit of the reverse biased photodiode and the input stage of a low noise amplifier, characterized by a noise bandwidth B, an input resistance R_L and equivalent noise temperature T_L. Assuming a 100 percent amplitude modulated optical signal, the peak ac current will be the same as the (average) dc value I_L. There will be two noise sources due to the diode: a thermal noise from the series resistance (surface contacts and bulk N^+ region) and a shot noise due to the reverse leakage current (thermal generation). It is important to note that the dc photocurrent will itself contribute to a further shot noise. The signal-to-noise ratio may therefore be written as

$$\frac{S}{N} = \frac{I_L^2 R_L/2}{2q(I_s + I_L)BR_L + kT_L B} \tag{6.7}$$

where I_s is the reverse "dark" current of the diode. For the case where the light generated current $I_L \gg kT_L/(2qR_L)$, and where the reverse leakage current is negligible compared to the photocurrent, this simplifies to

$$\frac{S}{N} = \frac{I_L}{4qB} = \frac{\eta_q W_{opt}}{4hvB} \tag{6.8}$$

This depends only on the quantum efficiency, the optical frequency, and the signal bandwidth; it is referred to as the quantum noise limit for the signal-to-noise ratio. For example, a detected current of 1 nA will give a unity signal-to-noise ratio for $B \sim 1$ GHz. This would correspond to an optical power of approximately 1.5 nW at a wavelength of 0.85 μm.

This quantum noise limit is of course present in conventional radio frequency systems; however, in this case the frequency v is several orders of magnitude lower and the quantum noise limited signal-to-noise ratio is so high for normal power levels that it does not represent an achievable limit for any practical system at radio frequencies. For example, a one microwatt signal with a 1 MHz bandwidth has a quantum noise limited S/N of about 60 dB at one micron wavelength and 120 dB at a one meter (RF) wavelength; detector and

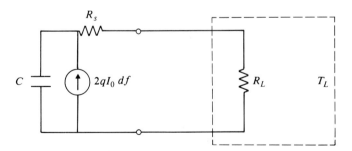

FIGURE 6.4
Equivalent circuit of reverse biased photodiode and input stage of amplifier.

amplifier noise would make it virtually impossible to get anywhere near this latter value.

6.2.4   The Avalanche Photodiode or APD

In most situations, it is the amplifier noise which sets the limit on signal-to-noise ratio. Currently, one of the best "front ends" is the GaAs MESFET, which has low noise properties and a capacitive input comparable to the PIN photodiode. An alternate solution is to use the property of avalanche multiplication to increase the gain before the first conventional amplifier stage. We saw in Chap. 3 that if a diode is biased close to its avalanche breakdown voltage, the reverse current is multiplied by a factor M (see Fig. 3.13). In Eq. (6.7), if the photocurrent I_L is multiplied by M, the effect of the amplifier noise term $kT_L B$ is reduced, as can be seen by dividing the numerator and denominator by I_L^2.

However, the avalanche process itself generates noise because the avalanche gain for each particle is different at any given time and the mean square value of the gain is greater than the square of the mean; the overall signal-to-noise ratio is maximum for a particular value of avalanche gain M. The general expression for signal-to-noise ratio is [5, 6]:

$$\frac{S}{N} = \frac{q^2\eta_q^2\Phi_0^2 A^2/2}{2qB(q\eta_q\Phi_0 A + I_s)M^{d-2} + 2qI_x BM^{-2}} \qquad (6.9)$$

where I_x is the equivalent noise current of the following amplifier, including thermal noise and possible guard ring leakage noise of the photodiode. The factor d represents the excess noise introduced by the avalanche process. A value $d = 2$ would correspond to noise-free multiplication. Values in practice lie in the range $2.3 < d < 4$.

For equal electron and hole ionization multiplication coefficients, the value of d tends to the value 3. For cases where one coefficient is much greater than the other, low values of d (approaching 2.3) are obtained [6].

For silicon, the ratio of electron avalanche multiplication coefficient α to hole coefficient β is greater than 100. This gives rise to excellent APD performance (values of d close to 2). However, for Ge (usable at longer wavelengths such as 1.3 or 1.55 microns), $\alpha/\beta \sim 2$ and poor APD performance is obtained (large values of d).

For InP, the ratio is reported to be 0.3 to 0.4. The best α/β ratio obtained using heterojunction structures is reported as $\alpha/\beta \sim 8$ for a GaAlAs/GaAs structure with 45% Al content. The value of d is still far from that obtained with silicon.

The optimum multiplication factor for maximum signal-to-noise ratio may be estimated by taking the derivative of Eq. (6.9) [7], assuming d is constant

$$M_{opt} = \left\{ \frac{2\left[\int_0^B I_x\, df\right]\Big/ B}{(d-2)(q\eta_q\Phi_0 A + I_s)} \right\}^{1/d} \qquad (6.10)$$

If the diode is biased to yield this optimum value of S/N, the actual value is given by

$$\left[\frac{S}{N}\right]_{\text{opt}} = \frac{(q\eta_q \Phi_0 A)^2/2}{dqB(q\eta_q \Phi_0 A + I_s)M_{\text{opt}}^{d-2}} \tag{6.11}$$

For low bias, where M is less than M_{opt}, the amplifier noise I_x limits the performance. For higher than optimum bias voltage, $M > M_{\text{opt}}$ and the avalanche multiplication noise limits the performance.

6.2.4.1 AVALANCHE PHOTODIODE STRUCTURE. The requirement for depletion-layer thickness for the APD is the same as for the PIN photodiode and is determined by the required quantum efficiency at the given wavelength. The avalanche region is most conveniently created by using a profile similar to the hyperabrupt varactor diode discussed in Chap. 5, the difference being that in the avalanche photodiode, a thick lightly doped drift region is required for photon absorption. In other words, the depletion layer must pass the moderately doped layer into the lightly doped layer at moderate voltages. Figure 6.5(a) illustrates a typical APD structure, with a guard ring to reduce surface edge effects. Figure 6.5(b) and (c) illustrate the ideal step profile and the diffused or implanted profile.

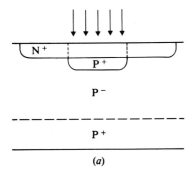

(a)

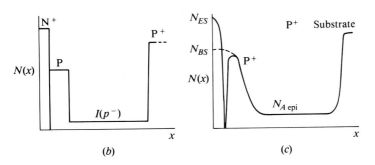

(b) (c)

FIGURE 6.5
Avalanche photodiode structure details. (a) Sectional view; (b) ideal impurity profile; (c) diffused or ion implanted profile.

If no moderately doped region exists at the start of the drift region, a very high bias voltage is required to achieve the required avalanche multiplication (the diode will always be operating close to breakdown for useful gains). If, on the other hand, the moderately doped region has too high a doping level, breakdown will occur before the depletion layer even reaches the start of the lightly doped layer. The design problem is thus similar to that of the hyperabrupt varactor but with quite different end results.

Figure 6.6 shows the manner in which the reach-through and breakdown voltages vary with surface doping level for a typical avalanche photodiode. The structure analyzed has a lightly doped π region of 10^{13} cm^{-3} extending from the surface to the P$^+$ substrate, the latter being at a depth of 120 μm from the surface [8]. The N$^+$ gaussian profile has a surface concentration of 10^{21} cm^{-3} with a characteristic length of 2 μm. The moderately doped P region is gaussian with a (fixed) characteristic length of 10 μm. Figure 6.6 shows the effect of varying the surface concentration N_{BS} of this P diffused region.

At low values of N_{BS}, the breakdown voltage tends to the value in excess of 1000 volts associated with a 120 μm lightly doped layer. Note that the reach-through voltage V_{rt} only has a value of about 100 volts. For $N_{BS} = 4 \times 10^{15}$ cm^{-3}, the breakdown voltage decreases to 500 volts, with a slight increase in reach-through voltage. Values of N_{BS} in excess of 5×10^{15} give a breakdown voltage less than the reach-through value—i.e., the depletion layer extends only partly through the 120-μm π region before breakdown occurs. This condition is undesirable for an avalanche photodiode since a significant fraction of the light is absorbed outside the depletion layer, giving poor quantum efficiency. A reason-

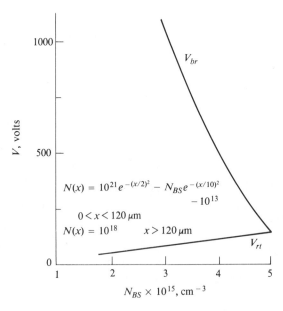

FIGURE 6.6
Effect of peak doping level on performance of an avalanche photodiode. Breakdown and reach-through voltages versus surface P region doping level, N_{BS}, computed using the BIPOLE program (see Chap. 15).

able choice of N_{BS} for this diode is in the range of 4.2 to 4.6 × 10^{15} cm^{-3}; the whole of the π region is then occupied by the depletion layer and the breakdown voltage V_{br} is low enough (400 to 250 volts) so that useful avalanche gain is achieved at moderate bias (close to V_{br}).

It is clear that APD design and fabrication conditions are quite critical, specially for the P diffused region. The requirement for high purity silicon for the background material is an additional factor leading to the high cost of APDs.

6.3 SOLAR CELLS

6.3.1 General *I-V* Considerations

The solar cell is designed to provide maximum electrical power into a resistive load from direct (or concentrated) sunlight. Figure 6.7 shows the *I-V* characteristics of an illuminated PN junction diode; this characteristic applies also to the PIN diode discussed above, but for signal detection applications, as we have seen, the diode is always reverse biased. In the solar cell application, to extract electrical power, operation is confined to the "forward voltage, reverse current" quadrant. The general *I-V* relation in the presence of an optical generated current I_L is, to an excellent approximation, given by superposition of the constant current term to the normal *I-V* equation for the diode

$$I = I_0\left[\exp\left(\frac{V_a}{mV_t}\right) - 1\right] - I_L \tag{6.12}$$

In fact, I_L is slightly bias-dependent. In reverse bias more current is generated from absorption within the depletion region of the junction than for forward bias

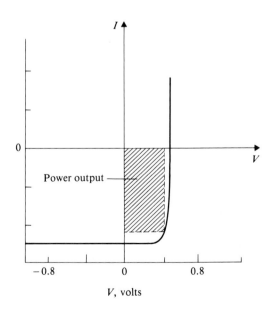

FIGURE 6.7
I-V characteristic of a PN junction solar cell.

(where the depletion layer is narrower). In the neutral region on either side of the depletion layer, the opposite occurs, with more neutral region current begin generated in forward bias. However, to a good approximation, these effects cancel, and the superposition principle can be applied for most practical purposes. Strictly, I_L is then defined as the light generated current for $V_a = 0$. The power output for a specified terminal voltage V_a is given by

$$P = V_a \left\{ I_L - I_0 \left[\exp \left(\frac{V_a}{mV_t} \right) - 1 \right] \right\} \tag{6.13}$$

The maximum power output $V_m I_m$ (where $dP/dV_a = 0$), together with I_L and the open circuit voltage, defines the *fill factor* (FF)

$$FF = \frac{I_m V_m}{I_L V_{oc}} \tag{6.14}$$

where V_{oc} is the open circuit voltage ($I = 0$). This is found from Eq. (6.12) to be

$$V_{oc} = mV_t \ln \left[1 + \frac{I_L}{I_0} \right] \tag{6.15}$$

The power rectangle is shown in Fig. 6.7. Differentiating Eq. (6.13) and rearranging terms to obtain the terminal current for maximum output power gives

$$I_m = I_L \left[1 - \left(\frac{mV_t}{V_m} \right) \right] \tag{6.16}$$

The corresponding voltage is found to be

$$V_m = V_{oc} - mV_t \ln \left[1 + \frac{V_m}{mV_t} \right] \tag{6.17}$$

The optical generated current I_L in amperes is comparable to the incident optical power in watts (as we saw above, $hv/q \sim 1$ at the long wavelength end of the visible spectrum). The power available from the sun is of order 100 mW/cm^2 (see below). The diode extrapolated reverse saturation current density, $J_0 = I_0/\text{area}$, is dominated by the moderately doped thick region, similar to the epitaxial layer of P^+NN^+ diodes

$$J_0 = q \left(\frac{n_i^2}{N_A} \right) \frac{W_p}{\tau_n} \tag{6.18}$$

Typical values are: $N_A = 10^{16} \text{ cm}^{-3}$, $W_p = 200 \text{ }\mu\text{m}$, $\tau_n = 10 \text{ }\mu\text{s}$. This gives $J_0 \sim 3 \times 10^{-12} \text{ A/cm}^{-2}$. The value of open circuit voltage from Eq. (6.15) is seen to be close to 0.6 V for an ideal diode ($m = 1$). If space-charge recombination dominates, the value of I_0 is much higher, but so is mV_t. The net result is always an open circuit voltage ~ 0.6 V for bright sunlight. This tells us from Eq. (6.16) that the current I_m for maximum power output is within 5 to 10 percent of the I_L value and the corresponding voltage V_m is $\sim 3mV_t$ below the open circuit value.

6.3.2 Carrier Concentration Distribution

Figure 6.8 shows the impurity profile of a typical solar cell. Notice the shallow diffused region at the back; this type of structure is referred to as a *back surface field* (BSF) cell. From Eq. (6.15) we know that for a high value of V_{oc} a low value of extrapolated reverse saturation current I_0 is required. From our knowledge of the narrow base diode and the P^+NN^+ structures in Chap. 3, we know that I_0 will be lower for the BSF solar cell than for an equivalent structure with no P^+ region, particularly if the diffusion length is long compared to the device thickness W_t.

Figure 6.9 shows the computed carrier distributions at four wavelengths, each corresponding to a particular value of α. At the long wavelength (low α) the current is generated primarily in the thick base region, whereas at short wavelengths (high α), the current is generated closer to the surface.

Figure 6.10 shows the current densities in each region versus α. As the wavelength decreases, more of the light is absorbed at the surface and eventually the base contribution decreases. The emitter and space-charge layer contributions rise as more of the photons are absorbed in these regions near the surface, until at very short wavelengths, the generated electron hole pairs recombine close to the surface and the total current becomes very small. This is because of high surface recombination and low bulk lifetime associated with the high doping level near the surface of a diffused diode. These results indicate the fact that design of a solar cell to match the sun's spectrum is not trivial. Figure 6.11 shows the ideal spectrum, based on blackbody radiation for a temperature of 5800°C. The actual

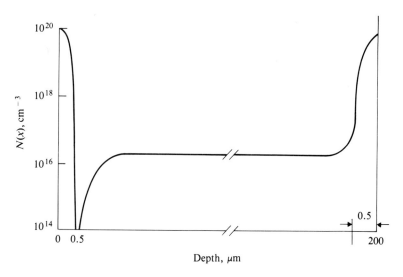

FIGURE 6.8
Impurity profile with values for a typical solar cell. The N diffused layer has a sheet resistance of 18 Ω/sq.

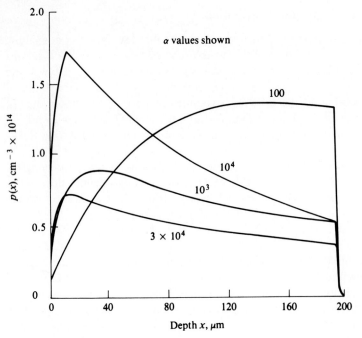

FIGURE 6.9
Carrier distributions for four different values of α for the solar cell of Fig. 6.8 for $V = 0.4$ V, computed using the BIPOLE program. The incident photon flux density is constant at 10^{18} cm^{-2}.

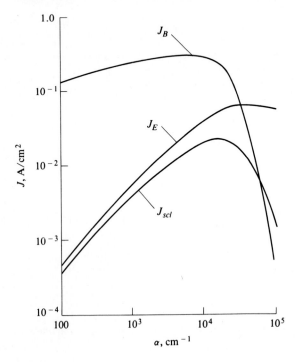

FIGURE 6.10
Current densities in each region: J_B (base), J_E (emitter), J_{scl} (space-charge layer) versus absorption coefficient for the solar cell of Fig. 6.8, computed using the BIPOLE program.

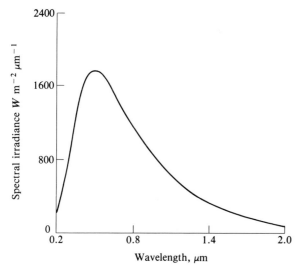

FIGURE 6.11
The spectrum corresponding to blackbody radiation at a temperature of 5800 K for a power density of 1353 W/m². Actual spectra for AM0 or AM1 conditions are slightly different due to absorption, etc, after Sze [3].

spectrum is slightly different. AM0 is defined to be the *air mass 0* condition existing outside the earth's atmosphere [1]. The power density for this case is 135 mW/cm². The conditions on the earth's surface are not unique, but AM1 is taken to be the corresponding spectrum at sea level at midday when the sun's rays travel vertically; the corresponding power density is close to 100 mW/cm².

6.3.3   Factors Determining Solar Cell Efficiency

The power conversion efficiency, η, neglecting series resistance losses and losses due to reflection at the surface, is defined as

$$\eta = \frac{P_{elect}}{P_{sun}}$$

$$= F_{il} I_L \frac{V_{oc} - K_{vt} V_t}{P_{sun}} \tag{6.19}$$

where P_{elect} is the electrical power output, P_{sun} is the sunlight power incident on the surface, K_{vt} is a factor which may be deduced from Eq. (6.17) to lie between 3 and 5. The factor F_{il} is the amount by which the current for maximum power is reduced from the short circuit output value. We saw above that this is in the range 0.9 to 0.95.

Determination of I_L depends upon several factors. Firstly, due to the finite thickness of the diode, some of the incident light will not be absorbed and will pass through the diode. Figure 6.12 shows the percentage of current available for

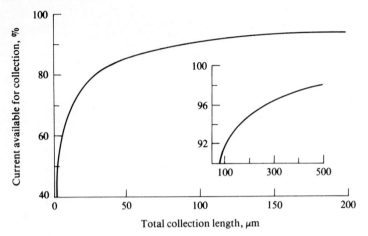

FIGURE 6.12
Percentage of spectral current available for collection as a function of diode total thickness for silicon.
(*After Dunbar and Hauser* [10]. *Reprinted with permission from Solid State Electronics, Pergamon Press PLC.*)

a given total diode thickness. For 100 μm, approximately 90 percent of the photons are absorbed before passing out the far side.

The bandgap effectively determines a long wavelength cutoff for the light. From Fig. 6.11, it is clear that for silicon ($E_g = 1.1$ eV) a non negligible part of the infrared tail of the sun's spectrum is "lost." However, for GaAs ($E_g = 1.43$ eV) an even greater component is "lost." The value of I_L/P_{sun} may now be estimated using Eq. (6.6) by taking an effective value of hv/q over the spectrum. The peak of the spectrum occurs at about 0.6 μm. This corresponds to $hv/q \sim 2.0$. The ratio I_L/P_{sun} can therefore be estimated from Eq. (6.6) as approximately 0.5.

The open circuit voltage can be estimated approximately using Eq. (6.15) and assuming that the base region component of I_0 dominates. From Eq. (6.18), substituting for $n_i^2 = N_c N_v \exp(-E_g/kT)$ gives

$$V_{oc} = V_t \left[\ln (I_L) - \ln (K_{bg}) + \frac{E_g}{kT} \right] \tag{6.20}$$

where, from Eq. (6.18), $K_{bg} = (qW_p/N_A \tau)N_c N_v$. Because the logarithm removes the sensitivity of this result to minor variations in values, we can obtain a very good estimate of open circuit voltage using Eq. (6.20). For AM0 conditions, with the reduction factors discussed above, $J_L \sim 0.06$ A/cm^2 (published values [10] give a more exact value of 0.053 A/cm^2); using a doping level $N_A = 10^{16}$ cm^{-3}, a base width $W_p = 100$ μm, and a carrier lifetime $\tau_n = 10$ μs, we find $K_{bg} \sim 1.6 \times 10^6$ for silicon. The resulting open circuit voltage becomes using $V_t = 0.025$ V:

$$V_{oc} = V_t[-2.8 - 14.2 + 44] \sim 0.67 \text{ V}$$

Note the dominance of the bandgap term.

We can see at a glance the effect of parameter changes on the open circuit voltage. Increasing the base doping level by a factor of 10, or the carrier lifetime by a factor of 10, will have the effect of increasing V_{oc} by approximately 60 mV. If N_A is increased above 10^{18} cm^{-3}, Auger recombination will start to intervene and the carrier lifetime will decrease as N_A^{-2}, thus decreasing V_{oc} rather than increasing it. Note that a reduction in center layer width will improve V_{oc} but only so long as the value of I_L is not adversely affected. From Fig. 6.12, it can be seen that below about 50 μm it could be expected that the reduction in I_L would be significant.

The bandgap of the semiconductor material clearly has an important effect on the open circuit voltage (it is important to note that to a first approximation, the value of I_L is *not* affected by the bandgap). Using GaAs instead of silicon will increase the value of V_{oc} in the above example from 0.67 to 0.97 V, if we neglect the reduction in I_L due to the "lost" near infrared component of absorbed photons from the sun's spectrum. From Eq. (6.19) it is clear that the efficiency of a GaAs cell should be substantially higher than a similar silicon cell.

We can now estimate the probable efficiencies obtainable. Substituting the above results into Eq. (6.19), using $K_{vt} = 4$ and $F_{il} = 0.9$ gives $\eta = 0.45 \ (0.67 - 0.1) = 0.25$. This is very close to values obtained from exact computer solutions. Note that in an ideal diode, with the K_{bg} term in Eq. (6.20) reduced to 0, the maximum possible efficiency would still be limited to a value of order 30 percent. M. Wolf [11] quotes an efficiency of 25 percent in 1980. T. Tiedje et al. [12] report 29.8 percent in 1984.

We see from the above that the power conversion efficiency of solar cells is theoretically limited to a value of order 30 percent. The choice of device parameters will be such as to approach the theoretical limit as closely as possible. For each diode, corresponding to a particular set of parameter values, the open circuit voltage and short circuit current will be different, and these values will determine the "dc operating point" for the cell in order to obtain maximum efficiency for a given sunlight condition. Figure 6.13 shows the computed curve of efficiency versus output voltage for the cell discussed above (Fig. 6.8). The voltage will in practice be selected by choice of the load resistance. Figure 6.14 shows the effect of altering the junction depth. Clearly, a very shallow junction depth is desirable for efficient operation. This is because the heavily doped diffused layer has significant recombination, both in the bulk (Auger and SRH) and at the surface (where the effective recombination velocity at the Si-SiO$_2$ interface can vary from 10 to 10^4 cm/s). In fact a simple calculation shows that the reverse saturation current for this region is comparable to that for the thick base region and therefore affects the overall value of J_0.

Figure 6.15 shows the effect of varying the base region doping level N_A. At low doping levels the efficiency falls off because of the reduced open circuit voltage as seen from the above discussion. At high doping levels, Auger recombination makes the lifetime fall off faster than the doping level increases, thus again confirming the conclusions outlined above.

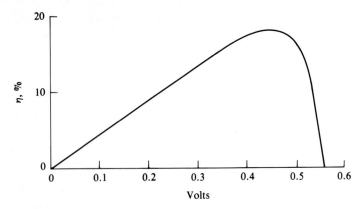

FIGURE 6.13
Efficiency versus operating voltage for the silicon solar cell of Fig. 6.8, computed using the BIPOLE program.

6.3.4 Technology Aspects of Solar Cells

Apart from the vertical impurity profile influence on efficiency, there are a number of important effects related to the surface and contacts. Figure 6.16 shows a typical contact arrangement. The goal is to cover as little of the surface as possible by the opaque metal contacts; however, in the extreme case of, for example, one narrow contact along the edge of the surface, a large series resistance is added, in this case equal to one square of sheet resistance. Since the

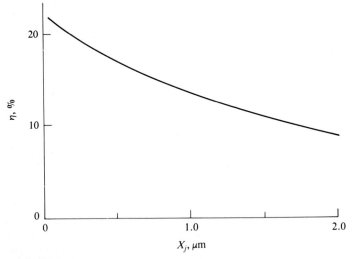

FIGURE 6.14
Efficiency versus junction depth for the silicon solar cell described by Fig. 6.8, computed using the BIPOLE program.

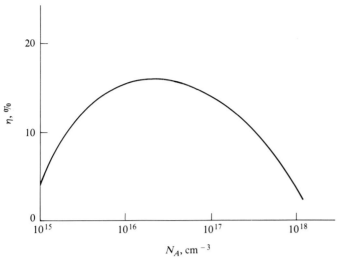

FIGURE 6.15
Efficiency versus thick base doping level N_A for the solar cell described by Fig. 6.8, computed using the BIPOLE program.

emitter is shallow, sheet resistance values will be in the range of a few tens of ohms per square. This is one reason why a low doped diffused layer is not desirable. It is obvious that power lost in the series resistance will degrade the efficiency. This wili be particularly true for cells operating at high power densities—the concentrated cell.

Figure 6.17 shows the I-V characteristics in the presence of series resistance.

In order to reduce loss by light reflected from the surface of the cell, anti-reflection coatings are used. For the AM0 spectrum, 36 percent of the incident light is reflected in the case of a silicon surface [10]. This is reduced to 15.6 percent loss if a 0.08 μm SiO layer is used, and 17.6 percent if a 0.11 μm SiO$_2$ coating is used.

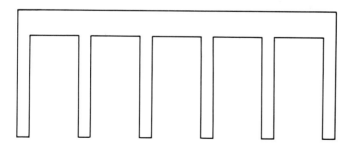

FIGURE 6.16
Typical contact arrangement for solar cell.

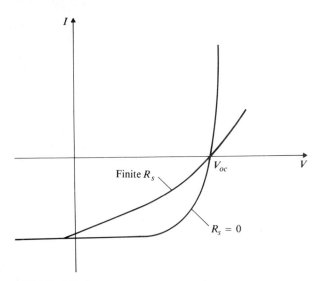

FIGURE 6.17
I-V characteristics in presence of series resistance R_s.

An enormous amount of effort has been devoted to solar cell research over the past several years for both space and terrestrial applications. The focus of the work has been in two main directions: firstly cells for maximum possible efficiency using advanced technology; secondly low efficiency cells using techniques suitable for mass production. In the first category, gallium arsenide and heterojunctions have received considerable attention, as have configurations for increasing the amount of light penetrating through the surface (antireflection coatings and "textured" surfaces [2]). For low cost applications using mass production methods, the emphasis is on thin film cells in which the active layers are polycrystalline or amorphous material deposited on suitable substrates such as glass or plastic. CdS is a popular material for such low cost cells. The silicon solar cell is a (very good) compromise between high efficiency and low cost and has been available commercially for several years. Finally, it should be noted that use of concentrated sunlight can improve the efficiency. Solar cells operating at 100 or 1000 times normal sunlight intensities (using concentrating lenses) will, from Eq. (6.15), have a higher I_c and hence a higher V_{oc} value (although high level injection means that m tends to a value of 2). Auger recombination becomes important but the overall efficiency is improved. M. A. Green et al. [13] report a maximum possible value of 34 percent using concentrated sunlight with a thin base structure.

PROBLEMS

6.1. It is useful to work with exponential impurity profiles for hand analysis. By equating gradients and values of profiles at the junction X_j for a gaussian profile: $N(x) =$

$N_g \exp - (x/x_g)^2 - N_{\text{epi}}$ and an exponential: $N(x) = N_e \exp - (x/x_e)$, show that $x_e = x_g^2/2X_j$ and $N_e/N_g = \exp (X_j/x_g)^2$. Using a logarithmic vertical scale and linear distance scale, draw these two profiles for $N_g = 10^{21}$, $N_{\text{epi}} = 10^{16}$ cm^{-3}, $x_g = 0.1$ μm; neglect bandgap narrowing effects.

6.2. Consider a PIN photodiode with a gaussian profile as above, with $N_g = 10^{20}$ cm^{-3}, $N_{\text{epi}} = 10^{14}$ cm^{-3}, $X_j = 0.3$ μm, $W_{\text{epi}} = 50$ μm. Calculate the appropriate constants for the exponential profile approximation. Assuming that the depletion layer for reverse bias extends far into the epitaxial layer, derive an approximate expression for the depletion-layer thickness on both sides of the junction and determine the values for a bias of 5 and 50 V.

6.3. For the diode of Prob. 6.2, calculate the internal quantum efficiency for high speed applications (include both sides of the depletion layer) at wavelengths of 0.4, 0.6, 0.85, and 1.0 μm (use α values of 10^5, 5×10^3, 7×10^2, and 50 cm^{-1}, respectively).

6.4. For the diode used in Probs. 6.2 and 6.3, with a diameter of 100 μm, calculate the cutoff frequency at both 5 and 50 V bias, due to:
(i) internal transit time
(ii) diode RC time constant
(iii) diode capacitance and amplifier input resistance of 50 Ω and 1 kΩ.

6.5. For the diode used in the above problems, with a carrier lifetime in the epitaxial layer and space-charge layer of 1 μs, and using an amplifier with an equivalent noise temperature of 300 K and a bandwidth of 1 MHz, calculate, using a reverse bias of 50 V, (i) the incident optical power (assume 100% modulation) for which the quantum noise contributes equally to the sum of all the other noise sources for $R_L = 50$ Ω, (ii) the signal-to-noise ratio for this condition, (iii) the value of load resistance R_L for which the diode dark current noise contributes equally to the amplifier noise, and (iv) the RC limited bandwidth for case (iii); assume $\lambda = 0.85$ μm.

6.6. Consider an avalanche photodiode with $d = 2.3$ and where necessary assume the diode and load characteristics of Probs. 6.2 through 6.5, with $R_L = 50$ Ω. Calculate and plot on a sketch, S/N versus M for I_L equal to 10^{-10} and 10^{-7} A, $B = 1$ MHz. Assuming $V_{br} = 200$ V and using Eq. (3.123) for M with $n = 3$, determine the values of bias voltage corresponding to maximum S/N.

6.7. Consider an idealized avalanche photodiode impurity profile of the type shown in Fig. 6.5(b) with a P region consisting of a 3 μm layer doped 0.5×10^{16} cm^{-3} and a 50 μm layer doped 10^{13} cm^{-3}. Calculate the breakdown voltage assuming $E_{br} = 3 \times 10^5$ V/cm and note the relative amount of voltage developed across the two layers. Observe also the criticalness of the total dopant integral in the thin moderatey doped layer.

6.8. A BSF solar cell has a "base" region 100 μm wide doped 10^{16} cm^{-3} with a 1 μs carrier lifetime in this region and an equivalent lifetime in the (forward biased) space-charge layer of 0.1 μs. The junction depth is at 0.5 μm and the surface concentration of the gaussian diffused layer is 10^{20} cm^{-3}. Assume an area of 1 cm^2. Using the results of Chap. 3 and Prob. 6.1, calculate, making suitable approximations, the values of saturation currents I_0 ($m = 1$) and I_{scr} ($m = 2$) and corresponding open circuit voltage for $I_L = 100$ mA. Using the complete diode equation given by $I = I_0 \exp (V_a/V_t) + I_{scr} \exp (V_a/2V_t)$ calculate and sketch the output power versus voltage. Compare the maximum output power to that obtained using Eqs. (6.16) and (6.17) assuming first no space-charge recombination current, second only space-charge recombination current.

6.9. Use Eq. (6.20) to calculate approximately the ratio of power conversion efficiency of a GaAs solar cell to that of a silicon solar cell. You can estimate the approximate difference in I_L due to the long wavelength cutoff from Fig. 6.11.

6.10. The effect of the thickness of the solar cell may be deduced from Eq. (6.20). Calculate the maximum power output under ideal conditions [using Eqs. (6.16) and (6.17) as well] as the thickness W_p is varied from 500 to 10 μm. Note that Fig. 6.12 may be used to find the reduction in I_L. Assume AM0 conditions.

REFERENCES

1. S. C. Jain, E. L. Heasell, and D. J. Roulston, "Recent advances in the physics of silicon PN junction solar cells including their transient response," *Progress in Quantum Electronics*, vol. 11, pp. 105–204 (1987).
2. *IEEE Transactions on Electron Devices Special Issue on Photovoltaics* (May, 1984).
3. S. M. Sze, *Physics of Semiconductor Devices*, 2d ed., Wiley, New York (1981).
4. W. W. Gartner, "Depletion layer photoeffects in semiconductors," *Phys. Rev.*, vol. 116, pp. 84–87 (October 1959).
5. R. J. McIntyre, "Multiplication noise in uniform avalanche photodiodes," *IEEE Trans. Electron Devices*, vol. ED-13, p. 164 (January 1966).
6. J. R. Biard and W. W. Schaunfield, "A model of the avalanche photodiode," *IEEE Trans. Electron Devices*, vol. ED-14, pp. 233–238 (May 1967).
7. J. C. Tandon and D. J. Roulston, "Comparison of avalanche photodiode and photoparametric diode signal to noise ratio," *Solid State Electronics*, vol. 16, pp. 1503–1505 (December 1973).
8. R. Kumar, S. G. Chamberlain, and D. J. Roulston, "Two-dimensional computer simulation of the breakdown characteristics of a multi-element avalanche photodiode array," *IEEE Trans. Electron Devices*, vol. ED-31, pp. 928–933 (July 1984).
9. M. P. Thekaekara, "Data on incident solar energy," *Suppl. Proc. 20th Annual Meet. Inst. Environ. Sci.*, p. 21 (1974).
10. P. M. Dunbar and J. R. Hauser, "A study of efficiency in low resistivity silicon solar cells," *Solid State Electronics*, vol. 19, pp. 95–102 (February 1976).
11. M. Wolf, "Updating the limit efficiency of silicon solar cells," *IEEE Trans. Electron Devices*, vol. ED-27, pp. 751–760 (April 1980).
12. T. Tiedje, E. Yablonovitch, G. D. Cody, and B. G. Brooks, "Limiting efficiency of silicon solar cells," *IEEE Trans. Electron Devices*, vol. ED-31, pp. 711–761 (May 1984).
13. P. Campbell and M. A. Green, "The limiting efficiency of silicon solar cells under concentrated sunlight," *IEEE Trans. Electron Devices*, vol. ED-33, pp. 234–239 (February 1986).
14. W. C. Dash and R. Newman, "Intrinsic optical absorption in single crystal germanium and silicon at 77 and 300 K," *Phys. Rev.*, vol. 99, pp. 1151–1155 (August 1955).
15. H. R. Philipp and E. A. Taft, "Optical constants of germanium in the region 1 to 10 eV," *Phys. Rev.*, vol. 113, pp. 1002–1005 (February 1959).
16. H. R. Philipp and E. A. Taft, "Optical constants of silicon in the region 1 to 10 eV," *Phys. Rev. Lett.*, vol. 8, p. 13 (1962).
17. D. E. Hill, "Infrared transmission and fluorescence of doped gallium arsenide," *Phys. Rev.*, vol. 133A, pp. 866–872 (February 1964).
18. H. C. Casey, D. D. Sell, and K. W. Wecht, "Concentration dependence of the absorption coefficient for n and p type GaAs between 1.3 and 1.6 eV," *Jnl. Appl. Phys.*, vol. 46, pp. 250–257 (January 1975).

CHAPTER
7

BIPOLAR TRANSISTORS— BASIC STRUCTURE AND THEORY

7.1 INTRODUCTION

In this chapter we will examine from a simple theoretical standpoint the funda-mental operation of the bipolar transistor, sometimes referred to as the *bipolar junction transistor* or BJT. The structure of the standard discrete NPN device will be described briefly in terms of its geometry and impurity profile. We then derive the basic terminal characteristics using the simplifying approach of uniform doping in each region. The small signal equivalent circuit is discussed including both low frequency and capacitive components. This leads to a discussion of base resistance, transition frequency, and maximum oscillation frequency. A discussion of the Ebers-Moll large signal model and breakdown voltages follows. This enables the reader to grasp the essential features of transistor operation and also some limits of the elementary theory. Subsequent chapters will build upon the results of the present chapter to consider modifications to the simple theory to account for real structures with non constant doping levels, followed by a detailed study of operation at high current levels and an introduction to various "real" devices including a range of modern IC transistors and high voltage power switching transistors.

7.2   THE VERTICAL NPN STRUCTURE

Here we discuss from a very simple basic viewpoint the transistor structure. More advanced structural details will be considered in Chaps. 12 and 13. Figure 7.1 shows the essential features of the conventional NPN transistor [1]: (a) is a diagram of the geometry, or mask layout; (b) shows a cross section through the device, with (c) the impurity profile on the same vertical axis. In order to understand the structure, it is useful to follow the fabrication sequence in outline. The starting material for the discrete transistor is normally a wafer with a thin epitaxial layer grown on a heavily doped N^+ substrate for mechanical support. The substrate is of order 200 or more microns thick. The epitaxial layer thickness and doping level are chosen as a function of the desired base-collector breakdown voltage, according to the basic diode properties discussed in Chap. 3. It is typically a few microns thick.

Two diffusions are required to complete the device. These are carried out using standard masking and photolithographic techniques (see Chaps. 12 and 13 for more details). Firstly a base diffusion is performed, normally using the column III element boron. The furnace temperature and diffusion time are chosen to give the required base-collector junction depth—typically 1 to 3 microns for a small analog or digital device. In order to maintain the required breakdown voltage, the residual epitaxial-layer thickness (between the base-collector junction and the N^+ substrate) must be sufficient. The surface concentration of boron will be of order 10^{19} cm^{-3}.

During the boron diffusion, an oxide layer is grown; the wafer is thus ready for the second diffusion step. This is normally phosphorus or arsenic (column V

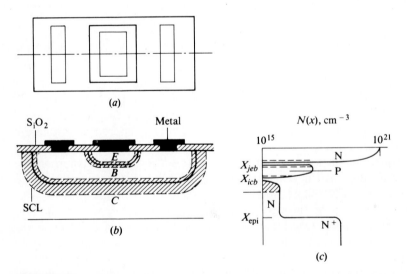

FIGURE 7.1
NPN transistor structure made using the double diffused process. (a) Geometry based on mask layout; (b) cross section; (c) impurity profile.

elements). The combination of furnace temperature and time are chosen such that
the N type emitter diffuses typically to a depth of order one-half that of the base.
The surface concentration of the N type emitter is of order 10^{21} cm^{-3}, limited
usually by the solid solubility of the donor atoms in the silicon.

The final processing steps consist of two masks and corresponding photolithographic steps—for the creation of contact windows for the base and emitter
contacts and for the metallization pattern to determine the final layout of base
and emitter connections to suitable bonding pads (not shown on Fig. 7.1, but see
later chapters for details in various transistor structures).

The theory of operation of the transistor hinges upon the doping levels and
vertical distances of the three regions: emitter, base, and collector [2]. The lateral
dimensions (geometry) determine essentially only the active operating area, A, of
the device and hence the actual values of current at a given base-emitter forward
bias voltage. Some additional parameters, specifically base resistance, are also
determined by the horizontal geometry. The impurity profile shown in Fig. 1(c) is
redrawn in Fig. 7.2 with depth, x, shown as the horizontal axis. In order to
assimilate more readily the basic operation of the transistor, we will initially

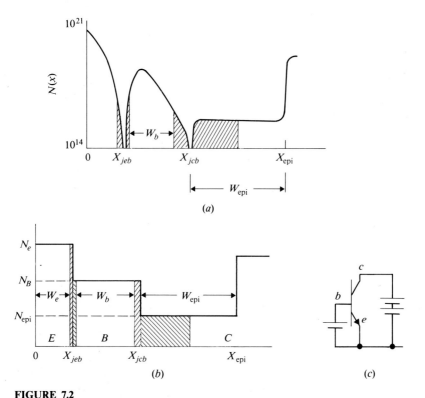

FIGURE 7.2
(a) Impurity profile for a double-diffused NPN transistor; (b) simplified impurity profile with constant
doping levels (space-charge regions are hatched); (c) simplified common emitter bias arrangement; V_c
is several volts and V_{BE} is approximately 0.7 V.

replace the actual impurity profile by regions of constant doping level, as shown in Fig. 7.2(*b*). The bias arrangement for normal amplifier operation is a forward bias on the base-emitter junction and a reverse bias on the base-collector junction as shown in Fig. 7.2(*c*). This means that the emitter-base space-charge region will be narrow, and can usually be neglected to a first approximation in considering the widths of the neutral emitter and base regions. The collector is always more lightly doped than the base (otherwise the transistor could not have been fabricated as discussed above, with creation of the base-collector junction by diffusion into the epitaxial layer). The base-collector depletion layer will therefore extend more into the collector side than into the base side [see Eq. (2.31)]; once again, to a first approximation, we can neglect the width of this depletion layer on the base side, compared to the neutral region base width. This parameter, of crucial importance to the operation of the transistor, will henceforth be denoted by W_b.

7.3 ANALYSIS OF THE SIMPLIFIED NPN STRUCTURE

Under the normal "active region" bias arrangement discussed above, holes will be injected into the N type emitter from the base; electrons will be injected from the N type emitter into the P type base. This is similar to an NP diode. Figure 7.3 shows diagrams of resulting minority carrier concentration versus distance and the currents. Since the base-collector junction is reverse biased, according to the law of the junction, the minority carrier concentrations at each side of this depletion layer are zero. The collector thus acts as a "sink" for electrons diffusing across the base. The electric field within the base-collector space-charge layer is such as to carry the electrons through to the N type collector (the reader should think about this by writing down the appropriate equation for electron drift current and electric field determined from the solution of Poisson's equation). In fact, because of the relatively high field within this space-charge layer, the electrons rapidly attain their saturated drift velocity. The electron current I_{nc} is thus given by [3]:

$$I_{nc} = qAv_d n_c \tag{7.1}$$

where v_d is the saturated drift velocity, and n_c is the electron concentration within the space-charge layer. Clearly, there is a contradiction between the result of zero electron concentration based on the law of the junction, and the finite value given from Eq. (7.1)

$$n_c = \frac{I_{nc}}{qv_d A} \tag{7.2}$$

A simple calculation [see below after Eq. (7.5)] suffices to show that for most practical cases, however, this value is close enough to zero to be neglected, and for the rest of this chapter we consider the reverse biased base-collector

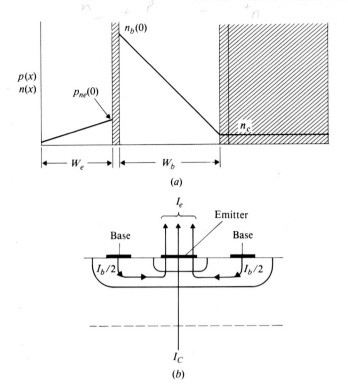

FIGURE 7.3
(a) Minority carrier concentration versus distance in NPN transistor; (b) current components.

junction to be a perfect sink for electrons. The effect of n_c becomes very important at high currents, and this will be studied in Chap. 10.

Referring to Fig. 7.3, the hole current into the emitter, assuming the narrow base diode theory to apply with $W_e \ll L_p$, is given by [see the diode result Eq. (3.9)]

$$I_{pe} = qAD_p \frac{p_{ne}(0) - p_{ne0}}{W_e} \qquad (7.3)$$

where p_{ne0} is the thermal equilibrium hole concentration in the emitter and D_p is the hole diffusion constant. $p_{ne}(0)$ is the injected hole concentration given by

$$p_{ne}(0) = p_{ne0} \exp\left[\frac{V_{BE}}{V_t}\right] \qquad (7.4)$$

as in Eq. (2.71), where V_{BE} is the forward base-emitter bias.

The electron current flowing across the base from the collector is given by:

$$I_{nc} = \frac{qAD_n n_b(0)}{W_b} \qquad (7.5)$$

Note the minor difference between this ideal expression and that for a narrow base diode—the absence of the $-n_{bo}$ thermal equilibrium term, since the electron concentration at $x = W_b$ is now assumed zero due to the reverse biased collector.

We can now check the fact that n_c given by Eq. (7.2) is small enough to be neglected. Equating the expression for collector current given by Eq. (7.1) and by Eq. (7.5) gives the result $n(0)/n_c = v_d W_b/D_n$. For $v_d = 10^7$ cm/s and $D_n = 10$ cm^2/s, we see that for a 1 micron base width the ratio is 100 and for a 0.1 micron base width, the ratio is 10. These results imply a 1 and 10 percent error, respectively, in neglecting n_c as being small compared to $n(0)$.

In Eq. (7.5) the value of $n_b(0)$ is given by

$$n_b(0) = n_{bo} \exp \left[\frac{V_{BE}}{V_t} \right] \tag{7.6}$$

We are now in a position to examine the most fundamental terminal characteristics of the transistor. The total emitter and collector currents are given (following conventional sign notation) from Fig. 7.3(b) by:

$$I_e = -(I_b + I_C)$$

$$I_C = I_{nc}$$

$$I_b = I_{pe} \tag{7.7}$$

The current flowing into the base terminal is, in this simple case, due only to the holes injected into the emitter and must therefore be equal to I_{pe}. Note that this current comes in "from the side" as is evident from the discussion on the transistor structure and Fig. 7.3. The most frequently used transistor configuration for both analog and digital applications is that in which the input signal is applied to the base terminal and the output is taken from the collector, with the emitter being "common" to the input and output ports. This common emitter circuit is shown in Fig. 7.4. Considering the practical situation where $V_{BE} \gg V_t$, we obtain the expression for dc current gain

$$\beta_0 = \frac{I_C}{I_b} = \frac{I_{nc}}{I_{pe}} \tag{7.8a}$$

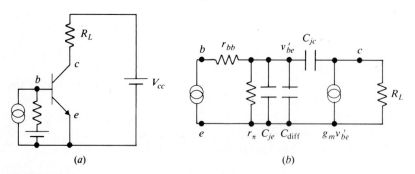

(a)   *(b)*

FIGURE 7.4
(a) The common emitter transistor circuit; (b) small signal equivalent circuit.

or substituting from Eq. (7.3) to Eq. (7.6):

$$\beta_0 = \frac{W_e D_n n_{b0}}{W_b D_p p_{e0}} \tag{7.8b}$$

In terms of doping levels, substituting from the mass action law for n_{b0} and p_{e0}, this becomes

$$\beta_0 = \frac{W_e D_n N_{De}}{W_b D_p N_{Ab}} \tag{7.9a}$$

where N_{De} and N_{Ab} are the doping levels in the emitter and base regions, respectively.

Substituting ball-park numbers, assuming comparable values for emitter and base region widths, D_n about twice D_p, a base doping level $N_{Ab} = 10^{17}$ cm^{-3} and an emitter doping level with an effective value (taking bandgap narrowing into account) $N_{De} = 10^{18}$ cm^{-3} gives a common emitter current gain $\beta_0 \sim 20$. We know from our previous work on diffused regions in diodes that this result is likely to be far from the true value for gain, but it will serve for the purposes of this discussion, and it certainly indicates the trends necessary for high gain: heavily doped emitter, lightly doped base, thick (wide) emitter, thin (narrow) base. Note that, because of our assumption used in Eq. (7.5) of $W_b \ll L_n$, we have neglected recombination in the base. The reader may show very simply that using base minority carrier charge divided by base lifetime τ_b for this component of base current gives $\beta_0 = \tau_b/t_{bb}$ where t_{bb} is the base transit time $W_b^2/2D_n$ [see below Eq. (7.15)]. The total gain is therefore

$$\beta_0 = \frac{1}{(W_b D_p N_{Ab}/W_e D_n N_{De}) + t_{bb}/\tau_b} \tag{7.9b}$$

Neglecting the recombination effect on gain is therefore valid in nearly all real transistors, where base lifetimes are of order 1 μs and base transit times t_{bb} are of order 1 ns for 1 micron base widths.

7.4   SMALL SIGNÁL EQUIVALENT CIRCUIT ANALYSIS

Figure 7.4(b) shows the small signal equivalent circuit of the transistor in the common emitter configuration. It is clear that the ac or small signal current gain dI_c/dI_b will be identical to the above derived dc gain, for the case considered of substantial forward base-emitter bias. The small signal input conductance is given by (neglecting the voltage drop in the base resistance)

$$g_\pi = \frac{dI_b}{dV_{BE}} \tag{7.10}$$

This may be written as

$$g_\pi = \frac{dI_b}{dI_C}\frac{dI_C}{dV_{BE}}$$

The first term in parentheses is simply $1/h_{fe}$ where h_{fe} is the ac or incremental current gain and the second term is, by definition (change in output current with respect to change in input voltage), the transconductance g_m. We thus obtain the expression for input conductance as

$$g_\pi = \frac{g_m}{h_{fe}} \qquad (7.11)$$

For the conditions considered so far, the gain is constant and therefore $h_{fe} = \beta_0$. From the equations for I_C, Eqs. (7.5) and (7.6), we see that

$$g_m = \frac{I_C}{V_t} \qquad (7.12)$$

7.4.1 Capacitance Terms

The equivalent circuit may be extended to include high frequency components by simple addition of the junction (depletion-layer) and diffusion capacitances. Following the diode treatment of Sec. 2.5, the depletion-layer capacitances may be written as

$$C_{je} = \frac{C_{je0}}{(1 - V_{BE}/V_{bie})^{\gamma_e}} \qquad (7.13a)$$

$$C_{jc} = \frac{C_{jc0}}{(1 - V_{BC}/V_{bic})^{\gamma_c}} \qquad (7.13b)$$

where V_{bie} and V_{bic} are the built-in barrier potentials of the emitter-base and collector-base junctions, determined by the doping levels as per Eq. (2.24); C_{je0}, C_{jc0} are the depletion-layer capacitances of the two junctions at zero bias; γ_e, γ_c are the capacitance-voltage nonlinear coefficients discussed in Eqs. (2.53) to (2.59). Although for the assumptions used in the present chapter, the constant doping levels imply γ values of $1/2$, the emitter-base junction is more closely approximated by a linearly graded junction with $\gamma = 1/3$; see Eq. (2.59) and corresponding discussion in Chap. 9.

The diffusion capacitance in this simplified treatment is due to the minority carrier (electron) charge stored in the base region, Q_b. From Fig. 7.3(a), this is given by the area under the $n(x)$ distribution over the base width W_b. For normal forward bias, $n_b(0) \gg n_{b0}$ and we can write

$$Q_b = \frac{qAn_b(0)W_b}{2} \qquad (7.14a)$$

The diffusion capacitance $C_{\text{diff}} = dQ_b/dV_{BE}$ is therefore

$$C_{\text{diff}} = \frac{dQ_b}{dI_C}\frac{dI_C}{dV_{BE}} \tag{7.14b}$$

Because both the charge and the current are proportional to $n_b(0)$, the ratio dQ_b/dI_c is the same as the ratio Q_b/I_C. By combining Eqs. (7.5) and (7.14a), this ratio is seen to be the diffusion transit time t_{bb} and is given by

$$t_{bb} = \frac{dQ_b}{dI_C} = \frac{W_b^2}{2D_n} \tag{7.15}$$

The second term in Eq. (7.14b) is the transconductance. The final expression for the diffusion capacitance is therefore

$$C_{\text{diff}} = g_m t_{bb} \propto I_C \tag{7.16}$$

We will consider other terms in the diffusion capacitance in Chap. 9 (due mainly to hole charge in the neutral emitter and free electron charge in the base-collector space-charge layer).

7.4.2 Base Resistance

To complete the high frequency equivalent circuit, the base resistance must be added. This resistance is due to the horizontal flow of base current under the emitter in the active base region (refer to Fig. 7.5) [2]. Its value is given by the doping level and thickness of the neutral base region, together with the horizontal dimensions of the active base region (which are the same as those of the emitter)

$$r_{bb} = \left[\frac{1}{F_b}\right]\left[\frac{L/B}{q\mu_p N_{Ab} W_b}\right] \tag{7.17}$$

The factor F_b requires some discussion. Because the discrete transistor under consideration has two base contacts disposed on either side of the emitter running along its full B direction as shown in Fig. 7.5, the base current enters from both sides and is zero at the centre, $x = L/2$. The total path length is therefore $L/2$ instead of L and because both base contacts are in parallel the resistance

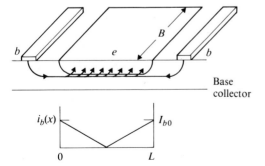

Base
collector

FIGURE 7.5
Origin of base resistance.

"seen" at the base terminal is reduced further by a factor of one-half. Further-more, the base current is due to injection of holes into the emitter and this will occur uniformly from 0 to $L/2$ and $L/2$ to L as shown in Fig. 7.5. This distributed nature of the current means that the effective path length will be further reduced. The determination of the reduction factor may be shown as follows.

Base Resistance Reduction Factor

The uniform injection of hole current into the emitter gives a linear decrease in hole current in the base with distance x

$$i_b(x) = I_{b0} \frac{1-x}{L'} \tag{7.18}$$

where for simplicity in presentation we have introduced $I_{b0} = I_b/2$, $L' = L/2$. The resistance in an incremental distance dx is $dR = (\rho/W_b B)\,dx$, where the resistivity of the base region between emitter and collector is $\rho = 1/q\mu_p N_{Ab}$. The value of base resistance may be found by evaluating the total power dissipated:

$$W = \int_0^{L'} i_b^2 \, dR \tag{7.19a}$$

$$= I_{b0}^2 \frac{\rho}{W_b B} \int_0^{L'} \left(1 - \frac{x}{L'}\right)^2 dx \tag{7.19b}$$

$$= \frac{I_{b0}^2 \rho}{W_b B} \frac{L'}{3} \tag{7.20}$$

We thus see that the additional factor due to the distributed nature of the base current is 1/3. The final value of F_b in Eq. (7.17) is therefore 12 for the discrete transistor.

7.4.3 Transition Frequency f_t

The single most important parameter used to characterize the high frequency behavior of a transistor is the common emitter current gain bandwidth product, sometimes called the transition frequency and always denoted by f_t [2]. This is the frequency at which the short circuit common emitter current gain extrapo-lates to unity. Consider this gain at high frequencies based on the equivalent circuit of Fig. 7.4(b). Defining the output current as that flowing in a short circuit from the output (collector) terminal to ground (the standard test condition), means that the collector-base capacitance C_{jc} is effectively in parallel with the base-emitter junction capacitance C_{je} and diffusion capacitance C_{diff}. The ratio of output to input current is therefore

$$h_{fe}(\omega) = \frac{i_c}{i_b} = \frac{g_m v_{be}'}{i_b} \tag{7.21a}$$

Since v'_{be}/i_b is the impedance of g_π is in parallel with the total capacitance $C_{\pi t}$, we obtain

$$h_{fe}(\omega) = \frac{g_m/g_\pi}{1 + j\omega C_{\pi t}/g_\pi} = \frac{h_{fe}(0)}{1 + j\omega h_{fe}(0)C_{\pi t}/g_m} \tag{7.21b}$$

where we define $C_{\pi t}$ as the sum of all three capacitances C_{je}, C_{jc}, and C_{diff}. $h_{fe}(\omega)$ is sketched in Fig. 7.6. It is customary to use h_{fe} as the low frequency gain and henceforth we will not distinguish between $h_{fe}(0)$ and h_{fe}. The 3 dB cut-off frequency or beta cut-off frequency of current gain, $f_{c\beta}$ is given from Eq. (7.21b) by

$$f_{c\beta} = \frac{g_\pi}{2\pi C_{\pi t}} = \frac{g_m}{2\pi h_{fe} C_{\pi t}} \tag{7.22}$$

The frequency at which the above gain [Eq. (7.21b)] extrapolates to unity, bearing in mind that $h_{fe} \gg 1$, is given by

$$f_t = \frac{g_m}{2\pi C_{\pi t}} \tag{7.23}$$

The behavior of f_t versus collector current can best be seen by examining the expression

$$\frac{1}{2\pi f_t} = \frac{C_{\pi t}}{g_m}$$

$$= \left(\frac{V_t}{I_c}\right)(C_{je} + C_{jc}) + t_{bb} \tag{7.24}$$

where we have substituted for C_{diff} from Eq. (7.16). Since the two depletion-layer capacitances are, for all practical purposes, independent of collector current, it is clear that at low currents f_t is proportional to I_c, tending at high values of I_c to a

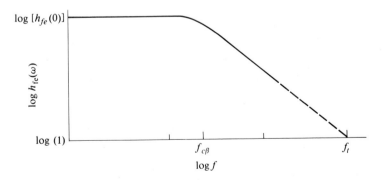

FIGURE 7.6
Common emitter current gain versus frequency, defining the beta cut-off frequency $f_{c\beta}$ and the f_t of the transistor.

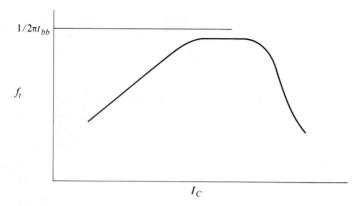

FIGURE 7.7

Transition frequency f_t versus collector current I_C.

value $1/2\pi t_{bb}$. This behavior is illustrated in Fig. 7.7. Note the direct relationship between maximum possible f_t (never actually attained before high current falloff effects come into play) and base transit time, which is proportional to W_b^2. Once again, the importance of achieving a narrow base width is apparent. It must be noted that the small signal equivalent circuit of Fig. 7.4(b) becomes invalid at frequencies of order f_t; the 6 dB per octave slope shown in Fig. 7.6 will only be observed in most devices well below the f_t frequency, since for frequencies above f_t the vertical delay in the base, t_{bb}, becomes a significant fraction of one period. The concept of f_t is nevertheless an extremely useful one, provided it is used only to characterize the transistor at frequencies well below f_t, but typically can be up to about ten times the beta cut-off frequency, $f_{c\beta}$. We shall return in more detail to an examination of maximum f_t in Chap. 9, including the other components of charge due to free carriers, and in Chap. 10 we discuss high current f_t falloff effects.

7.4.4   Maximum Oscillation Frequency $f_{m\,osc}$

Another very important parameter which characterizes the high frequency operation of the transistor is the maximum oscillation frequency, or unity power gain frequency $f_{m\,osc}$. In fact, this is often more important than the f_t and is a good measure of transistor performance not only for power gain in small signal and large signal amplifiers, but also for wideband analog amplifiers and even (to a certain extent) for nonsaturating logic gates.

Consider the small signal equivalent circuit with a source v_s, series resistance R_s and load resistance R_L shown in Fig. 7.8. Assuming that the frequency is sufficiently high such that the capacitive current through $C_\pi = C_{je} + C_{\text{diff}}$ is much greater than the current through the resistance r_π, a simple nodal analysis

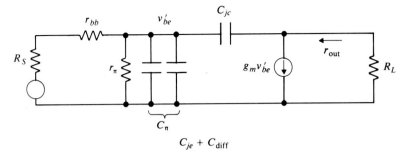

FIGURE 7.8
Circuit for determination of maximum oscillation frequency $f_{m\,osc}$.

gives the output resistance:

$$r_{out} = \frac{C_\pi}{C_{jc}\,g_m} \tag{7.25}$$

Since the input resistance at high frequencies (where the reactance of C_π is much less than the value of r_{bb}) is r_{bb}, the high frequency power gain G_p is given by

$$G_p = \frac{i_c^2 R_L}{i_b^2 r_{bb}}$$

$$= \frac{g_m^2 R_L}{4 r_{bb}\,\omega^2 C_\pi^2} \tag{7.26}$$

Substituting the above value r_{out} for R_L (i.e., considering matched output load conditions for maximum power gain) gives

$$G_{p\,max} = \frac{g_m}{4\omega^2 C_\pi C_{jc}\,r_{bb}} \tag{7.27}$$

The maximum oscillation frequency is that for which the power gain is equal to unity. Using this condition, and substituting f_t for $g_m/2\pi C_\pi$ gives the final result

$$f_{m\,osc} = \sqrt{\frac{f_t}{8\pi C_{jc}\,r_{bb}}} \tag{7.28}$$

This result indicates that it is not sufficient to obtain a high value of f_t (for example, by decreasing the base width W_b), but that r_{bb} and C_{jc} must also be kept low. The presence of the r_{bb} term is particularly important, since one way of reducing the value of r_{bb} would be to have an increased base width as can be seen from Eq. (7.17). There is thus clearly some compromise to be adopted for a "good" transistor design which is not equivalent to designing solely for a maximum f_t. This point will be addressed in more detail in Chap. 9.

7.5 GENERAL DC OPERATION— THE EBERS-MOLL EQUATIONS

Let us now return to the original equations for the currents flowing in the transistor, but instead of assuming the base-collector junction to be reverse biased, we will allow it to become forward biased. The total emitter current is given as before by

$$I_E = -(I_{pe} + I_{nc}) \tag{7.29}$$

where

$$I_{pe} = qAD_p p_{ne0} \left[\exp\left(\frac{V_{BE}}{V_t}\right) - 1 \right]$$

and the collector electron current flowing across the base region is now given by including the law of the junction at both sides of the base, width W_b (see Fig. 7.9)

$$I_{nc} = \frac{qAD_n}{W_b} [n_b(0) - n_b(W_b)]$$

$$= \frac{qAD_n n_i^2}{W_b N_{Ab}} \left[\exp\left(\frac{V_{BE}}{V_t}\right) - \exp\left(\frac{V_{BC}}{V_t}\right) \right] \tag{7.30}$$

The second term is due to injection of electrons from the N type collector into the base and has the effect of decreasing the concentration gradient in the base.

A final term defines the magnitude of the hole current flowing into the N collector due to the forward bias created by V_{BC}. This may be obtained by recollecting the behavior of the P^+NN^+ diode and equating the N type epitaxial collector layer to the center layer of the diode. Using the result of Chap. 3, Eq. (3.55), assuming that the NN^+ interface reduces the hole current essentially to zero, gives

$$I_{pc} = \left(\frac{qAp_{nc0} W_{epi}}{\tau_{epi}}\right) \exp\left(\frac{V_{BC}}{V_t}\right) \tag{7.31}$$

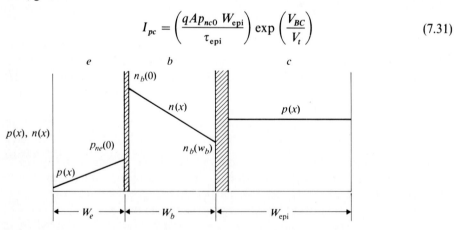

FIGURE 7.9
Minority carrier distributions for forward biased base-emitter and base-collector junctions.

The total collector current I_C is

$$I_C = I_{nc} + I_{pc} \tag{7.32}$$

Let us now define four parameters which describe the dc operation of the transistor; firstly forward and reverse common base current gains:

$$\alpha_F^{-1} = \frac{(qAD_n n_{b0}/W_b) + (qAD_{pe} p_{e0}/W_e)}{qAD_n n_{b0}/W_b} \tag{7.33}$$

$$\alpha_R^{-1} = \frac{(qAD_n n_{b0}/W_b) + (qAp_{nc0} W_{epi}/\tau_{epi})}{qAD_n n_{b0}/W_b} \tag{7.34}$$

secondly, emitter and collector "diode" reverse saturation currents:

$$I_{ES} = \frac{qAD_n n_{b0}}{W_b} + \frac{qAD_p p_{e0}}{W_E} \tag{7.35}$$

$$I_{CS} = \frac{qAD_n n_{b0}}{W_b} + \frac{qAp_{nc0} W_{epi}}{\tau_{epi}} \tag{7.36}$$

From the above four equations, it can readily be seen that a "reciprocity condition" applies

$$\alpha_R I_{CS} = \alpha_F I_{ES} \tag{7.37}$$

Noting the conventional current flow shown in Fig. 7.10, where I_C is normally positive in an NPN transistor, combining Eqs. (7.29) and (7.32), we can now write the emitter and collector currents in the form:

$$I_E = -I_{ES}\left[\exp\left(\frac{V_{BE}}{V_t}\right) - 1\right] + \alpha_R I_{CS}\left[\exp\left(\frac{V_{BC}}{V_t}\right) - 1\right] \tag{7.38}$$

$$I_C = \alpha_F I_{ES}\left[\exp\left(\frac{V_{BE}}{V_t}\right) - 1\right] - I_{CS}\left[\exp\left(\frac{V_{BC}}{V_t}\right) - 1\right] \tag{7.39}$$

These are the well-known Ebers-Moll equations [5]. Within the simplifications made in their derivation, they describe the operation of the transistor in any bias configuration. Moreover, because of the reciprocity condition given by Eq. (7.37) there are only three parameters required to completely describe the dc operation of the transistor with the Ebers-Moll model.

Figure 7.10 shows the two diode circuit representation of this model. For the normal active region, where the base-emitter junction is forward biased and the base-collector junction is reverse biased, the Ebers-Moll equations simplify to the form

$$I_E = -I_{ES}\left[\exp\left(\frac{V_{BE}}{V_t}\right) - 1\right] - \alpha_R I_{CS} \tag{7.40}$$

$$I_C = \alpha_F I_{ES}\left[\exp\left(\frac{V_{BE}}{V_t}\right) - 1\right] + I_{CS} \tag{7.41}$$

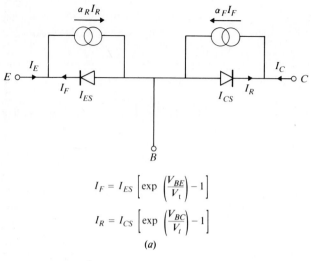

$$I_F = I_{ES}\left[\exp\left(\frac{V_{BE}}{V_t}\right) - 1\right]$$

$$I_R = I_{CS}\left[\exp\left(\frac{V_{BC}}{V_t}\right) - 1\right]$$

(a)

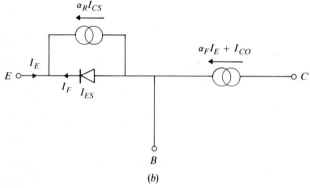

(b)

FIGURE 7.10

(a) Two diode circuit representation of the Ebers-Moll NPN model; (b) simplified model valid for normal active region.

Or, using the notation in Fig. 7.10,

$$I_C = -\alpha_F I_E - I_{CO} \tag{7.42}$$

where

$$I_{CO} = (1 - \alpha_F \alpha_R)I_{CS} \tag{7.43}$$

When both junctions are reverse biased, the equations simplify further to the form

$$I_E = -(1 - \alpha_F)I_{ES} \tag{7.44}$$

$$I_C = -(1 - \alpha_R)I_{CS} \tag{7.45}$$

The Ebers-Moll equations are widely used in nonlinear circuit analysis programs such as SPICE [6] and WATAND [7]. They give sufficiently accurate results for many applications, particularly if the base resistance and a collector resistance are added to the model. The major approximations of the model are not due to the assumed constant doping levels in each region (the four Ebers-Moll parameters can be defined for any impurity profile), but rather are due to the fact that variations in the space-charge layer widths with bias voltage have been neglected. Furthermore, variations from the ideal theory occur at high current levels and these effects are not represented. We shall discuss extensions to the model after these effects have been dealt with in Chaps. 8, 9, and 10.

A further difference between the model and a real device behavior lies in the fact that the reverse saturation currents are assumed in the above derivation to be due to diffusion in the neutral regions. We already saw in Chap. 3 that under reverse bias, the currents in junctions are dominated by generation within the space-charge region. This need not create a problem for most intended uses of the Ebers-Moll model as long as the terms I_{ES} and I_{CS} are determined from measurements using values extrapolated (on a log-linear plot of I versus V) from forward bias measurements in the range 0.5 to 0.7 V, for typical silicon transistors. Of course, the increase in base current due to space-charge recombination in the emitter-base space-charge layer will affect the current gain, particularly at low bias; even this effect is often not too important in many circuit applications. However, we shall include the effect in subsequent treatments in Chap. 9. Figure 7.11 shows the general dc characteristics for common base and common emitter configurations, as predicted by the Ebers-Moll equations and the theory presented in this chapter. In the next section we consider the voltage limitation due to breakdown and in Chaps. 8, 9, and 10 we discuss various effects which make the characteristics depart from the rather idealized behavior predicted by the above theory.

7.6 VOLTAGE LIMITATIONS IN THE TRANSISTOR

It is relatively simple to extend the results for avalanche multiplication in diodes to the case of the transistor. The formula for breakdown of a one-sided abrupt junction [Eq. (3.125)] may be applied directly to the base-collector junction

$$V_{cbr} = \frac{\varepsilon}{2q} \frac{E_{br}^2}{N_{epi}} \tag{7.46}$$

where N_{epi} is the doping level of the collector epitaxial layer. This assumes that its thickness W_{epi} is sufficient for the space-charge not to extend to the heavily doped N^+ substrate, and also neglects the fact that the breakdown voltage will be reduced because of the radius of curvature of the sidewall region. It also assumes implicitly that space-charge penetration into the moderately doped base can be neglected.

For the emitter base junction, the problem is more complicated, because the one-sided abrupt junction approximation is very poor in this case. A crude estimate for the emitter-base breakdown voltage can be made by using an "average"

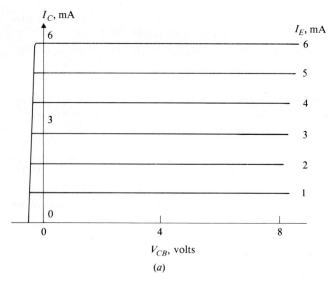

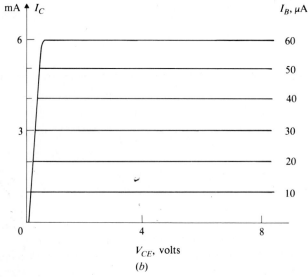

FIGURE 7.11
Direct current transistor characteristics based on the simplified theory of this chapter. (*a*) Common base configuration; (*b*) common emitter configuration.

doping for the base. However, a much better approximation is to assume a linearly graded emitter-base junction with a gradient "*a*" and use Eq. (3.126), with Eq. (2.50), to give the value of "*a*."

An effect special to the transistor (not present in diodes) can occur if the base-collector space-charge layer extends right across the neutral base; this happens if the collector epitaxial layer doping is sufficiently high and leads to the "base punch-through" condition. This effect is not normally present but is discussed further in Chap. 9 in connection with Early voltages.

The most important breakdown condition in transistors is the collector-emitter breakdown voltage, often referred to as the sustaining voltage or BV_{CEO}. It is always less than the value of base-collector breakdown voltage discussed above. Let us examine this limit. In Sec. 3.8 we discussed avalanche multiplication. Equation (3.123) provided a simple empirical expression which we reproduce here:

$$M = \frac{1}{1 - (V_{CB}/V_{cbr})^n} \tag{7.47}$$

We can modify the Ebers-Moll equations for the normal active region by multiplying all the current crossing the base-collector junction by this factor. Thus Eqs. (7.40) to (7.43) become

$$I_E = -I_F - M\alpha_R I_{CS} \tag{7.48}$$

where
$$I_F = I_{ES} \left[\exp \left(V_{BE}/V_t \right) - 1 \right]$$

$$I_C = M(\alpha_F I_F + I_{CS})$$

$$= M\alpha_F I_F + M I_{CS} \tag{7.49}$$

If the base is open-circuited, the emitter and collector currents become equal ($I_E = -I_C$) and therefore, combining the above expressions,

$$I_C = \frac{M I_{CS}(1 - \alpha_F \alpha_R)}{1 - M\alpha_F} \tag{7.50}$$

$$= \frac{M I_{CO}}{1 - M\alpha_F} \sim \frac{I_{CO}}{1 - M\alpha_F} \tag{7.51}$$

When the term $M\alpha_F$ becomes equal to unity, the emitter and collector currents tend to infinity. Note that since α_F lies in the range 0.95 to 0.99, only a very small amount of multiplication M is required to create this "breakdown" situation. The corresponding collector-emitter breakdown voltage BV_{CEO} will therefore be considerably less than the collector-base breakdown voltage V_{cbr} and is given by the condition

$$M\alpha_F = 1 = \frac{\alpha_F}{1 - (BV_{CEO}/V_{cbr})^n} \tag{7.52}$$

or

$$BV_{CEO} = V_{cbr}(1 - \alpha_F)^{1/n}$$

$$= \frac{V_{cbr}}{\beta_0^{1/n}} \tag{7.53}$$

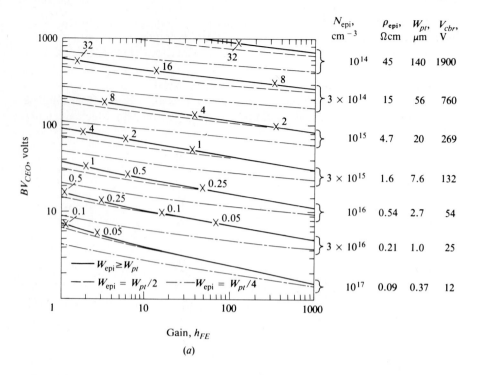

N_{epi}, cm^{-3}	ρ_{epi}, Ωcm	W_{pt}, μm	V_{cbr}, V
10^{14}	45	140	1900
3×10^{14}	15	56	760
10^{15}	4.7	20	269
3×10^{15}	1.6	7.6	132
10^{16}	0.54	2.7	54
3×10^{16}	0.21	1.0	25
10^{17}	0.09	0.37	12

Gain, h_{FE}

(a)

FIGURE 7.12

Computed BV_{CEO} versus h_{FE} curves, obtained by numerical solution of the ionization integral, for various collector doping levels using the BIPOLE program: (a) uses ionization data of Crowell, Sze, and Lee et al. [8, 9]; (b) uses ionization data of Van Overstraeten and De Man [10, 11, 12]. The solid

Since the current gain β_0 is typically of order 100 and since n is typically 2 to 3, it is clear that the value of BV_{CEO} can be less than V_{cbr} by a factor 2 to 4. This open-circuit base condition is often used as a safe upper limit on collector voltage. It should be realized, however, that if the emitter current is fixed by the circuit (as in common-base operation) the voltage may be increased toward the value V_{cbr}.

It is relatively simple to compute exactly the values of BV_{CEO} for various doping levels since the collector is normally uniformly doped and a one-sided abrupt junction approximation may be used in the solution of Poisson's equation. From Chap. 3, the result for the ionization integral may be used and numerically integrated from the $E(x)$ solution to Poisson's equation. From Eq. (7.51), the condition for BV_{CEO} occurs when $M\alpha_F = 1$. Since the ionization integral value is $S_e = 1 - 1/M$ and since $\alpha_F = \beta_0/(1 + \beta_0)$ we obtain the relation for collector-emitter breakdown:

$$\beta_0 = \frac{1 - S_e}{S_e} \tag{7.54}$$

Figure 7.12 shows computed curves of BV_{CEO} for various collector epitaxial

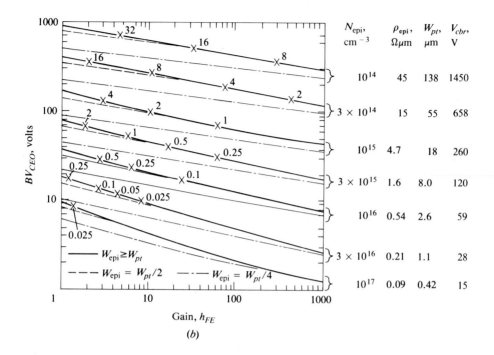

curves are for epitaxial layer thickness greater than the reach-through value W_{pt}; the broken curves are for thicknesses equal to one-half and one-quarter W_{pt}, respectively. The numbers on each curve represent junction radius of curvature in microns for cylindrical breakdown equal to the value on the vertical axis. (*After Roulston and Depey* [13]. *Reproduced with permission IEE.*)

layer doping levels as a function of the gain β_0. The numerical results also indicate the breakdown limits due to the radius of curvature effect and reach-through. Reach-through is the condition which occurs when the collector-base space-charge layer extends from the base-collector junction to the NN^+ low-high junction.

For example, from Eq. 7.12(*a*), consider a doping level of 10^{15} cm^{-3}. For an epitaxial layer thicker than or equal to 4.7 microns, the collector-base breakdown voltage is 269 V. For a gain of 100, the value of BV_{CEO} is 110 V. This requires a radius of curvature of almost 4 microns if the collector-base breakdown voltage is to be greater than 110 V. We also deduce from the broken curve for $W_{epi} = W_{pt}/2$ that the value of BV_{CEO} is only very slightly reduced if the epitaxial layer thickness is reduced by a factor of two from 4.7 to 2.35 microns.

It is significant that even numerical computations must rely on imperfectly known physical data—in this case the ionization coefficients. These appear to be dependent on the material, given the spread in published values versus electric field [13]. It is because of this fact, combined with the reduction in breakdown voltage due to sidewall radius of curvature effects, that for many purposes the semi-empirical equations (7.46) and (7.53) are adequate.

7.7 CONCLUSIONS

In this chapter, we have introduced the bipolar transistor structure in simplified form, but including all the essential features fundamental to a basic understanding. The following chapters will deal with the more realistic case of non constant doping levels in the base and emitter regions, and with the effects of non constant depletion-layer widths; high current effects are then dealt with.

PROBLEMS

7.1. Show that the current gain, when limited only by transport across the base, including base recombination, is τ_b/t_{bb}.

7.2. A transistor has an emitter 20 μm × 500 μm, and a collector 80 μm × 500 μm, with base contacts running the full length of the 500 μm emitter. The junction depths are 1 and 2 μm from the surface and the N^+ collector region starts at 8 μm from the surface. The (constant) doping levels in the emitter (N type) base and collector are 10^{18}, 10^{17}, and 10^{16} cm^{-3}, respectively. Using this information, calculate the gain β, the emitter-base and base-collector junction capacitances, and the f_t for $V_{CB} = 10$ V and for two values of collector current I_c: 1 mA and 10 mA.

7.3. Using the information in Prob. 7.2, calculate the base resistance and the maximum oscillation frequency.

7.4. As V_{CB} increases in the reverse bias direction, so does the depletion-layer thickness in the base. Using Eq. (2.16) applied to each side of the b-c junction, and Eq. (2.17) on the collector side (assuming the latter to be at a significantly lower doping level than the base), derive an expression for the neutral base width as a function of the metallurgical base width ($X_{jcb} - X_{jeb}$). Calculate this value of W_b for the above transistor for $V_{CB} = 10$ and 20 V. Hence calculate the change in I_C and find the corresponding small signal (or incremental) output resistance $\Delta V_{CB}/\Delta I_C$. This phenomenon is known as *base width modulation* and will be treated further in connection with Early voltage in Chap. 9.

7.5. For the transistor data given above, plus the carrier lifetime in the collector epitaxial layer $\tau_{epi} = 0.1$ μs, calculate the values of the four Ebers-Moll model parameters, using Eqs. (7.33) and (7.34).

7.6. By rearranging Eqs. (7.33) to (7.36) to take account of the fact that the emitter area is not normally the same as the collector area, show that the reciprocity condition [Eq. (7.37)] remains valid.

7.7. Using the breakdown formula in Chap. 7, calculate the base-collector and collector-emitter breakdown voltages for the transistor described above.

7.8. Using Fig. 7.11(b), determine the collector doping level and thickness (assume non-reach-through design), for $BV_{CEO} = 50$ V for two values of gain β equal to 10 and 100. From the corresponding V_{cbr} values given by this figure, use Eq. (7.53) to determine the equivalent values of n in the avalanche multiplication law.

REFERENCES

1. S. Ghandhi, *VLSI Fabrication Principles*, Wiley, New York (1983).
2. P. E. Gray, D. DeWitt, A. R. Boothroyd, and J. F. Gibbons, "Physical electronics and circuit models of transistors," *SEEC*, vol. 2, Wiley, New York (1964).

3. D. J. Roulston, "Low current base-collector boundary conditions in GHz frequency transistors," *Solid State Electronics*, vol. 18, pp. 845–847 (October 1975).

4. R. C. Pritchard, J. B. Angell, R. B. Adler, J. M. Early, and W. M. Webster, "Transistor internal parameters for a small-signal representation," *Proc. IRE*, vol. 49, pp. 725–739 (April 1961).

5. J. J. Ebers and J. C. Moll, "Large-signal behavior of junction transistors," *Proc. IRE*, vol. 42, p. 1761 (1954).

6. L. W. Nagle and D. O. Pederson, "Simulation program with integrated circuit emphasis (SPICE)," *Proc. 16th Midwestern Symp. Circuit Theory*, Waterloo, Ontario (1973).

7. I. Hajj, K. Singhal, J. Vlach, and P. Bryant, "WATAND—a program for the analysis and design of linear and piecewise linear networks," *Proc. 16th Midwestern Symp. Circuit Theory*, Waterloo, Ontario (1973).

8. R. van Overstraeten and H. de Man, "Measurement of the ionization rates in diffused silicon PN junctions," *Solid State Electronics*, vol. 13, pp. 583–608 (May 1970).

9. A. G. Chynowith, "Ionization rates for electrons and holes in silicon," *Phys. Rev.*, vol. 109, pp. 1537–1540 (1958).

10. C. R. Crowell and S. M. Sze, "Temperature-dependence of avalanche multiplication in semiconductors," *Appl. Phys. Lett.*, vol. 9, pp. 242–244 (1966).

11. C. A. Lee et al., "Ionization rates of holes and electrons in silicon," *Phys. Rev.*, vol. 134, pp. A761–A773 (1964).

12. J. Conradi, "Temperature effects in silicon avalanche diodes," *Solid State Electronics*, vol. 17, pp. 99–106 (1974).

13. D. J. Roulston and M. Depey, "Emitter-collector breakdown voltage BV_{CEO} versus gain h_{fe} for various NPN collector doping levels," *Electronics Letters*, vol. 16, pp. 803–804 (October 1980).

CHAPTER
8

ANALYSIS WITH REAL IMPURITY PROFILES— THE EMITTER

8.1 INTRODUCTION

Real bipolar transistors are made, as described briefly in the last chapter, using either diffusion of P and N type impurities or ion implants followed by some thermal processing. In general, the final impurity profiles of the base and emitter dopants are often approximated by gaussian or erfc type or, in some cases, exponential functions. In this and the subsequent chapter, we consider how the electrical characteristics of the device are related to the actual impurity profile.

The nonuniformly doped emitter is perhaps the most difficult region of the transistor to examine using analytic methods. Unfortunately, it is also a very critical region from a transistor design viewpoint, since it is this region which determines the current gain β_0. The analysis is complicated because of the simultaneous presence of the following physical mechanisms, of similar quantitative importance: heavy doping bandgap narrowing, Shockley-Read-Hall recombination, Auger recombination, built-in retarding field. In this chapter we present a theoretical study of this region which indicates quite clearly the trends for different emitter doping profiles and which is sufficiently accurate for estimating the gain (in conjunction with the base region parameters) for a particular choice of emitter parameters.

A separate section deals with the modern polysilicon emitter, with theoretical explanations of its operation for various cases, including tunneling and thermionic emission across the very thin interfacial layer between the polysilicon and monosilicon regions.

8.2 APPROXIMATIONS TO REAL IMPURITY PROFILE

Figure 8.1(a) shows a typical emitter impurity profile. The effective doping level is shown on the same figure. Figure 8.1(b) shows an approximation to the real effective doping profile. This consists of a retarding field region, thickness W_{e1}, near the emitter-base space-charge layer, with a region, thickness W_{e2}, of constant doping near the surface. Figure 8.1(c) shows the corresponding minority carrier distribution. If we represent the retarding field region by an exponential impurity profile

$$N(x) = N_0 \exp\left(\frac{x}{x_{0e}}\right) \tag{8.1}$$

the analytic solution becomes quite tractable. An exponential profile gives an excellent fit to a gaussian law over a few decades of $N(x)$ provided the depth x under consideration is a couple of characteristic lengths below the surface, $x = 0$. We shall first examine the case where no recombination is present. This analysis is based on the treatment given in Chap. 3 for the diffused region of a diode, but with the refinement of the addition of the constant doping region. We then extend the result to include both SRH and Auger recombination and show that, for all except the most modern very shallow emitter devices, it is essential to include recombination if the results are to be meaningful.

8.2.1 The Analytic Solution, No Recombination

Following the treatment of Sec. 3.4.2, where the retarding field was considered in some detail, it is clear that the minority carrier distribution will fall for increasing x ($x = 0$ is here defined to be the space-charge layer boundary for convenience). The hole distribution may be well approximated by an exponential decay as was

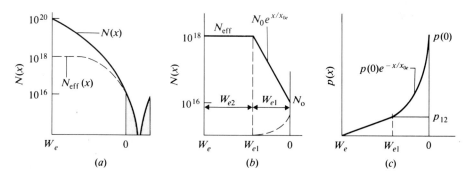

FIGURE 8.1
(a) Typical diffused emitter profile with effective profile shown by the broken curve—log vertical scale; (b) approximation to real effective profile—log vertical scale; (c) minority carrier distribution—linear vertical scale.

shown by Eq. (3.37), although now we have a different boundary condition. The original "sink" condition used in the derivation of Eq. (3.37) is replaced by a carrier concentration determined by transport through the uniformly doped region thickness W_{e2} in Fig. 8.1. Since transport through this W_{e2} region is by diffusion, with no retarding field, it is a fair approximation to argue that the hole concentration p_{12} at $x = W_{e1}$ is determined primarily by the properties of the retarding field region. Based on our previous work, the hole distribution will be exponential:

$$p(x) = p(0) \exp\left(-\frac{x}{x_{0e}}\right) \tag{8.2}$$

The value at $x = W_{e1}$ is given by

$$p_{12} = \frac{p(0)}{r_e} \tag{8.3}$$

where $p(0)$ is the injected hole concentration

$$p(0) = \frac{n_i^2}{N_0} \exp\left(\frac{V_{BE}}{V_t}\right) \tag{8.4}$$

N_0 is the doping level at the space-charge layer boundary and the parameter

$$r_e = \frac{N_{\text{eff}}}{N_0} \tag{8.5}$$

where N_{eff} is the "saturation" level of the effective doping level $\sim 10^{18}$ cm^{-3}. Having thus obtained the value of p_{12}, the current flowing through the W_{e2} region is (neglecting recombination) given by

$$J_h = \frac{qD_{p2}\,p_{12}}{W_{e2}} \tag{8.6}$$

Here we use D_{p2} to denote the average hole diffusion coefficient in the heavily doped region W_{e2}; its value will be quite low (of order 1 cm^2/s). This expression for current may be rewritten in terms of the applied base-emitter voltage by using Eqs. (8.2) to (8.5):

$$J_h = \frac{qD_{p2}\,n_i^2}{N_{\text{eff}}\,W_{e2}} \exp\left(\frac{V_{BE}}{V_t}\right) \tag{8.7}$$

Notice the rather surprising result that it is only the parameters describing the heavily doped region which enter into this equation. This is true only in the case where recombination in the W_{e1} region is neglected. Let us now proceed to modify the above result by including the effect of recombination in both the retarding field and uniformly doped regions. We will, as usual, make the simplifying assumption that the shape of the distribution remains unaltered in the presence of recombination. This can be shown to be an excellent approximation for the retarding field region.

8.2.2 Solution in the Presence of Recombination

The current in region W_{e1} due to recombination with an effective average lifetime τ_{e1} is given by

$$J_{re1} = \frac{Q_{e1}}{\tau_{e1}} \tag{8.8}$$

where Q_{e1} is the minority carrier charge from $x = 0$ to $x = W_{e1}$. If W_{e1} is at least two or three times the value x_{0e}, that is, if the value of r_e is at least ten, the charge may be well approximated as

$$Q_{e1} = qp(0)x_{0e} \tag{8.9}$$

The recombination current may thus be written as:

$$J_{re1} = \frac{qx_{0e}\,n_i^2}{\tau_{e1}\,N_0} \exp\left(\frac{V_{BE}}{V_t}\right) \tag{8.10}$$

In the region with constant effective doping level, the recombination current is given by

$$J_{re2} = \frac{Q_{e2}}{\tau_{e2}} = \frac{qp_{12}\,W_{e2}}{2\tau_{e2}} \tag{8.11}$$

This may be written in terms of the applied bias voltage using Eqs. (8.3) to (8.5) as

$$J_{re2} = \frac{qW_{e2}\,n_i^2}{2\tau_{e2}\,N_{\text{eff}}} \exp\left(\frac{V_{BE}}{V_t}\right) \tag{8.12}$$

Combining the three components of hole current [Eqs. (8.7), (8.10), and (8.12)] gives the total current injected into the emitter

$$J_{p\,\text{tot}} = \frac{qD_{p2}\,n_i^2}{N_{\text{eff}}\,W_{e2}} K_e \exp\left(\frac{V_{BE}}{V_t}\right) \tag{8.13}$$

where

$$K_e = 1 + \left[\frac{x_{0e}\,W_{e2}}{L_{e1}^2}\right]\frac{N_{\text{eff}}}{N_0} + \left[\frac{W_{e2}^2}{2L_{e2}^2}\right]$$

and where we have substituted $L_{e1} = \sqrt{D_{p2}\tau_{e1}}$ and $L_{e2} = \sqrt{D_{p2}\tau_{e2}}$.

Let us now examine the relative magnitudes of the terms contributing to the total injected current. For a typical diffused or implanted emitter, $W_{e1} \sim W_{e2} \sim W_e/2$. Since the doping ratio $N_{\text{eff}}/N_0 \sim 100$, the value of x_{0e} is of order $W_{e1}/4$. The above expression for K_e may thus be written for typical profiles as

$$K_e = 1 + \left[\frac{W_e}{L_{e1}}\right]^2\frac{1}{1600} + \left[\frac{W_e}{L_{e2}}\right]^2\frac{1}{8} \tag{8.14}$$

In region W_{e1} the doping level changes from a value of order 10^{16} cm^{-3} at the space-charge layer boundary for typical bias, to a value determined by bandgap narrowing at $x = W_{e1}$. The recombination will be primarily due to

Shockley-Read-Hall mechanisms and will be doping-dependent. Figure 1.11 shows a recent collection of measured values of lifetime versus doping. It is seen that the value of minority carrier lifetime for this doping range lies in the range 1 to 100 ns or more.

In region W_{e2} it may safely be assumed that Auger recombination will dominate. Using the range of published values of Auger recombination coefficient C_n [1], the lifetime in this region will lie in the range 0.01 to 100 ns.

Taking typical average values for lifetimes in the two regions, $\tau_{e1} = 100$ ns and $\tau_{e2} = 0.1$ ns gives $L_{e1} = 3$ μm and $L_{e2} = 0.1$ μm; we thus obtain the following results for K_e in Eq. (8.14):

$$\text{For } W_e = 1.0 \text{ } \mu\text{m} \qquad K_e = 1 + 0.8 + 12$$

$$\text{For } W_e = 0.2 \text{ } \mu\text{m} \qquad K_e = 1 + 0.03 + 0.5$$

$$\text{For } W_e = 0.1 \text{ } \mu\text{m} \qquad K_e = 1 + 0.01 + 0.125$$

It is quite clear that even if substantially different values are used for the profile and lifetime parameters, the behavior changes dramatically from the case for deep (1 micron) emitters, where recombination increases the injected current by an order of magnitude, to shallow (0.1 micron) emitters, where recombination is virtually negligible. In the latter case, the hole current has the same value at the space-charge layer boundary as at the ohmic contact; the emitter is referred to as "transparent."

Figure 8.2 shows computed minority carrier distributions and current densities for different conditions, from the emitter-base space-charge layer boundaries to the surface/metal contact. Note the relative insensitivity of the minority carrier distribution to the value used for recombination lifetime. Note also the drastic decrease in hole current density for the deep emitter case (due to recombination) and the relatively constant current density for the shallow emitter structure, where recombination is almost negligible.

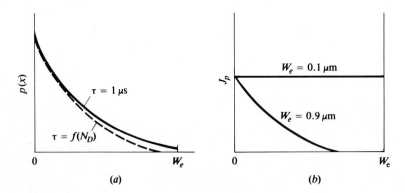

(a)   *(b)*

FIGURE 8.2
Results for a gaussian emitter profile: (*a*) Computed minority carrier concentration for two lifetime conditions, $W_e = 0.9$ μm; (*b*) current density versus distance in the emitter for two different thicknesses, $\tau = f(N_D)$.

8.2.3 A Single Region Analytical Model for the Emitter

In Fig. 8.3 we have included curves to model the combined SRH and Auger lifetime-dependence on doping level from Fig. 1.11 based on the empirical formulation of the form proposed by Selvakumar [2] and given by Eq. (1.39)

$$\frac{1}{\tau_p} = \frac{C_s^2 N^2}{D_p(N)n_{ie}^4} \tag{8.15}$$

where N is the doping level, D_p is the hole diffusion coefficient, and n_{ie} is the intrinsic carrier concentration as modified from the value n_i due to bandgap narrowing. It is seen that a reasonably good fit can be obtained for the various sets of available data.

Selvakumar [3] showed that using this form of lifetime versus doping dependence, the following analytical expression for hole saturation current

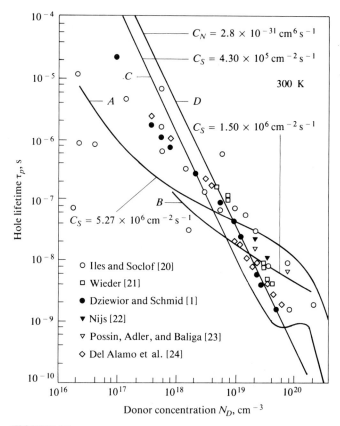

FIGURE 8.3

Measured minority carrier lifetimes versus doping levels for N type silicon, with curves fitted using Eq. (8.15). (*After Selvakumar and Roulston [4]. Reprinted with permission from Solid State Electronics, Pergamon Press PLC.*)

density injected into the emitter is obtained:

$$J_p = qC_s \frac{\cosh [Ki\,(W_E)] + S_f \sinh [Ki\,(W_E)]}{\sinh [Ki\,(W_E)] + S_f \cosh [Ki\,(W_E)]} \qquad (8.16)$$

where

$$Ki\,(W_E) = C_s \int_0^{W_E} \frac{N(x)}{D_p(x)n_{ie}^2}\, dx$$

and

$$S_f = \frac{C_s N(0)}{S_p n_{ie}^2(0)}$$

S_p is the effective hole recombination velocity at the surface. For the case under consideration, with metal contacts, S_p is so high that S_f is very small.

Figure 8.4 shows the normalized current density versus the Ki number for a range of S_f values.

The Ki number contains the minority carrier transport properties associated with bulk transport, while the S_f term contains surface recombination effects. This is a rather attractive formulation for studying emitter behavior and has been shown [4] to give good agreement with numerical simulations. For thin emitters the value of Ki is small. For example, a silicon emitter with $W_E = 0.25 \ \mu m$ with a gaussian impurity profile having a surface concentration

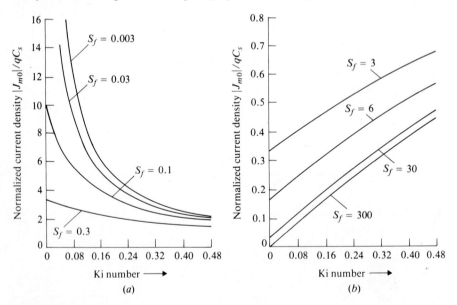

FIGURE 8.4
Normalized emitter current density versus Ki number for various values of S_f using Eq. (8.16). (*After Selvakumar and Roulston* [4]. *Reprinted with permission from Solid State Electronics, Pergamon Press PLC.*)

$N(0) = 10^{20}$ cm^{-3} has $Ki = 0.173$ (assuming a Slotboom bandgap narrowing model [5], the mobility data given in Chap. 1 [6], and with $C_s = 3.16 \times 10^6$ cm^{-2} s^{-1}). In this case the injected current depends strongly on both the bulk and surface properties, as can be seen from the figure. The importance of reducing the surface recombination velocity is apparent from the figure—this leads us to consideration of the polysilicon contacted emitter to be discussed in the next section. For this case, and for $S_f > 1$, Eq. (8.16) can be simplified for a very thin emitter to give

$$\frac{J_p}{qC_s} = \frac{1}{S_f} + Ki\,(W_E)$$

In the limit of zero thickness, giving $Ki = 0$, the current is determined entirely by S_f.

8.3 THE POLYSILICON EMITTER

8.3.1 Structural Details

Figure 8.5 shows the cross section of the simplest structure currently in widespread use with a polysilicon emitter contact. This arrangement was originally used because of the self-alignment possible between the emitter contact and the emitter "diffusion" [7]. The details of the process are given in Sec. 14.3. It very soon became apparent, however, that the polysilicon emitter process yielded higher current gain, thus providing a double advantage [8]. Two slightly different fabrication processes are available. An N$^+$ polysilicon layer in close contact with the N$^+$ monocrystalline emitter is obtained in practice by using a dip etch (HF) prior to the polysilicon deposition. Measurements [19] indicate that there exists in this case an oxide layer of about 7 Å thickness. A device in which a slightly (but significantly from an electrical point of view) thicker (of order 14 Å) interfacial oxide layer exists between the polysilicon and the monocrystalline regions

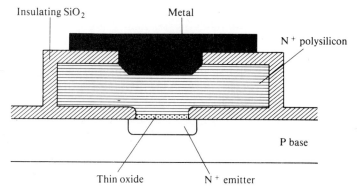

FIGURE 8.5
Basic polysilicon emitter structure.

is obtained when a standard RCA clean is used prior to polysilicon deposition, and has quite a profound effect on the electrical characteristics.

In both cases the dc current gain is considerably higher than would be expected using standard current transport equations with the same mobility and lifetime data as for monocrystalline silicon.

De Graaf and De Groot [9] studied the gain mechanism on the basis of the interfacial oxide layer being the controlling factor. They provided algebraic relations for tunneling and showed that with layer thickness of order 20 Å and effective barrier heights for hole-tunneling current of about 1 eV the theory could explain measured results. Subsequently Ning and Isaac [10] used a model for transport in the polysilicon and concluded that measured results were satisfactorily explained if the hole mobility in the polysilicon layer was considerably lower than in a monocrystalline layer of the same doping level. Their measured mobility lent support to this possibility. Other workers [11] considered trapping and recombination at grain boundaries, and reduced mobility in a 0.05 μm region of the polysilicon layer as being the explanation for the increased current gain of polysilicon emitter transistors.

We shall now proceed to examine two cases from a theoretical standpoint. The treatment can be simplified by considering only the case where there is one grain in the vertical direction. Earlier models [12, 13] took multigrain structures into account, but experimental evidence shows that grain dimensions are such that often only one to three grains exist in the vertical direction and that for typical lifetime and mobility values the hole current is determined primarily by transport in the first grain. Even if this is not the case, the detailed grain structure is seldom, if ever, completely characterized, so use of a single polysilicon region characterized by its effective doping level, mobility, and carrier lifetime is convenient. This is the approach which we shall use here [16].

8.3.2 Case Where No Interfacial Layer is Present

In this situation, the polysilicon layer is treated almost the same as any silicon layer of the same doping and thickness. Solution of the current transport equations for a hole concentration p_s at the polysilicon/monocrystalline silicon interface (shown in Fig. 8.6) will give the same result as that obtained in Chap. 3 for a diode of arbitrary region thickness. Let this thickness be W_{pol}. The current is:

$$J_{ps} = \frac{qD_{\text{pol}}p_s}{L_{\text{pol}}}\frac{1}{\tanh\left(W_{\text{pol}}/L_{\text{pol}}\right)} \tag{8.17}$$

Experimental evidence [10] indicates that the hole mobility (and hence diffusion coefficient) in some N type polysilicon may be less than the mobility in monocrystalline silicon with the same doping level by a factor r_μ where $r_\mu \sim 0.07$ [17]. This is favorable to device characteristics since it gives a lower current, and hence a higher current gain, than would otherwise be the case.

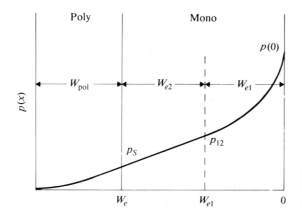

FIGURE 8.6
Minority carrier distribution in polysilicon emitter.

The results obtained for the standard emitter may now be extended to this new situation. Consider the overall carrier distribution from emitter-base space-charge layer boundary to the metal contact at the surface shown in Fig. 8.6.

It will be convenient to introduce the concept of recombination velocity S_p at the polysilicon/monosilicon interface where

$$S_p = \frac{J_{\text{pol}}}{q p_s} \tag{8.18}$$

The current density in the highly doped W_{e2} part of the emitter is given in the absence of recombination by

$$J_p = q D_{p2} \frac{(p_{12} - p_s)}{W_{e2}} \tag{8.19}$$

Using the above definition of surface recombination velocity, and recognizing the equality of the currents in the mono and poly regions at the interface, we can express the value of p_s as

$$p_s = \frac{p(0)}{r_e[1 + S_p W_{e2}/D_{p2}]} \tag{8.20}$$

It follows that the current density is given in terms of $p(0)$ by

$$J_p = q p_s S_p = \frac{q p(0) S_p}{r_e[1 + S_p W_{e2}/D_{p2}]} \tag{8.21}$$

Note that we have derived this result for the case of no recombination in the monocrystalline emitter. This is often the situation in practice, because in polysilicon contacted emitters the depth of the N^+ region is usually very shallow, i.e., less than 0.2 μm.

When recombination cannot be neglected, an approach similar to that used previously [Eq. (8.13)], where the minority carrier distribution is assumed to be unaltered when recombination is present, yields a result of the form given by Eq.

(8.13) but with K_e given by

$$K_e = \left[\frac{1}{1 + D_{p2}/S_p W_{e2}}\right]\left\{1 + \frac{W_{e2}(2 + S_p W_{e2}/D_{p2})}{2S_p \tau_{e2}} + \frac{x_{0e} r_e(1 + S_p W_{e2}/D_{p2})}{\tau_{e1} S_p}\right\} \quad (8.22)$$

It may be noted that as S_p decreases due to the properties of the polysilicon layer and interface, the terms involving τ_{e1} and τ_{e2} increase in value. This means that an emitter which is "transparent" (i.e., no change in hole current over distance W_e) for an ohmic contact [Eq. (8.13)] may become "opaque" (i.e., large change in hole current over distance W_e) for a polysilicon contact.

8.3.3 Device With Thin Interfacial Oxide Layer

Figure 8.7 shows the simplified band diagram from metal to monosilicon for the case where a thin interfacial oxide layer, thickness δ, exists between the poly and mono regions. Because of the fact that this layer is very thin, carriers can tunnel through. Band bending [13] at the polysilicon and monosilicon interfaces with the thin oxide is assumed to be negligible, as is the effect of traps—reasonable assumptions for the high doping levels ($\sim 10^{20}$ cm^{-3}) used in emitters. It is also possible that thermionic emission occurs at high temperatures.

The hole current tunneling through the thin oxide layer is given by [13]

$$J_{pt} = B_h\left[\exp - \frac{E_g + E_{sn}}{kT}\right]\left[\exp\left(\frac{q\phi_R}{kT} + \frac{qC_h V_{ox}}{2}\right) - \exp\left(\frac{q\phi_L}{kT} - \frac{qC_h V_{ox}}{2}\right)\right] \quad (8.23)$$

where $E_{sn} = E_{Fn} - E_c$ is the spacing between the (electron) Fermi levels, assumed constant, and the conduction band edge; ϕ_R and ϕ_L are the respective quasi-

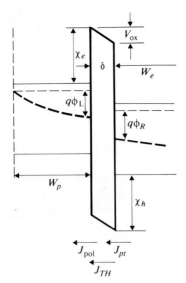

FIGURE 8.7
Simplified band diagram for a polysilicon emitter transistor with a thin interfacial layer between the poly and mono regions. (*After Eltoukhy and Roulston* [13]. *Reproduced with permission © 1982 IEEE.*)

Fermi level separations on the right and left sides of the oxide layer; V_{ox} is the voltage drop across the oxide layer. The constants C_h and B_h are given by

$$C_h = \frac{2\pi\delta}{h} \sqrt{\frac{2m_i^*}{\chi_h}}$$

$$B_h = A_h T^2 \frac{\exp(-b_h)}{1 - C_h kT}$$

where h is Planck's constant, m_i^* the effective mass of holes in the oxide layer, χ_h is the barrier height for holes tunneling through the oxide layer, thickness δ, A_n is the effective Richardson constant for holes, and b_h is given by

$$b_h = \frac{4\pi\delta}{h} \sqrt{2m_i^* \chi_h}$$

The above result is based on quantum mechanical tunneling using the Wentzel-Kramers-Brillouin (WKB) approximation [14, 15], with the linear term in the Taylor expansion of the tunneling probability included. Although an approximate result, the representation of the interfacial oxide layer as a rectangular barrier is already a simplification and it is consistent with the rather sparse knowledge available for a given process.

At moderate current densities, the voltage across the oxide, V_{ox}, is small and may be neglected, to obtain a simplified expression for hole tunneling current

$$J_{pt} = \frac{B_h}{N_V} (p_R - p_L) \tag{8.24a}$$

$$= q S_{TU}(p_R - p_L) \tag{8.24b}$$

where p_R and p_L are the hole concentrations on the right and left sides of the barrier, and S_{TU} is an effective recombination velocity for tunneling.

If thermionic emission is present, the corresponding current is given by

$$J_{TH} = \frac{A_h T^2}{N_V} \left[\exp\left(-\frac{\chi_h}{kT} \right) \right] (p_R - p_L) = q S_{TH}(p_R - p_L) \tag{8.25}$$

The total current density at the interface is:

$$J_{pol} = J_{pt} + J_{TH}$$

and the combined recombination velocity at the monocrystalline surface due to injection into the poly, tunneling, and thermionic emission across the interface is given by

$$S_{eff} = S_p \frac{S_{TH} + S_{TU}}{S_p + S_{TH} + S_{TU}} \tag{8.26}$$

Figure 8.8 shows computed curves of S_{eff} versus oxide thickness for various hole barrier heights. As noted previously, the oxide thickness for the RCA clean process has been found to be approximately 14 Å. Barrier heights for holes have

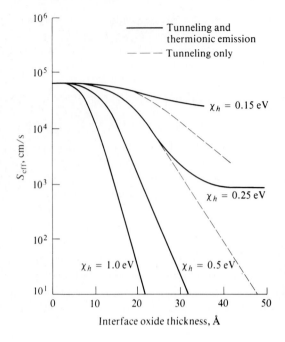

FIGURE 8.8
Effective recombination velocity at the monocrystalline to oxide interface for a polysilicon emitter, versus oxide thickness, in angstroms. The polysilicon doping level is taken to be 10^{20} cm^{-3} and its thickness is 0.3 μm. The Auger process is assumed to be the only recombination mechanism present in the polysilicon layer. (*After Bakker, Roulston, and Eltoukhy* [16]. *Reproduced with permission © 1984 IEE.*)

been deduced [17] using this model to be approximately 1.1 eV. It is thus seen that recombination velocities of order 10^3 cm/s are possible. Figure 8.9 shows results for recombination velocity for various doping levels. Normally the polysilicon doping level is the same as at the surface of the monocrystalline region since the dopant is introduced after polysilicon deposition. 10^{20} cm^{-3} is a typical value.

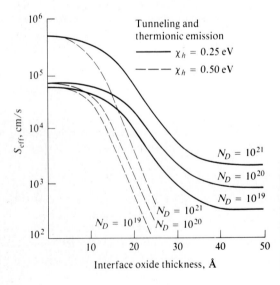

FIGURE 8.9
Effective recombination velocity versus interfacial oxide thickness (angstroms) for various polysilicon doping levels; data not given is the same as for Fig. 8.8. (*After Bakker, Roulston, and Eltoukhy* [16]. *Reproduced with permission © 1984 IEE.*)

8.3.4 Emitter Resistance Due to Oxide Voltage Drop

The majority carrier electron current which tunnels through the thin interfacial oxide layer creates the voltage drop V_{ox}. The expression for electron current density may be written in the absence of band-bending adjacent to the interfacial layer as [13]

$$J_{nt} = A_e T^2 \exp\left[\frac{E_c - E_F}{kT}\right] \frac{\exp(-b_e)}{1 - C_e kT} \, 2 \sinh\left[\frac{qC_e V_{ox}}{2}\right] \qquad (8.27)$$

where

$$b_e = \frac{4\pi\delta}{h} \sqrt{2m_i^* \chi_e}$$

$$C_e = \frac{2\pi\delta}{h} \sqrt{2m_i^* / \chi_e}$$

A_e is the modified Richardson constant for electrons, χ_e the effective barrier height for electrons, and m_i^* the effective mass in the oxide layer. For interfacial layer thicknesses of practical interest (< 20 Å) the argument of the sinh term is small with respect to unity. In this case, the sinh can be approximated by its argument and the above equation simplifies to [17]

$$J_{nt} = A_e T^2 \exp\left(\frac{E_c - E_F}{kT}\right) \frac{\exp(-b_e)}{1 - C_e kT} \, qC_e V_{ox} \qquad (8.28)$$

Figure 8.10 shows the computed voltage drop across the oxide layer divided by the collector current, versus base-emitter voltage for different oxide layer thicknesses. The locus showing the approximation used in Eq. (8.28) is also shown.

From Eq. (8.28), the emitter resistance can be obtained [17]

$$R_e = \frac{1 - C_e kT}{AqC_e T^2 A_e} \exp(+b_e) \exp\left[-\frac{E_c - E_F}{kT}\right] \qquad (8.29)$$

This is plotted in Fig. 8.11 versus both the oxide thickness and doping level, using the approximate relation between $(E_c\text{-}E_F)$ and doping level N_d given by Mertens et al. [18].

The values of resistance obtained with a 14 Å layer (corresponding to the RCA clean) can be sufficiently high to seriously degrade the performance in some applications. For this reason, and for reasons of reproducibility, current trends in industry are to use only the HF dip etch method of surface preparation prior to polysilicon deposition. This has been shown [19] to give an interfacial layer of about 7 Å and presents only a small barrier for tunneling.

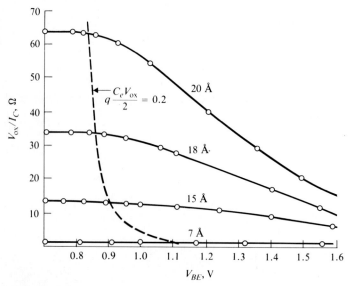

FIGURE 8.10

The computed voltage drop V_{ox} normalized with respect to collector current, plotted versus base emitter voltage V_{BE} for a range of oxide thicknesses, for an emitter area 6 μm × 6 μm. The dashed line shows the conditions under which the expression $qC_e V_{ox}/2 = 0.2$ is satisfied. (*After Ashburn et al.* [17]. *Reproduced with permission © 1987 IEEE.*)

8.3.5   Base Current Dependence on Emitter Data

It is instructive to see how variations in the polysilicon layer and in the mono-crystalline emitter affect the base current, and hence the gain. We will use, for illustrative purposes, computed results for a polysilicon emitter transistor with the following characteristics (after Eltoukhy and Roulston [13]):

Polysilicon layer doping level 10^{20} cm^{-3}

Polysilicon layer thickness   100 nm

Base surface doping level 2×10^{18} cm^{-3}

Emitter base junction depth in monocrystalline region   0.2 μm

Barrier height for electrons   0.6 eV

Barrier height for holes   1.1 eV

Emitter-base space-charge layer lifetime   100 ns

Other data is given in [13]. Figure 8.12 shows the hole current density at the base emitter junction versus V_{BE}. The broken curve is a reference for a metal contact structure. Note that even for the case where there is no interfacial oxide,

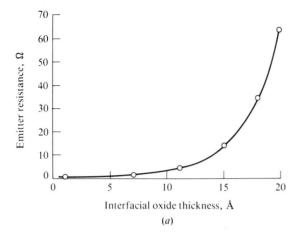

(a)

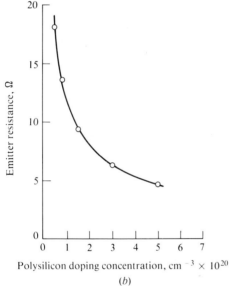

(b)

FIGURE 8.11
(a) Interfacial oxide resistance versus oxide thickness for $\chi_e = 0.4$ eV; (b) oxide layer resistance versus poly-silicon doping level $\chi_e = 0.4$ eV. Emitter area = 6 μm × 6 μm. (*After Ashburn et al.* [17]. *Reproduced with permission © 1987 IEEE.*)

a reduction in hole current occurs, according to Eq. (8.22). As the oxide layer increases in thickness, the hole current is reduced. However, at the same time, for large oxide thicknesses (note that SiO_2 could be intentionally grown), the base emitter voltage increases due to the voltage drop across the oxide layer necessary to maintain electron tunneling current as per Eq. (8.28). In order to see the behavior of current gain, the data of Fig. 8.13 is presented.

The gain for the metal contact reference case is 40. A factor of 2 improvement occurs with the 100 nm thick polysilicon layer and no interfacial oxide. This factor increases to over 4 as the oxide increases to 40 Å. Note that at low collector current densities, the current gain becomes independent of the poly-

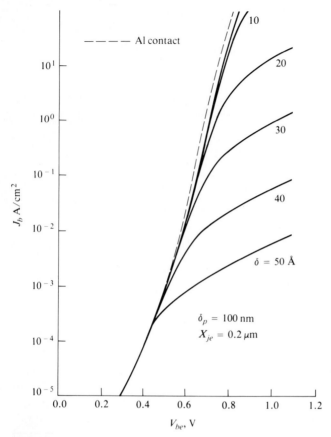

FIGURE 8.12

Hole current density versus base-emitter voltage for an aluminum and a polysilicon contacted emitter for various interfacial oxide layer thicknesses. (*After Eltoukhy and Roulston* [13]. *Reproduced with permission* © *1982 IEEE.*)

silicon and interfacial layer because of emitter-base space-charge recombination current.

Figure 8.14 shows the hole current density versus V_{BE} for two different emitter-base junction depths. It is revealing to note that for the case of the metal contact, increasing the junction depth gives a decrease in hole current injected into the emitter. This is perfectly understandable because at these quite shallow depths there is little recombination (the emitter is almost transparent, see Sec. 8.2.2 and Fig. 8.2); the behavior can best be visualized by imagining a "narrow base diode" emitter—the thicker the "base" of the diode, the lower is the injected current.

However, when the polysilicon layer is incorporated, with a 30 Å interfacial oxide, the reverse behavior occurs. Increasing the monocrystalline junction depth from 0.1 to 0.2 μm actually increases the hole current. This is because now the

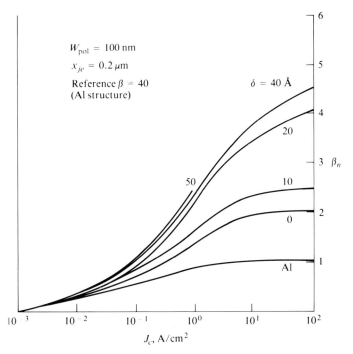

FIGURE 8.13

Normalized current gain versus collector current density for an aluminum contact and for a poly-silicon contacted emitter with various oxide thicknesses. (*After Eltoukhy and Roulston* [13]. *Reproduced with permission* © *1982 IEEE.*)

effective recombination velocity at the monocrystalline surface is so low that recombination in the monocrystalline region dominates [refer to Eq. (8.22)].

In spite of this rather complex dependence on emitter-base junction depth, it is clear from Fig. 8.14 that for a fixed junction depth (refer to the curves for either 0.1 or 0.2 μm), the addition of the polysilicon and interfacial layer always yields a reduction in hole current and hence an improvement in gain.

8.3.6 Experimental Polysilicon Model Verification

Figure 8.15 shows a series of measured and computed results on a test device made with 0.4 micron LPCVD polysilicon, implanted with a dose of 1×10^{16} cm^{-2} [17]. This was followed by a 30 min drive-in at 900°C for the arsenic case and 10 min at the same temperature for the phosphorus-doped device. The oxide thickness for both the HF dip etch and the RCA clean device was obtained by High Resolution Electron Microscopy measurements as 8 ± 3 Å and 14 ± 2 Å respectively [19]. The unknown model parameters (electron tunneling barrier height χ_e, hole tunneling barrier height χ_h, and mobility reduction

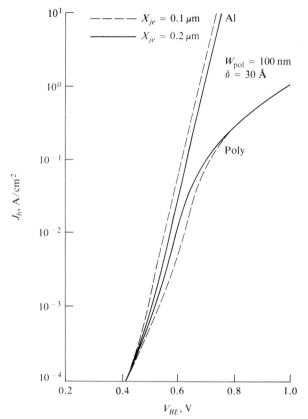

FIGURE 8.14
Hole current density at the emitter-base junction versus base-emitter voltage for an aluminum and a polysilicon contacted emitter for two different values of emitter-base junction depth. (*After Eltoukhy and Roulston* [13]. *Reproduced with permission © 1982 IEEE.*)

parameter r_μ for hole mobility in the polysilicon compared to the mobility in the monocrystalline material for the same doping level) were obtained by fitting as illustrated in the figure. It is seen that the model gives excellent agreement with measured results. It should be noted that there is still some doubt about the true nature of hole transport, and recombination in the doped polysilicon layer and the r_μ mobility parameter should be considered purely as a convenient means of modeling a more complex phenomenon. The modeling of the silicon-polysilicon interface, covering a few atomic layers, is still somewhat simplified because of the lack of reliable information about the precise nature of the SiOx region. It is possible that the equivalent hole barrier height with the HF dip etch treatment is higher than the above-mentioned 1.1 eV value, or that the equivalent thickness is higher than 8 Å. Interpretation of values for this type of situation is clearly not easy and it is possible that the mobility reduction is only a modeling artifice masking other effects. Indeed, recent results [25] indicate that the polysilicon grain size can be of order 0.1 μm. It is thus likely that the effective minority carrier mobility is the same as in monocrystalline silicon. The HF dip etch results

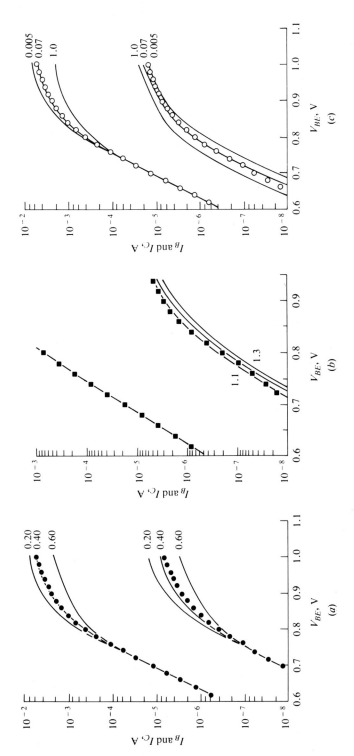

FIGURE 8.15
Measured and computed Gummel plots on a test polysilicon emitter device. (*After Ashburn et al.* [17].) (*a*) Comparison of computed and measured base-emitter voltage for an arsenic-doped RCA device (the effective barrier height for electrons is the variable parameter); (*b*) computed and measured base and collector current versus V_{BE} for phosphorus-doped RCA device (effective barrier height for holes is the parameter); (*c*) computed and measured base and collector current versus V_{BE} for the arsenic-doped HF dip etch device. The hole mobility ratio is the variable parameter. (*Reproduced with permission* © *1987 IEEE.*)

207

can then be explained with the above tunneling model if an oxide thickness of 11 Å is used, with the same barrier height, and the increased mobility value.

Polysilicon emitter transistors have become almost standard in all VLSI and similar high speed applications today. Although the technological improvements have been driven by the need for smaller geometries (achievable with the self-alignment properties of the polysilicon process—see Chap. 14) the enhanced gain is a decided benefit. The higher gain may be "traded" at the design/fabrication stage for a lower gain but an increased base doping, giving a lower value of base resistance. According to Eq. (7.28), this ensures improved maximum oscillation frequency and, by extrapolation, generally improved high speed performance.

Current technology trends are to use HF dip etch followed by thermal annealing sufficient to cause the oxide to break up. For sufficient thermal processing the emitter can then behave almost as if no thin oxide layer were present [19, 26]. In this case one would expect the results of Sec. 8.3.2 to apply. The situation is further complicated by the fact that some epitaxial realignment of the polysilicon layer occurs.

8.4 CONCLUSIONS

In this chapter we have studied the "real emitter" structure. It has been shown that the uniform doped approximation used in Chap. 7 is almost useless in terms of device design and that the variation in doping level, plus high doping effects, must be taken into account. The polysilicon emitter was seen to provide not only a method of self-aligning the metal contact to the emitter but to yield improved gain. In Chap. 14 we discuss the more practical details of polysilicon emitter transistors.

PROBLEMS

8.1. The impurity profile of a bipolar transistor is given by the double gaussian

$$N(x) = N_e \exp -\left(\frac{x}{x_e}\right)^2 - N_b \exp -\left(\frac{x}{x_b}\right)^2 + N_{epi}$$

where $N_e = 10^{21}$, $N_b = 10^{19}$, $N_{epi} = 10^{16}$ cm^{-3}, $x_e = 0.3$ μm, and $x_b = 0.6$ μm.

Plot, using a log-linear scale, this profile and the effective doping profile using Eqs. (1.34) and (1.35). Determine graphically the surface doping level and characteristic length of the exponential which represents a good fit to the effective doping profile near the emitter-base junction, and estimate the widths W_{e1} and W_{e2} defined in Fig. 8.1.

8.2. Using Eq. (2.50), find the linear gradient of the above profile in the vicinity of the emitter-base junction. Hence, using Eq. (2.40), calculate the value of depletion-layer width d_n and the doping level $N(d_n)$ for a total voltage (barrier less applied bias voltage) of 0.1 V.

8.3. From the above results, use Eq. (8.7) to calculate J_h for $V_{BE} = 0.6$ V.

8.4. Calculate the recombination currents J_{re1} and J_{re2} given by Eqs. (8.10) and (8.12) using the typical carrier lifetime data given in the text (0.1 and 100 ns for the heavily and moderately doped regions, respectively).

8.5. Consider the approximation used in writing Eq. (8.2) and the effect of recombination on the shape of the minority carrier distribution in region W_{e1}. The calculation was based on the fact that hole diffusion and drift currents almost cancel [see Eqs. (3.29) and (3.40)]. One can argue that the shape of $p(x)$ will deviate from the assumed exponential decay when the recombination current is equal to the diffusion current.
  (i) Show that this leads to the conditions $L_p = x_0$ where L_p is the hole diffusion length in this region W_{e1} and discuss the implications.
  (ii) Use Eq. (8.6) to determine an expression for the ratio of $J_h/J_{h\,\text{diff}}$ in the absence of recombination and estimate typical values.
  (iii) Derive an expression for the ratio of recombination current to original (no recombination) current for the condition used in part (i) above considering only W_{e1}.
  (iv) Calculate typical values of this ratio for typical emitter impurity profiles.

8.6. This question concerns the hole distribution and current in the uniformly doped polysilicon region of a poly emitter device. Use the Auger recombination formula [Eq. (1.36)] with $C_n = 10^{-31}$ cm^6 s^{-1} with a polysilicon doping level of 10^{20} cm^{-3} and calculate L_p in Eq. (8.17) using (i) the low mobility given in Table 1.2; (ii) the mobility reduced by the factor $r_\mu = 0.07$. Hence discuss the shape of the hole distribution for each case in a 0.3 μm thick polysilicon emitter, and calculate the two values of the recombination velocity S_p at the poly/mono interface.

8.7. Use Eq. (8.23) to derive Eq. (8.24a) for the case where the voltage drop across the oxide layer is negligible.

8.8. Show that the effective recombination velocity given by Eq. (8.26) reduces to the S_p value if the recombination velocity for tunneling is very high, and to the S_{TU} value if the latter is very low (assume $S_{TH} = 0$).

8.9 Examine the contributions to hole current in Eq. (8.22) using the same data as in the discussion of Eq. (8.14), and show that the reduction of the surface recombination velocity to values in the range 10^2 to 10^5 cm/s (due to the replacement of a metal contact by a polysilicon contact) can have the effect of increasing significantly the importance of recombination in the monosilicon emitter region.

8.10. With reference to Fig. 8.15, suggest reasons for the tendency of the I_C and I_b curves to saturate at high V_{BE} values.

REFERENCES

1. J. Dziewior and W. Schmid, "Auger coefficients for highly doped and highly excited silicon," *App. Phys. Lett.*, vol. 31, pp. 346–348 (September 1977).
2. C. R. Selvakumar, "A new minority carrier lifetime model for heavily doped GaAs and InGaAsP to obtain analytical solutions," *Solid State Electronics*, vol. 30, pp. 473–474 (July 1984).
3. C. R. Selvakumar, "Simple general analytical solution to the minority carrier transport in heavily doped semiconductors," *Jnl. Appl. Phys.*, vol. 56, pp. 3476–3478 (December 1987).
4. C. R. Selvakumar and D. J. Roulston, "A new simple analytical emitter model for bipolar transistors," *Solid State Electronics*, vol. 30, pp. 723–728 (July 1987).
5. J. W. Slotboom and H. C. de Graaff, "Measurements of bandgap narrowing in Si bipolar transistors," *Solid State Electronics*, vol. 19, pp. 857–862 (1976).

6. N. D. Arora, J. R. Hauser, and D. J. Roulston, "Electron and hole mobilities in silicon as a function of concentration and temperature," *IEEE Trans. Electron Devices*, vol. ED-29, pp. 292–295 (February 1982).

7. M. Takagi, K. Nakayama, Ch. Tevada, and H. Kamiko, "Improvement of shallow base transistor technology by using a doped polysilicon diffusion source," *J. Jap. Soc. Appl. Phys. (Suppl.)*, vol. 42 (1972).

8. J. Graul, A. Glasl, and H. Murtman, "High-performance transistors with arsenic-implanted polysil emitters," *IEEE Jnl. Solid State Circuits*, vol. SC-11, pp. 491–495 (August 1976).

9. H. C. de Graaff and J. G. de Groot, "The SIS tunnel emitter: a theory for emitters with thin interfacial layers," *IEEE Trans. Electron Devices*, vol. ED-26, pp. 1771–1776 (November 1979).

10. T. H. Ning and R. D. Isaac, "Effect of emitter contact on current gain of silicon bipolar devices," *IEEE Trans. Electron Devices*, vol. ED-27, pp. 2051–2055 (November 1980).

11. A. Neugroschel, M. Arienzo, Y. Komen, and R. D. Isaac, "Experimental study of the minority-carrier transport at the polysilicon-monosilicon interface," *IEEE Trans. Electron Devices*, vol. ED-32, pp. 807–816 (April 1985).

12. A. A. Eltoukhy and D. J. Roulston, "Minority carrier injection into polysilicon emitters," *IEEE Trans. Electron Devices*, vol. ED-29, pp. 961–964 (June 1982).

13. A. A. Eltoukhy and D. J. Roulston, "The role of the interfacial layer in polysilicon emitter bipolar transistors," *IEEE Trans. Electron Devices*, vol. ED-29, pp. 1862–1869 (December 1982).

14. A. T. Fromhold, Jr., *Quantum Mechanics for Applied Physics and Engineering*, Academic Press, New York (1981).

15. R. Stratton, "Volt-current characteristics for tunneling through insulating films," *Jnl. Phys. Chem. Solids*, vol. 33, p. 1177 (1962).

16. G. W. Bakker, D. J. Roulston, and A. A. Eltoukhy, "Effective recombination velocity of poly-silicon contacts for bipolar transistors," *Electronics Letters*, vol. 20, pp. 622–624 (July 1984).

17. P. Ashburn, D. J. Roulston, and C. R. Selvakumar, "Comparison of experimental and computed results on arsenic and phosphorus doped polysilicon emitter bipolar transistors," *IEEE Trans. Electron Devices*, vol. ED-34, pp. 1346–1353 (June 1987).

18. R. P. Mertens, J. L. Van Meerbergen, J. F. Nijs, and R. J. Van Overstraeten, "Measurement of the minority carrier transport parameters in heavily doped silicon," *IEEE Trans. Electron Devices*, ED-27, pp. 949–955 (1980).

19. G. R. Wolstenhome, M. Jorgensen, P. Ashburn, and G. R. Booker, "An investigation of the thermal stability of interfacial oxide in polycrystalline silicon emitter bipolar transistors by comparing device results with high-resolution electron microscopy observations," *Jnl. App. Phys.*, vol. 61, pp. 225–233 (January 1987).

20. P. A. Iles and S. I. Soclof, in Photovoltaic Specialists Conf. 1975.

21. A. W. Wieder, "Emitter effects in shallow bipolar devices: measurement and consequences," *IEEE Trans. Electron Devices*, vol. ED-27, pp. 1402–1408 (August 1980).

22. J. F. Nijs, Ph.D. Thesis, Katholieke Universiteit Leuven, Belgium (1982).

23. G. E. Possin, M. S. Adler, and B. J. Baliga, "Measurement of the PN product in heavily doped epitaxial emitters," *IEEE Trans. Electron Devices*, vol. ED-31, pp. 3–17 (January 1984).

24. J. del Alamo, S. Swirhun, and R. M. Swanson, "Simultaneous measurement of hole lifetime, hole mobility, and bandgap narrowing in heavily doped n-type silicon," *1985 IEDM Tech. Digest*, pp. 290–293 (December 1985).

25. D. Gold and D. J. Roulston, Dept. of Metallurgy and Science of Materials, Oxford University, unpublished results (1988).

26. J. C. Bravman, G. L. Patton, and J. D. Plummer, "Structure and morphology of polycrystalline silicon–single crystal silicon interfaces," *J. Appl. Phys.*, vol. 57, pp. 2779–2782 (April 1985).

CHAPTER
9

REAL IMPURITY PROFILES— THE BASE AND COLLECTOR REGIONS

9.1 INTRODUCTION

In this chapter, we continue our study of the "real impurity profile" and focus on the base and collector regions. Firstly, we take a look at the emitter-base and collector-base space-charge layer properties (capacitance, breakdown voltages, and recombination current). Then follows a detailed study of the base quasi-neutral region. This includes consideration of the retarding and accelerating field region in the base and their effect on base transit time; a full analysis of the I_C versus V_{BE} characteristics using the Gummel integral approach is then carried out, leading to a rigorous yet simple study of Early voltages (a widely used means of characterizing the small signal output conductance).

The total charge components contributing to the f_t of the transistor are then examined in detail, including the charge in the neutral emitter and the free carrier charge in the emitter-base space-charge layer. This leads to a discussion of maximum oscillation frequency and the "optimum" transistor.

Sidewall effects are considered where appropriate, for example, in the determination of space-charge recombination currents (where surface recombination is also included), and in the determination of junction capacitances. All of the theory in this chapter is restricted to low level injection conditions. High current effects will be considered in the next chapter.

9.2   SPACE-CHARGE REGIONS

9.2.1   Emitter-Base Space-Charge Layer Breakdown Voltage and Capacitance

We have already pointed out in Chaps. 3 and 7 that in the vicinity of the junction, a double-diffused gaussian type impurity profile can be approximated by a linearly graded junction. Let us recall the results of Eqs. (2.55) and (2.58). For a double gaussian given by

$$N(x) = N_e \exp\left[-\left(\frac{x}{x_e}\right)^2\right] - N_b \exp\left[-\left(\frac{x}{x_b}\right)^2\right] + N_{\mathrm{epi}} \qquad (9.1)$$

the equivalent linearly graded profile in the vicinity of the emitter-base junction X_{jeb}, $N(x) = ax$ is given by

$$a = N_b \frac{2X_{jeb}}{x_b^2} \exp\left[-\left(\frac{X_{jeb}}{x_b}\right)^2\right]\left[1 - \left(\frac{x_b}{x_e}\right)^2\right] \qquad (9.2)$$

In Chap. 2 we showed in Fig. 2.8 how this linear graded law fitted two typical double gaussian impurity profiles—one a shallow junction device, the other a deep junction device. Figure 9.1 shows the gradient a computed for a range of double gaussian impurity profiles at zero bias and at breakdown (in each case, a is computed from the actual doping levels and corresponding depletion-layer edges). Surface concentrations are $N_e = 10^{21}$ cm^{-3}, $N_b = 3 \times 10^{19}$ cm^{-3}. Results are given for two cases: $x_b/x_e = 1.6$ and 2.0. The lower value is typical for real transistor profiles. Notice the difference between the zero bias and the breakdown

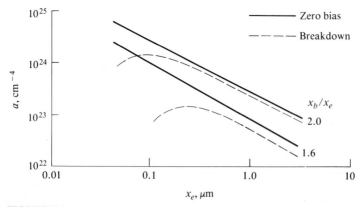

FIGURE 9.1
Computed results for linear gradient a of the e-b junction at zero bias and at breakdown versus characteristic length of emitter gaussian impurity profile, for the net profile given by Eq. (9.1).

values of "linear" gradient a. The downward curvature at breakdown for shallow junction depths corresponds to the onset of base punch-through; in this case the depletion layer extends well past the peak in the base doping profile toward the base-collector junction and the concept of a linear gradient is rather meaningless. It is clear, however, that for forward bias the linear graded approximation becomes increasingly valid, and for modeling purposes, such as depletion-layer capacitance versus voltage, the $\gamma = 1/3$ law will apply [see Eqs. (2.58) and (2.59)]. As an example, for a device with $x_b/x_e = 1.6$, and $x_e = 1.0\ \mu m$, the data corresponds to a zero bias gradient $a = 10^{23}\ cm^{-4}$.

Figure 9.2 gives computed results of breakdown voltage for gaussian profiles using both sets of ionization coefficients previously described (Chap. 3, Sec. 8). It should be noted that sidewall radius of curvature effects will further decrease the emitter-base breakdown voltage.

9.2.1.1 EMITTER-BASE SIDEWALL CAPACITANCE. Calculation of the sidewall component of emitter-base junction capacitance is complicated by the fact that the base impurity concentration varies along the sidewall in the vertical direction. The emitter diffusion in this region may be approximated by radial diffusion with a radial profile the same as the vertical profile provided that the location of the origin is shifted back into the oxide window opening by a distance $d \sim X_{jeb}/5$. Figure 9.3 shows this situation and the corresponding impurity profile. By considering a double gaussian vertical profile as given by Eq. (9.1) and a radial profile given by

$$N(r, \theta) = N_e \exp\left[-\left(\frac{r}{x_e}\right)^2\right] - N_b \exp\left[-\left(\frac{r\cos\theta}{x_b}\right)^2\right] + N_{epi} \qquad (9.3)$$

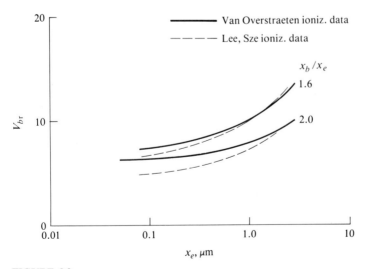

FIGURE 9.2
Computed emitter-base breakdown voltage for a double gaussian impurity profile [Eq. (9.1)] versus emitter gaussian characteristic length x_e. Results are given for two values of the ratio x_b/x_e.

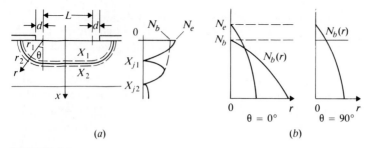

FIGURE 9.3
Cross section (a) and impurity profile (b) for sidewall capacitance analysis.

Poisson's equation may be solved in cylindrical coordinates for the depletion layer boundaries r_1 and r_2.

The sidewall capacitance per unit length of perimeter is given by

$$C_{j\,\text{edge}} = \varepsilon \int_0^{\pi/2} \left(\frac{r}{r_2 - r_1} \right) d\theta \tag{9.4}$$

It is clear from Fig. 9.3 that for $\theta = 90$ degrees, the capacitance per degree of arc is larger than for $\theta = 0$. This is because the doping level on the base side of the junction tends to a constant (N_b) high value along the surface ($\theta = 90$ degrees) and has a varying, but on average considerably lower, value $N_b \exp\left[-(x/x_b)^2\right]$, in the vertical ($\theta = 0$) direction. The ratio is typically between $2:1$ and $10:1$ depending on the profile. In order to facilitate presentation of the results, we have normalized them. Consider a uniform cylindrical junction whose capacitance per unit length of perimeter could be estimated as

$$C'_{j\,\text{edge}} = \frac{\pi}{2} C_{be\,\text{plane}} X_{jeb} \tag{9.5}$$

where $C_{be\,\text{plane}}$ is the capacitance per unit area of the plane region. We can then express the results in terms of the normalized factor F_c where

$$F_c = \frac{C_{j\,\text{edge}}}{C'_{j\,\text{edge}}} \tag{9.6}$$

Thus, for a transistor with emitter width L and length B, the total plane and sidewall capacitances are

$$C_{t\,\text{plane}} = BLC_{be\,\text{plane}} \tag{9.7}$$

$$C_{t\,\text{edge}} = \pi(B + L)F_c X_{jeb} C_{be\,\text{plane}} \tag{9.8}$$

In Fig. 9.4(a) $C_{be\,\text{plane}}$ is plotted versus emitter-base junction depth X_{jeb} for a range of values of X_{jcb}/X_{jeb}. The results are given for two surface base concentrations, and an emitter surface concentration $N_e = 10^{21}$ cm^{-3}. The bending and termination of the curves for low ratios of X_{jcb}/X_{jeb} are due to the approach

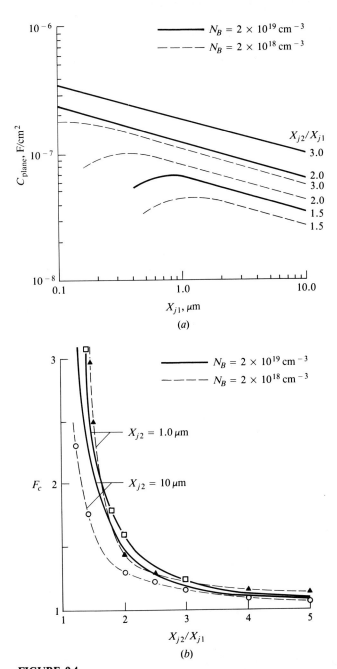

FIGURE 9.4

(a) Capacitance of plane region per unit area for two base surface concentrations N_b as a function of the junction depth X_{jeb} for various ratios of X_{jcb}/X_{jeb}; (b) normalized edge capacitance factor F_c versus X_{jcb}/X_{jeb} for two values of base surface concentration N_b and two junction depths. (*After Roulston and Kumar* [1]. *Reproduced with permission* © *1979 IEEE.*)

of base punch-through and would normally be outside the range of practical interest.

Figure 9.4(b) shows the factor F_c plotted versus the ratio of junction depths X_{jcb}/X_{jeb} for two values of base surface concentration. The advantage of plotting the results in this manner can be seen from the relative insensitivity to both junction depth and base surface concentration.

To illustrate the use of the curves, consider an RF power transistor with $B = 1$ cm, a stripe width $L = 2$ μm, an emitter-base junction depth $X_{jeb} = 0.5$ μm, with $N_b = 2 \times 10^{19}$ cm^{-3}. Figure 9.4(a) gives $C_{be \, plane} \sim 0.15 \times 10^{-6}$ F/cm^2. The total plane junction capacitance is thus, from Eq. (9.7),

$$C_{t \, plane} = 30 \text{ pF}$$

Figure 9.4(b) gives the value of $F_c = 1.5$. The total sidewall capacitance is thus, from Eq. (9.8),

$$C_{t \, edge} = 35 \text{ pF}$$

This example demonstrates the non-negligible contribution of the sidewall capacitance and the usefulness of the factor F_c.

9.2.2 Emitter-Base Recombination Current and Low Current Gain Falloff

Following the discussion for a diode in Sec. 3.6, it will be apparent that at low forward bias, there is likely to be an important additional component of base current due to recombination of excess carriers inside the emitter-base space-charge layer. This current I_{screb} is given by Eq. (3.106), repeated here for convenience:

$$I_{screb} = A_e \left[\frac{2qn_i}{\tau_{DE}} \right] \frac{4}{3} d_n \left[\frac{V_t}{V_{bie} - V_{BE}} \right] \exp \left(\frac{V_{BE}}{2V_t} \right) \tag{9.9}$$

where d_n is the depletion-layer width on the N side of the junction for the applied voltage V_{BE}, and τ_{DE} is the effective carrier lifetime in the emitter-base space-charge layer.

Because of the exp $(V_{BE}/2V_t)$ dependence of I_{screb}, the current gain β will be reduced at low forward bias of the base-emitter junction. The value of this reduced gain β_{screb} can be determined by using the expression for collector current [Eq. (7.5)] and dividing by I_{screb} given by Eq. (9.9). This may conveniently be expressed as follows:

$$\beta_{screb} = \left(\frac{3}{8} \right) \left[\frac{L_{DE}^2}{W_b d_n} \right] \left[\frac{n_i}{N_{Ab}} \right] \left[\frac{V_{bie} - V_{BE}}{V_t} \right] \exp \left(\frac{V_{BE}}{2V_t} \right) \tag{9.10}$$

We substituted L_{DE} for $\sqrt{D_n \tau_{DE}}$ to render each term in parentheses dimensionless. Figure 9.5 shows how the gain varies with collector current [$\propto$ exp (V_{BE}/V_t)].

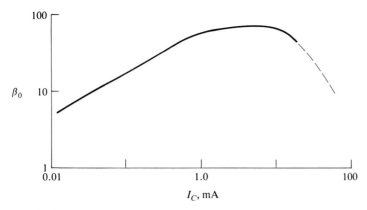

FIGURE 9.5
Typical current gain versus collector current.

Let us consider some typical values. d_n depends on the gradient a as plotted for diodes in Fig. 2.5 for zero bias.

Consider a double gaussian profile for a typical transistor with e-b and b-c junction depths of 0.5 and 1.0 μm, respectively, and with emitter and base surface doping concentrations $N_e = 10^{21}$ cm^{-3} and $N_b = 2 \times 10^{19}$ cm^{-3}. The value of a from numerical computations is found to be 3.6×10^{23} cm^{-4}, and the zero bias depletion-layer width for one side is $d_n = 0.06$ μm, with a built-in barrier $V_{bie} = 0.857$ V. For a forward bias of 0.6 V, the depletion-layer width decreases (approximately as the cube root of the total voltage $V_{bie} - V_{BE}$) to a value of 0.04 μm. This value will not be significantly different for a wide range of profiles. The remaining parameter which needs to be known for computation of this space-charge recombination current component is the effective carrier lifetime τ_{DE}. In Sec. 3.6 we have pointed out that according to Sah et al. [2] this lifetime is actually $2\sqrt{\tau_{p0}\tau_{n0}}$. Since these two quantities are very difficult to determine experimentally and are, in fact, position-dependent (both because of doping level changes and distance from surface [3]) the use of a single equivalent lifetime τ_{DE} is justified. Its value will be fixed for a given process and lies typically in the range 1 μs (for deep junction devices of order several microns) to 10 ns for shallow junction devices.

For a base width W_b of 1 μm and a constant base doping N_{Ab} of 10^{17} cm^{-3} the value of β_{screb} at $V_{BE} \sim 0.6$ V is about 20. This increases by a factor of 10 for a V_{BE} increase of about 100 mV. For bias voltages lower than the value for which $\beta_{screb} = \beta_0$ it is clear that the gain will be determined by the properties of the emitter-base space-charge layer and not by injection into the quasi-neutral emitter.

The above simple theoretical approach, as discussed in Sec. 3.6, corresponds with the results of Sah et al. for recombination centers at the intrinsic level [2]. Other recombination model parameters will give values of $I_{screb} \propto \exp(V_{BE}/mV_t)$ where $1 < m < 2$.

9.2.3 Emitter-Base Sidewall and Surface Recombination Currents

Figure 9.6 shows the three-dimensional nature of the emitter-base space-charge region. For modern small-dimension emitters, the volume of charge in the side-wall region can become comparable to that in the plane region of area A_e. The actual area to be used in Eq. (9.9) should therefore be approximately

$$A_{\text{eff}} = BL + \pi X_{jeb}(B + L) \tag{9.11}$$

In addition, the amount of recombination is increased due to surface effects. For a given process, this increase can be modeled conveniently by an equivalent extension of the space-charge layer vertically above the silicon surface over a distance X_{fs} as shown in the figure [4]. The value of X_{fs} can be obtained from measurements using two test structures with equal areas but different junction perimeters (i.e., two structures of different aspect ratios). Simple subtraction enables the value of X_{fs} to be determined. The final equivalent area is thus

$$A_{\text{eff}} = BL + (B + L)(\pi X_{jeb} + 2X_{fs})$$

The gain expression due to low current space-charge recombination [Eq. (9.10)] is thus reduced by the factor

$$\frac{1}{1 + \pi(X_{jeb} + 2X_{fs})(B + L)/BL} \tag{9.12}$$

Values of X_{fs} have been measured on a wide selection of devices and the results range from 20 μm to less than 1 μm for high quality surfaces.

The value of X_{fs} may be determined numerically if the relevant recombination data is known. Figure 9.7 gives results normalized with respect to surface oxide recombination velocity S_{ox} and bulk hole lifetime τ_{Dp}. The results are given versus surface state charge density Q_{ss} for several different junction depths and electron-to-hole lifetime ratios. It is apparent that in most cases the required data for the bulk SRH model parameters will not be readily available; use of a simple measurable parameter X_{fs} is thus preferable in practice.

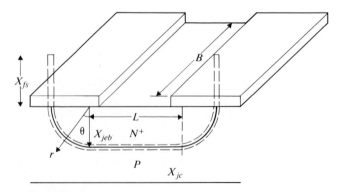

FIGURE 9.6
Three-dimensional diagram of the *e-b* space-charge layer.

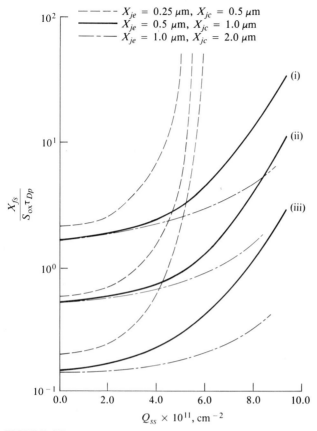

FIGURE 9.7

Normalized surface recombination parameter X_{fs} for an applied bias $V_{BE} = 0.4$ V, computed for a range of recombination parameters. Q_{ss} is the surface state charge. (i) $\tau_{Dn}/\tau_{Dp} = 10$; (ii) $\tau_{Dn}/\tau_{Dp} = 1.0$; (iii) $\tau_{Dm}/\tau_{Dp} = 0.1$. (*After Roulston and Eltoukhy* [5]. *Reproduced with permission* © *1985 IEE.*)

9.2.4 Base-Collector Space-Charge Layer

The base-collector junction may be represented accurately for most devices by the one-sided abrupt junction approximation, apart from small bias values (particularly in the range zero volts to forward bias), where a linearly graded approximation is more appropriate. Thus the capacitance and breakdown behavior is, for most devices, well modeled by the treatment used in the previous chapter. In particular the results contained in Fig. 7.12 may be used to determine the values of epitaxial layer doping and residual thickness (W_{epi} is defined from the base-collector junction to the NN$^+$ junction) for specified breakdown voltage conditions (V_{cbr} or BV_{CEO}). The radius of curvature effect of the base diffusion may also be deduced from this figure.

Note in particular that an optimum choice for residual epitaxial layer thickness may be considerably less than the value deduced solely from considerations of collector-base breakdown voltage. This is due to two reasons. Firstly, there is no advantage in having W_{epi} equal to the value for V_{br} deduced from one-dimensional or plane breakdown, if the radius of curvature of the diffused junction restricts the actual breakdown to a lower value. Figure 7.12 enables the optimum choice to be made so that no additional collector series resistance exists due to an "unused" epitaxial region near the NN^+ junction. Secondly, if BV_{CEO} is known to be the limiting factor in the intended application, there is no advantage to be gained by designing for a high V_{cbr}. In fact, in this situation, the "optimum" design would be such that the epitaxial layer just enters reach-through at the BV_{CEO} voltage, thus avoiding any "excess" series collector epitaxial resistance.

9.3 QUASI-NEUTRAL BASE REGION

Let us now turn our attention to the quasi-neutral base region. Figure 9.8 shows a typical impurity profile, with the boundaries determined by the *e-b* and *b-c* space-charge layers for given bias conditions. Some simple calculations based on typical bias values will show that normally the quasi-neutral base can be divided into a region W_{b1} from $x = 0$ to $x = x_{12}$ with an impurity profile which rises with increasing distance from the point of injection of the minority carriers, followed by a region W_{b2} from x_{12} to W_b with a decreasing profile. These two regions may be characterized by the respective electric fields; a retarding field in region W_{b1} and an accelerating field in region W_{b2}. The reader should refer back to Secs. 2.7.2 and 3.4.2 for the detailed discussion on the origin of the built-in electric field. Throughout the discussion on minority carrier transport in the base, it should be realized that use of the effective doping level (due to bandgap narrowing) is implied, although the difference between actual and effective doping profiles will not be nearly so significant as in the emitter region.

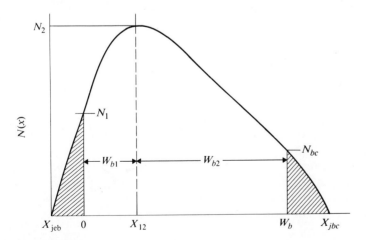

FIGURE 9.8
Typical base impurity profile with space-charge layers shown.

If we use the simplifying assumption (the only one which allows usable analytic results to be obtained) of exponential impurity profiles in each of the two regions, the amount of charge in each region (and hence the corresponding delay times) can be deduced.

Retarding Field Region

If the profile in region 1 is given by

$$N(x) = N_1 \, \exp\left(\frac{x}{x_{01}}\right) \tag{9.13}$$

the electric field is given by [see Eq. (3.32)]

$$E_1 = \frac{V_t}{x_{01}} \tag{9.14}$$

where

$$\frac{N_2}{N_1} = \exp\left(\frac{W_{b1}}{x_{01}}\right) \tag{9.15}$$

Let us denote the ratio of peak effective impurity concentration N_2 to the value at the e-b depletion-layer boundary N_1 by

$$r_1 = \frac{N_2}{N_1} \tag{9.16}$$

The minority carrier concentration can be shown to be [e.g., see Eq. (3.37) and omit recombination]

$$n(x) = n(0) \, \exp\left(-\frac{x}{x_{01}}\right) \tag{9.17}$$

Thus, in the retarding field region we have

$$n(x_{12}) = n(0) \, \exp\left(-\frac{W_{b1}}{x_{01}}\right) \tag{9.18}$$

Peak values of effective base doping are usually of order 10^{18} cm^{-3}. Values at the depletion-layer boundary depend on the amount of forward bias of the emitter-base junction. For a residual total junction voltage $V_{bie} - V_{BE}$ of order 0.1 V, this doping level is typically of order 10^{17} cm^{-3}. Thus the value of r_1 is of order 10 (but may be lower or higher depending on base profile and applied bias). To maintain analytic simplicity, let us limit the analysis to the case of large values of r_1 so that the carrier concentration at the position corresponding to peak base doping $n(x_{12})$ is much less than the injected value $n(0)$. The charge Q_1, due to the minority carriers, is therefore

$$Q_1 = qn(0)x_{01} = \frac{qn(0)W_{b1}}{\ln(r_1)} \tag{9.19}$$

Accelerating Field Region

We have not had occasion yet to encounter analysis of minority carrier flow in an accelerating field region. We would expect the converse of the result for the retarding field, but the result is not evident by inspection and we will now proceed to derive a result for this region, of width W_{b2}, extending from x_{12} to W_b. With the usual exponential approximation, the impurity profile is now given by (shifting the x reference for convenience to x_{12})

$$N(x) = N_2 \exp\left(-\frac{x}{x_{02}}\right) \tag{9.20}$$

The electric field is now in a direction to assist the flow of minority carriers by diffusion and by considering zero majority current flow conditions as in the derivation of Eq. (3.32) we find

$$E_2 = -\frac{V_t}{x_{02}} \tag{9.21}$$

where we may define an accelerating field factor

$$\eta = \frac{W_{b2}}{x_{02}} = \ln\left(\frac{N_2}{N_{bc}}\right) \tag{9.22}$$

where N_{bc} is the doping level at $x = W_b$. The electron current density (neglecting recombination which, as we saw in the last chapter, is negligible for nearly all real transistors in the base region) is given by

$$J_n = qD_n\left[\frac{dn}{dx} + \frac{nE}{V_t}\right] \tag{9.23}$$

or

$$\frac{dn}{dx} + \frac{nE}{V_t} - \frac{J_n}{qD_n} = 0$$

The solution to this equation is [6]

$$n(x) = \frac{-J_n W_{b2}}{qD_n} \frac{1 - \exp\left[-\eta(1 - x/W_{b2})\right]}{\eta} \tag{9.24}$$

where J_n is a negative quantity.

It is instructive to evaluate the individual diffusion and drift components of the electron current:

$$J_{n\,\text{diff}} = J_n \exp\left[-\eta\left(1 - \frac{x}{W_{b2}}\right)\right] \tag{9.25}$$

$$J_{n\,\text{drift}} = J_n\left\{1 - \exp\left[-\eta\left(1 - \frac{x}{W_{b2}}\right)\right]\right\} \tag{9.26}$$

It is apparent that for small values of x, that is, close to the point x_{12}, the drift component of current is much larger than the diffusion component of current for large values of the built-in field parameter η. On the other hand, at the right-hand side of the quasi-neutral base region x_{bc}, where the carrier concentration is assumed to be zero, the current is entirely diffusion current. Clearly, the electric field is assisting the flow of carriers across the base. Figure 9.9 shows some typical minority carrier-versus-distance plots for various values of η. Although for η values equal to about 8 the carrier distribution is almost horizontal, the maximum possible value of η is limited by the base impurity profile. Typically, the doping level at the point W_b is of order 10^{16} to 10^{17} cm^{-3}, so for the previously quoted value of maximum base doping N_2 of close to 10^{18} cm^{-3}, the value of η is limited to less than 4.

The value of carrier concentration at $x = x_{12}$ is found from Eq. (9.24) to be

$$n(x_{12}) = \frac{J_n W_{b2}}{qD_n \eta} \left[1 - \exp(-\eta)\right] \tag{9.27}$$

The current density is

$$J_n = \frac{qD_n n(x_{12})}{W_{b2}} \frac{\eta}{1 - \exp(-\eta)} \tag{9.28}$$

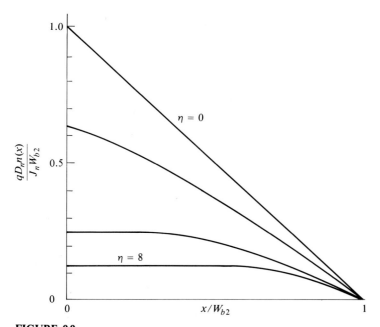

FIGURE 9.9
Minority carrier concentration versus normalized distance for various accelerating field factors. Note that the $x = 0$ reference corresponds to X_{12} of Fig. 9.8.

For η greater than about 3, this simplifies to

$$J_n = \frac{qD_n n(x_{12})\eta}{W_{b2}}$$ (9.29)

We note that this is a normal "narrow base" diode current increased by the accelerating field factor η. The minority carrier charge in the accelerating field region Q_{b2} is easily identified for large values of field factor as:

$$Q_{b2} = qn(x_{12})W_{b2}$$ (9.30)

Overall Base Current and Delay Times

Figure 9.10 shows a sketch of the idealized minority carrier-versus-distance plot. The exponential decrease in the retarding field region is followed by a more or less horizontal distribution of minority carriers until near the base-collector space-charge layer boundary, where a rapid decrease toward zero occurs. By using the above result for current [Eq. (9.29)] and $n(0)/n(x_{12})$ ratio r_1 [Eq. (9.16)] the final expression for current density may be written

$$J_n = F_{bb}\frac{qD_n n(0)}{W_b}$$ (9.31)

where

$$F_{bb} = \frac{\eta}{r_1}\frac{W_b}{W_{b2}}$$

For typical base profiles $\eta \sim 3$, $r_1 \sim 10$, and $W_b/W_{b2} \sim 2$; hence we see that F_{bb} is of order unity. This is equivalent to stating that the advantage of the accelerating

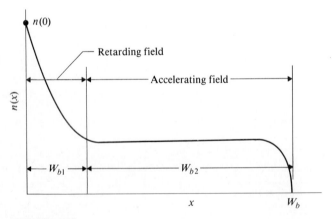

FIGURE 9.10
Minority carrier versus distance plot for the ideal retarding and accelerating field impurity profile.

field is more or less cancelled by the disadvantage of the retarding field in most typical diffused base regions.

The expressions for minority charge in each region, Eqs. (9.19) and (9.30) may be combined to give

$$Q_{b\,\text{tot}} = Q_{b1} + Q_{b2} = qn(0)\left[\frac{W_{b1}}{\ln (r_1)} + \frac{W_{b2}}{r_1}\right] \tag{9.32}$$

The total delay time due to this charge is given by

$$t_{bb} = \frac{Q_{b\,\text{tot}}}{J_n} = \left[\frac{W_{b1}}{\ln (r_1)} + \frac{W_{b2}}{r_1}\right]\left[\frac{W_{b2}}{D_n}\right]\left[\frac{r_1}{\eta}\right] \tag{9.33}$$

This expression may be compared to the value obtained previously [Eq. (7.15)] for a uniformly doped base $t_{bb} = W_b^2/2D_n$. Depending on the values of r_1 (typically from 1 to 10), W_{b1}/W_{b2} (from 0 to 1), and η (from 1 to 4), it is clear that the delay time may be either greater than or less than the value for a uniformly doped base.

The extreme values may be estimated by considering the base to be (i) entirely an accelerating field region $W_b = W_{b2}$, and (ii) entirely a retarding field region $W_b = W_{b1}$. For (i) the value of t_{bb} tends to $W_b^2/\eta D_n$. For (ii) the value can be deduced from Eqs. (3.39) and (9.19) to tend toward the value $[W_b^2/D_n][r_1/\ln r_1]$. This could be up to a factor of ten times higher than the $W_b^2/2D_n$ value for a uniformly doped base.

The apparent advantage of the so-called "drift" transistor (accelerating field only) is thus not readily obtained. In fact for most design purposes involving base delay times, it is a fair approximation to treat the base region as uniformly doped. Figure 9.11 shows computed minority carrier distributions for several different structures, for which the essential data is given in the accompanying table.

N_{peak} is the peak base doping level, R_{be}/sq is the sheet resistance of the active (pinched) base region, G_b is the base doping integral (Gummel integral—see next section). Device 4 is a VLSI polysilicon emitter structure; device 5 is a conventional integrated transistor; device 8 is a high voltage power switching device.

From the second diagram, with a linear vertical scale, it is evident that a triangular minority carrier distribution is quite a good approximation in many cases; this implies that the base region behaves roughly as for the uniformly doped case treated in Chap. 7, and confirms that the retarding and accelerating field regions have often, to a first approximation, a self-canceling effect. The computed results show that if the base transit time is expressed as $t_{bb'} = F_b W_b^2/2D_{n\,\text{av}}$ where $D_{n\,\text{av}}$ is the effective average value of electron diffusion coefficient, the values of F_b are equal to 0.81, 0.57, and 0.28 for devices 4, 5, and 8, respectively; this confirms the fact that the true "drift transistor" effect is only partially realized in a small number of cases. It is also of some interest to record the fact from [7] that for eight transistors studied, with f_t values ranging from 8 MHz to 5 GHz, the values of $D_{n\,\text{av}}$ were all within the range 15 to 22 cm²/s (effective

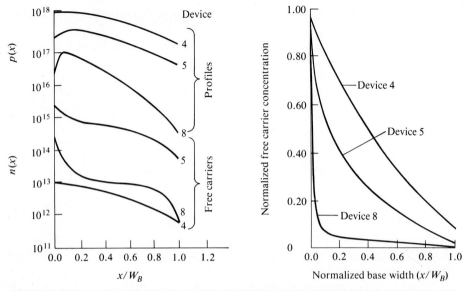

Device	X_{jeb}, μm	X_{jcb}, μm	W_b, μm	N_{peak}, $\times 10^{17}$	R_{be}/sq, Ω/sq	G_b, $\times 10^{10}$
4	0.02	0.23	0.12	19.0	3320	1.03
5	0.71	1.50	0.57	4.9	1990	1.50
8	10.0	25.0	12.0	1.1	455	4.23

FIGURE 9.11
Minority carrier distributions for some real NPN transistors computed using the BIPOLE program. The data for the devices is given in the accompanying table. (*After Zugelder and Roulston* [7]. *Reprinted with permission from Solid State Electronics, Pergamon Press PLC.*)

average mobility values from 600 to 845 cm^2/V·s). We thus conclude, somewhat surprisingly, that for approximate analysis, the uniformly doped base theory is very useful for estimating base related characteristics.

9.4 THE GENERALIZED I_C VERSUS V_{BE}, V_{BC} RELATION AND THE GUMMEL INTEGRAL

In the base region of any NPN transistor, the hole current is small (because of the high current gain) and, to a very good approximation, can be set to zero. We thus obtain

$$J_p = 0 = -qD_p \frac{dp}{dx} + q\mu_p pE \qquad (9.34)$$

$$E = \frac{V_t}{p} \frac{dp}{dx} \qquad (9.35)$$

The electron current density J_n given by

$$J_n = qD_n \frac{dn}{dx} + q\mu_n nE \tag{9.36}$$

may now be written in the form

$$J_n = qD_n\left[\frac{dn}{dx} + \frac{n}{p}\frac{dp}{dx}\right] \tag{9.37}$$

$$= \frac{qD_n}{p}\left[p\frac{dn}{dx} + n\frac{dp}{dx}\right] \tag{9.38}$$

The term in square brackets will be recognized as the derivative of the product $d(pn)/dx$. We may therefore rewrite Eq. (9.38) and integrate as follows:

$$\frac{J_n p}{qD_n} = \frac{d(pn)}{dx} \tag{9.39}$$

or

$$\frac{J_n}{q}\int_0^{W_b} \frac{p}{D_n}\,dx = (pn)_0 - (pn)_w \tag{9.40}$$

where

$$(pn)_0 = p(0)n(0) = n_i^2\, e^{V_{BE}/V_t} \tag{9.41}$$

$$(pn)_w = p(w)n(w) = n_i^2\, e^{V_{BC}/V_t} \tag{9.42}$$

So far we have made only one, very good, approximation, that of zero hole current (or constant quasi-Fermi level for the majority carriers). Note that in Eqs. (9.40) to (9.42) the boundary $x = 0$ is the boundary between the emitter-base space-charge layer and the quasi-neutral base region. The boundary W_b is the start of the collector-base space-charge layer. These are indicated in Fig. 9.12.

The left-hand side of Eq. (9.40) may be rewritten, without any loss of generality, by defining

$$G_b = \int_0^{W_b} p\,dx \tag{9.43}$$

Thus we have

$$(pn)_0 - (pn)_{W_b} = \frac{J_n}{qD_{n\,av}}G_b \tag{9.44}$$

where $D_{n\,av}$ is some average value of electron diffusion constant which, as already stated, is typically 15 to 22 cm^2/s, and G_b is the Gummel number or Gummel integral in the neutral base region [8]. For low level injection this is simply the integral of the base doping. Since we are dealing with minority carrier transport,

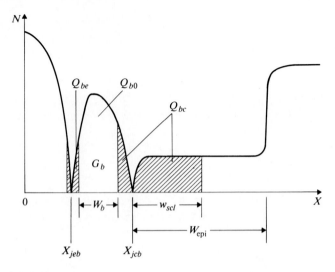

FIGURE 9.12
Impurity profile showing definitions used in the Gummel integral analysis.

the effective doping level $N_{A\,\text{eff}}(x)$ due to bandgap narrowing must be used, although the difference will be much less important than for the emitter, which is much more heavily doped than the base.

$$G_b = \int_0^{W_b} N_{A\,\text{eff}}(x)\,dx \tag{9.45}$$

This is a "constant" of the transistor and may be used to write the general $I_C = f(V_{BE}, V_{BC})$ relation as follows:

$$I_C = \left(\frac{qAn_i^2 D_{n\,\text{av}}}{G_b}\right)[e^{V_{BE}/V_t} - e^{V_{BC}/V_t}] \tag{9.46}$$

Notice that this is identical to the formula for the uniformly doped base transistor in which we have replaced $D_n/N_A W_b$ by $D_{n\,\text{av}}/G_b$. The significance of the generalized result using the Gummel integral lies in the fact that it may be determined easily from process fabrication data.

One of the most important parameters to be measured during fabrication of wafers is the pinched base sheet resistance R_{be}/sq. This is the sheet resistance of the quasi-neutral base under the emitter and is related to the actual impurity profile by the integral

$$R_{be}/\text{sq} = \frac{1}{q\displaystyle\int_0^{W_b}\mu_p N_A(x)\,dx} \tag{9.47a}$$

Taking the "average" mobility outside the integral enables us to write

$$R_{be}/\text{sq} = \frac{1}{q\mu_{p\,\text{av}}\,G_b'} \qquad (9.47b)$$

It is clear, therefore, that the sheet resistance is directly related to the Gummel integral and that, since base doping profiles nearly always cover the same range of doping levels, the average hole mobility varies only slightly from one device to another, as is the case for "average" electron diffusion constant. It is also significant to note that it is often convenient to obtain not the Gummel integral G_b but the quantity $G_b/D_{n\,\text{av}}$ [see Eq. (9.44)] from measurements since, as can be seen from Eq. (9.46), this only involves a knowledge of the area plus applied bias and temperature (for the value of n_i^2). Values of average hole and electron mobility have been determined for a wide range of devices in the base region by comparing computed results using the BIPOLE program with measured data [7]. This study, based on the devices used in Fig. 9.11 gave the following results:

$$\mu_{p\,\text{av}} = 277 \text{ cm}^2/\text{V}\cdot\text{s} \pm 20\%$$

$$D_{n\,\text{av}} = 18 \text{ cm}^2/\text{s} \pm 20\%$$

The Gummel integral as defined above will be slightly lower than the total integral from *e-b* to *c-b* junction because of the finite amount of charge inside both depletion layers. The integral of the effective doping level G_b is also smaller than the value of the integral of the actual doping level, G_b'—typically by a factor 0.9 to 0.3. As will be seen in the following section, the depletion layer components may be treated in a very general manner to yield important modeling parameters concerning the output conductance (Early voltage) in the normal mode of operation.

The current gain may now be defined for the general case of a transistor biased in the normal active region using Eq. (9.46) for I_C and Eqs. (8.7), (8.13), or (8.22) for I_b:

$$\beta_0 = \frac{D_{n\,\text{av}}/G_b}{K_e\,D_{p2}/N_{\text{eff}}\,W_{e2}} \qquad (9.48)$$

where, as we saw in Chap. 8, N_{eff} is the effective doping level in the heavily doped region, width W_{e2}, of the emitter and K_e in general is greater than unity and incorporates terms due to recombination [Eq. (8.13)] and injection into a poly-silicon layer [Eq. (8.22)].

9.4.1 The Normal Mode of Operation—
Early Voltage

Since increasing V_{bc} in the reverse bias direction (normal mode of operation) increases the charge in the *c-b* space-charge region, the value of G_b defined in the previous section will decrease. The collector current will therefore increase giving rise to a finite output conductance as illustrated in the $I_C - V_{CE}$ characteristics in Fig. 9.13. For modeling purposes this is sometimes represented by the Early voltage V_A [9], the intercept on the negative V_{CE} axis. Let us now evaluate V_A. It will be assumed that a constant base current I_B is equivalent to maintaining V_{BE} constant. This is a good approximation in most (but not all) transistors, since I_b is due primarily to current injected into the emitter and to recombination in the e-b space-charge layer. If neutral base region recombination is significant, the results will be different.

From Fig. 9.13 we see that

$$\frac{dI_C}{dV_{CE}} = \frac{I_C}{V_A + V_{CE}} \tag{9.49a}$$

where V_A and V_{CE} are defined as positive quantities. Thus,

$$V_A + V_{CE} = \frac{I_C}{dI_C/dV_{CE}} \tag{9.49b}$$

For simplicity, let us write

$$I_C = \frac{K}{G_b} \tag{9.50}$$

where, from Eq. (9.46), we may identify K for negative V_{bc} as

$$K = qAn_i^2 D_{n\,\text{av}}\, e^{V_{BE}/V_t} \tag{9.51}$$

Hence

$$\frac{dI_C}{dV_{CE}} = \frac{K}{G_b^2} \frac{dG_b}{dV_{CE}} \tag{9.52}$$

and therefore, from Eqs. (9.50) and (9.52)

$$\frac{I_C}{dI_C/dV_{CE}} = \frac{G_b}{dG_b/dV_{CE}} \tag{9.53}$$

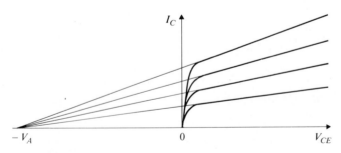

FIGURE 9.13
I_C versus V_{CE} characteristics showing finite output conductance and Early voltage definition.

We now relate the denominator of Eq. (9.53) to a very important and easily measured parameter, the collector-base junction capacitance C_{jc}. Relating the charge to the Gummel integral, $Q_b = qG_b$ (refer to Fig. 9.12), and since ΔQ_b in the neutral base due to a change $\Delta V_{CE} \approx \Delta V_{BC}$ is identical to ΔQ_b in the collector-base space-charge layer, it is obvious that we may write

$$\frac{dQ_b}{dV_{CE}} = \frac{dQ_{sc}}{dV_{CE}} = C_{jc'} \tag{9.54}$$

where $C_{jc'}$ is the capacitance per unit area of the base collector junction. Defining all quantities as magnitudes, we may thus combine Eqs. (9.49), (9.53), and (9.54) to write

$$V_A + V_{CE} = \frac{Q_b}{C_{jc'}} \tag{9.55}$$

From Fig. 9.12, it may be noted that

$$Q_b = Q_{b0} - Q_{be} - Q_{bc} \tag{9.56}$$

where Q_{b0} is the total Gummel integral charge between junctions, and Q_{be} and Q_{bc} are the components due to charge inside the two space-charge layers. It is also worth noting that for an abrupt junction (a good approximation to the reverse biased collector-base junction)

$$\frac{Q_{sc}}{V_{Gt}} = 2\left(\frac{dQ_{sc}}{dV}\right) = 2C_{jc'} \tag{9.57}$$

where V_{Gt} is the total voltage across the depletion layer, and for a linearly graded junction (a good approximation to the forward biased emitter-base junction)

$$\frac{Q_{sc}}{V_{tot}} = \frac{3}{2}\frac{dQ_{sc}}{dV} = \frac{3}{2}C_{jc'} \tag{9.58}$$

The value of Q_b in Eq. (9.56) may therefore be directly determined from the total integral Q_{b0} if the two junction capacitances per unit area are known (care must be exercised in taking account of the junction *areas*, particularly if sidewall and extrinsic base regions form an important part of the capacitance).

9.4.2   Variation of Early Voltage with V_{CE}

The integral Q_{bc} in Eq. (9.56) is identical on both sides of the collector-base junction. It may therefore be expressed as

$$Q_{bc} = qN_{epi}w_{scl} \tag{9.59}$$

where w_{scl} is the depletion-layer width in the epitaxial layer.

By neglecting the voltage drop across the base side of the collector-base space-charge layer, the total voltage drop across this reverse-biased junction can be written as

$$V_{bc} = \frac{1}{2\varepsilon} q N_{epi} w_{scl}^2 - V_{bic} \qquad (9.60)$$

where V_{bic} is the built-in barrier potential associated with this junction. Under normal operation, the base emitter junction is forward biased such that $V_{BE} \approx V_{bic}$. Thus, to a good approximation, Eq. (9.60) can be rewritten as

$$V_{CE} = \frac{1}{2\varepsilon} q N_{epi} w_{scl}^2 \qquad (9.61)$$

By defining the epitaxial layer reach-through voltage as

$$V_{pt\ epi} = \frac{1}{2\varepsilon} q N_{epi} W_{epi}^2 \qquad (9.62)$$

where W_{epi} is the width of the epitaxial layer between the collector-base junction and the N^+ substrate or buried layer, Eq. (9.61) can be expressed as

$$\sqrt{\frac{V_{CE}}{V_{pt\ epi}}} = \frac{w_{scl}}{W_{epi}} \qquad (9.63)$$

The base Gummel number can now be expressed using Eqs. (9.56) and (9.59) as

$$Q_b = Q_{b0}' - Q_{epi} \sqrt{\frac{V_{CE}}{V_{pt\ epi}}} \qquad (9.64)$$

where $Q_{b0}' = Q_{b0} - Q_{be}$ and $Q_{epi} = q N_{epi} W_{epi}$. Because of the exponential dependence of I_C upon V_{BE} the value of Q_{b0}' will be approximately constant over a large range of collector currents.

Punch-through of the base occurs when Q_b goes to zero. To avoid punch-through of the base, Q_{b0}' must be greater than Q_{epi}. If this condition is met, epitaxial layer reach-through will occur before base punch-through.

Using Eqs. (9.46) and (9.64), the collector current can now be written as [10]

$$I_C = \frac{const}{Q_{b0}'} \frac{e^{V_{BE}/V_t} - e^{V_{BC}/V_t}}{1 - \frac{Q_{epi}}{Q_{b0}'} \sqrt{\frac{V_{CE}}{V_{pt\ epi}}}} \qquad (9.65)$$

As well as providing a design criterion, Eq. (9.65) gives the complete dependence of I_C upon V_{CE}. In Fig. 9.14, the normalized collector current I_C is plotted as a function of V_{CE} for various values of Q_{epi}/Q_{b0}'. For values of Q_{epi} greater than 0.6 Q_{b0}', the dependence of I_C upon V_{CE} is quite pronounced and distinctly nonlinear.

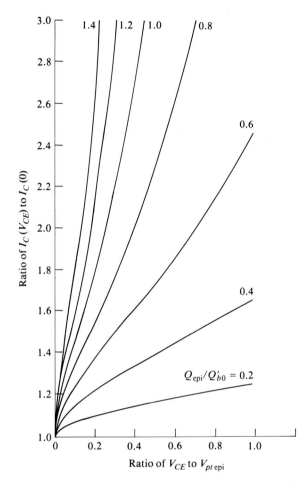

FIGURE 9.14
Normalized collector current versus collector emitter voltage for various values of Q_{epi}/Q'_{b0}. (*After Scott and Roulston* [10]. *Reprinted with permission from Solid State Electronics, Pergamon Press PLC.*)

Using Eq. (9.65) the Early voltage can be determined from the derivative of I_C with respect to V_{CE} and extrapolating back to zero collector current. In Fig. 9.15 we show the normalized Early voltage as a function of collector-emitter voltage for various values of Q_{epi}/Q'_{b0}. For low values of Q_{epi}/Q'_{b0} the Early voltage actually increases as the collector-emitter is increased. For high values of Q_{epi}/Q'_{b0}, the Early voltage increases to a maximum and then decreases. For the higher values of Q_{epi}/Q'_{b0} this decrease is quite dramatic.

Figure 9.16 shows a typical set of I_C versus V_{CE} characteristics for a transistor which encounters base punch-through at high V_{CE} values.

It is easy to see that as V_{CE} is increased from low to high values, the gradient dI_C/dV_{CE} decreases, levels off, then starts increasing again; this corresponds to values of Early voltage which increase, then peak, then start to decrease, eventually passing through 0 and changing sign. Users of CAD Gummel-Poon models in programs such as SPICE or WATAND must be careful in this situ-

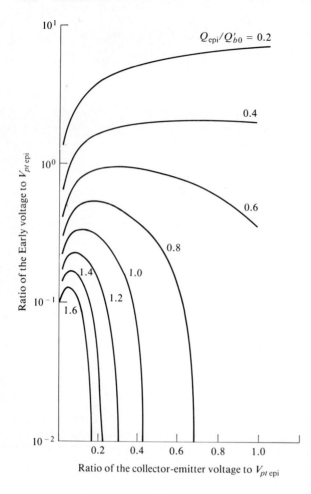

FIGURE 9.15
Early voltage versus V_{CE} for various values of the charge ratio Q_{epi}/Q_{b0}. (*After Scott and Roulston* [10]. *Reprinted with permission from Solid State Electronics, Pergamon Press PLC.*)

ation because the value of Early voltage used in the model is critical. This is discussed further in Chap. 15.

9.5   TOTAL DELAY TIME AND f_t

The concept of transition frequency, or common emitter current gain bandwidth product, was introduced in Chap. 7. We considered two major components of delay—base transit time t_{bb} and the RC time constant $(C_{je} + C_{jc})/g_m$; C_{je} and C_{jc} are the emitter-base and base-collector depletion-layer capacitances, respectively; g_m is the small signal transconductance $= I_C/V_t$.

Let us consider the various regions of the transistor as shown in the impurity profile diagram of Fig. 9.17. We recall that the base transit time t_{bb} is due to the electron charge Q_{n-ba} in the quasi-neutral base region. It is clear, however,

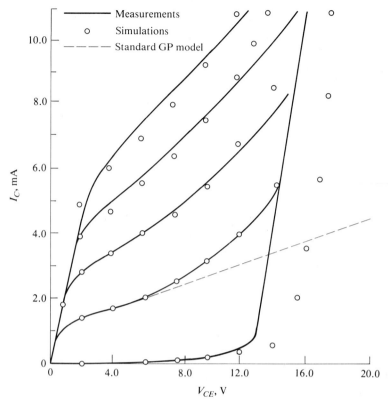

FIGURE 9.16
Common emitter characteristics for a transistor in which base punch-through occurs at high V_{CE} values. (*After Hebert and Roulston* [11]. *Reproduced with permission* © *1987 IEEE.*)

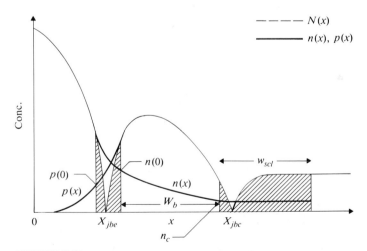

FIGURE 9.17
Minority carrier charge distribution in various regions.

that if the applied base-emitter voltage V_{BE} is increased by an amount ΔV_{BE} the hole charge in the emitter $Q_{p\text{-}em}$ will also increase. Furthermore, the finite amount of free electron charge $Q_{n\text{-}bcscl}$ in the base-collector space-charge layer discussed in Sec. 7.3 will also increase. Finally, the free carrier charge inside the emitter-base space-charge layer $Q_{pn\text{-}ebscl}$ (used in the determination of low forward bias recombination current [Eq. (3.106)] and current gain [Eq. (9.9)]) will increase due to the increase ΔV_{BE} in V_{BE}.

The analysis of Chap. 7 must therefore be extended to include these additional charges. Following the approach used in Sec. 7.4.1, we can define the diffusion capacitance as

$$C_{\text{diff}} = \frac{dQ_\Sigma}{dV_{BE}} \tag{9.66}$$

where Q_Σ is the sum of all the charges referred to above

$$Q_\Sigma = Q_{p\text{-}em} + Q_{n\text{-}ba} + Q_{n\text{-}bcscl} + Q_{pn\text{-}ebscl} \tag{9.67}$$

Proceeding as in Eq. (7.14b) we can write

$$\frac{dQ_\Sigma}{dV_{BE}} = \frac{dQ_\Sigma}{dI_C} \frac{dI_C}{dV_{BE}} \tag{9.68}$$

Since the charges $Q_{p\text{-}em}$, $Q_{n\text{-}ba}$, and $Q_{n\text{-}bcscl}$ are each proportional to $\exp{(V_{BE}/V_t)}$, that is, proportional to I_C, it follows that the derivatives dQ/dI_C will be the same as the ratios Q/I_C as was the case for the base charge discussed in Eqs. (7.14) and (7.15).

Let us now examine each of the above components of delay in turn.

Minority Carrier Charge in the Base

This has already been considered. As shown in Eq. (7.15) the base component of charge leads to the result

$$t_{bb} = \frac{Q_{n\text{-}ba}}{I_C}$$

This may be expressed as

$$t_{bb} = \frac{F_b W_b^2}{2D_n} \tag{9.69a}$$

where, from the discussion on the accelerating and retarding fields in the base region [Eq. (9.33)], F_b will lie typically in the range $0.5 < F_b < 2.0$

It should be noted that for very narrow base widths W_b, the ratio $n(0)/n_c = v_d W_b/D_n$ discussed in Chap. 7 in connection with Eqs. (7.1) to (7.6) is no longer so large that n_c can be neglected. Referring to Fig. 9.17, it is clear that an additional neutral region base charge $Q_{bc} = qn_c W_b$ exists in the case of a uniformly doped base. The effect of this charge may easily be shown [17, 18] to give a modified base delay time

$$t_{bb} = \frac{W_b^2}{2D_n}\left[1 + \frac{2D_n}{v_d W_b}\right] \tag{9.69b}$$

The second term in this equation will add approximately 10 percent to the delay time t_{bb} when the neutral base width $W_b = 0.3$ μm.

Charge in the Emitter

This hole charge is given by Eqs. (8.9) and (8.11) in the analysis of base current injected into the emitter region. Substituting from these two equations for charges using the two-region emitter model gives

$$Q_{p\text{-}em} = qAp(0)x_{0e} + \frac{qAp_{12}W_{e2}}{2}$$

Using the previously defined relations, this simplifies to

$$Q_{p\text{-}em} = F_e qAp(0)W_e$$

where

$$F_e = \frac{1}{\ln(r_e)}\frac{W_{e1}}{W_e} + \frac{1}{2r_e}\frac{W_{e2}}{W_e} \tag{9.70}$$

For typical emitter profiles, r_e is of order 10 and it is clear that the first term in F_e will normally dominate; i.e., most of the emitter hole charge is contained in the retarding field region of the emitter next to the emitter-base space-charge layer.

The delay time due to the above emitter charge is thus

$$t_{em} = \frac{Q_{p\text{-}em}}{I_C}$$

$$= \frac{F_e qp(0)W_e}{F_{bb}qD_n n(0)/W_b} \tag{9.71}$$

where the factor F_{bb} is given by Eq. (9.31) and is of order unity.

Under forward bias conditions, as already discussed, the emitter-base space-charge layer follows the symmetrical linearly graded law and $p(0) = n(0)$. The result for emitter delay thus simplifies to

$$t_{em} = \frac{F_e}{F_{bb}}\frac{W_e W_b}{D_n} \tag{9.72}$$

Since r_e is typically of order 10 and F_{bb} is invariably close to unity, this delay time is given approximately by

$$t_{em} \sim \frac{W_e W_b}{5D_n} \tag{9.73}$$

Comparing with Eq. (9.69) above, it is clear that for shallow junction devices the emitter component of delay is comparable to the base component and cannot in general be neglected. For example, a device with an emitter-base junction depth of 0.1 μm and a base-collector junction depth of 0.2 μm will have approximately $W_e = 0.1$ μm, $W_b = 0.1$ μm, and hence the emitter delay t_{em} will be approximately one-half the base delay t_{bb}.

In the case of a polysilicon contacted emitter, there will be an additional delay t_{pol} due to the charge stored in the polysilicon region. This may be calculated from the carrier distribution and will normally be less than the above delay t_{em}.

Base-Collector Space-Charge Layer Electron Delay Time

Figure 9.18 shows the details of the free electron charge within the base-collector space-charge layer. We recall that the value of n_c is related to the collector current density J_{nc} by Eq. (7.2)

$$n_c = \frac{J_{nc}}{qv_d}$$

Clearly, the presence of this electron concentration will modify the solution to Poisson's equation. For low currents, where $n_c \ll N_{epi}$ the actual values of depletion-layer boundaries are not changed significantly. However, for incremental changes in n_c due to changes in base-emitter voltage V_{BE}, there is an important effect which modifies the "expected" delay time. At first sight, we might imagine that the delay time associated with the electron concentration n_c is simply w_{scl}/v_d. In fact, due to the change in depletion-layer thickness when n_c

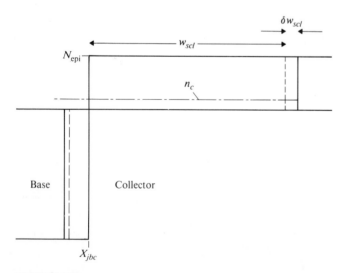

FIGURE 9.18
Details of the free electron charge in the collector-base space-charge layer.

changes, the resulting delay time is given by [12]:

$$t_{scl} = \frac{w_{scl}}{2v_d} \tag{9.74}$$

Exact Determination of Base-Collector Transit Time Delay

Let us define the net effective doping level on the uniformly doped collector side of the collector-base junction as $N_d' = N_{epi} - n_c$ and let us consider a change δn in n_c due to a change in collector current (or base-emitter voltage).

The free electron concentration now becomes $n_c + \delta n$ and the depletion-layer width changes from w_{scl} to a value $w_{scl} + \delta w_{scl}$. Applying Poisson's equation in the normal manner and integrating twice with the appropriate boundary conditions gives the total collector-base voltage (including the built-in barrier voltage) V_{cbt}

$$V_{cbt} = \frac{q}{2\varepsilon} N_d' w_{scl}^2$$

$$= \frac{q}{2\varepsilon} (N_d' - \delta n)(w_{scl} + \delta w_{scl})^2 \tag{9.75}$$

Thus we obtain

$$\frac{N_d' - \delta n}{N_d'} = \left[\frac{w_{scl}}{w_{scl} + \delta w_{scl}} \right]^2 \tag{9.76}$$

Using the fact that for small signals, $\delta n \ll n_c$ and $\delta w_{scl} \ll w_{scl}$, we obtain the approximation

$$\frac{\delta n}{N_d} = \frac{2\delta w_{scl}}{w_{scl}} \tag{9.77}$$

The change in free carrier charge within the space-charge layer is

$$\delta Q_n = -\delta n(w_{scl} + \delta w_{scl})$$

The change in fixed donor charge is

$$\delta Q_{N_d} = N_{epi}\, \delta w_{scl}$$

Hence we find

$$\frac{\delta Q_{N_d}}{\delta Q_n} = \frac{-N_{epi}\, \delta w_{scl}}{\delta n(w_{scl} + \delta w_{scl})}$$

$$= -\frac{\delta w_{scl}}{w_{scl}} \frac{N_{epi}}{\delta n} = -\frac{1}{2} \tag{9.78}$$

The total charge increment δQ_t due a change in current density J_{nc} and a corresponding change δn in electron concentration n is therefore

$$\delta Q_t = \delta Q_n + \delta Q_{N_d} = \delta Q_n - \frac{1}{2}\delta Q_n = \frac{\delta Q_n}{2} \tag{9.79}$$

Since

$$\frac{dQ_n}{dI_C} = \frac{d(qAn_c\,w_{scl})}{A\,dJ_{nc}} = \frac{w_{scl}}{v_d}$$

the corresponding overall delay time is

$$t_{scl} = \frac{dQ_t}{dI_C} = \frac{1}{2}\frac{dQ_n}{dI_C}$$

$$= \frac{w_{scl}}{2v_d}$$

as stated above in Eq. (9.74).

Delay Due to Free Carrier Charge in Emitter-Base Space-Charge Layer

This charge Q_{bescl} is, as we have already seen in Eq. (3.104), proportional to $\exp(V_{BE}/2V_t)$. A delay time t_{qbe} could be defined by

$$t_{qbe} = \frac{dQ_{bescl}}{dI_C}$$

but since Q_{bescl} is not proportional to collector current, the derivative does not therefore lead to a constant delay time, but rather a delay time which varies with V_{BE} bias. The effect may be considered as a modification of the emitter-base depletion-layer capacitance. Let us define a modified capacitance

$$C_{ebt} = C_{je}\,F_{ebscl} \qquad (9.80)$$

Figure 9.19 shows that the factor F_{ebscl} has a maximum value close to 2 at high forward bias. However, we have already seen in Chap. 7 that the diffusion capacitance dominates at high forward bias. This effect is therefore often not readily identifiable when modeling transistors and is often neglected, or combined either with a second-order modification to the value of C_{je} or base delay time t_{bb} [18].

Complete Expression for f_t

Using the above results and the definition for diffusion capacitance given by Eqs. (9.66) and (9.68) enables us to write

$$C_{\text{diff}} = g_m(t_{bb} + t_{scl} + t_{em} + t_{qbe}) \qquad (9.81)$$

This is the value which applies in the hybrid pi equivalent circuit of Fig. 7.8. The expression for f_t contains one additional delay due to collector series resistance R_c. This is the resistance of the collector region between the base-collector space-charge layer and the heavily doped substrate or buried layer. Because the capacitance C_{jc} is connected to the output collector terminal through this resistance,

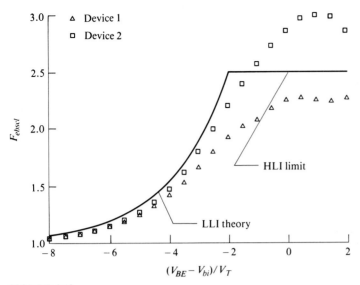

FIGURE 9.19
Correction factor F_{ebscl} for emitter-base capacitance due to free carriers within the space-charge layer. (*After Negus and Roulston* [13]. *Reprinted with permission from Solid State Electronics, Pergamon Press PLC.*)

an additional time constant $R_c\,C_{jc}$ exists in the short-circuit current gain, and thus in the expression for f_t. The reader may also consider the input Miller capacitance due to C_{jc} with a "load" resistance R_c and obtain the same result— an additional $R_c\,C_{jc}$ time constant in f_t. The final expression is, following a treatment similar to that of Chap. 7,

$$\frac{1}{2\pi f_t} = \frac{C_{je} + C_{jc}}{g_m} + t_{bb} + t_{scl} + t_{em} + t_{qbe} + R_c\,C_{jc} \qquad (9.82)$$

As suggested, it is sometimes more convenient to omit the t_{qbe} term and use a modified expression for C_{je} versus V_{BE}.

Use of the hybrid π small signal equivalent circuit is almost universal in both hand and computer analysis (the more general Ebers-Moll large signal representation reduces to the hybrid π model for small signal analysis). It is, however, important to recognize that lumping the diffusion capacitance across the (internal) base-emitter junction is based on the so-called charge-control approach, [14]. In reality, the change in total charge given by Eq. (9.67) is only partially "reclaimable" for a given change in voltage V_{BE}. It has been shown [15] that although the approximation is normally excellent in the common-emitter configuration, in common-base a time-dependent solution to the transport equation gives a base delay time (for high frequency collector current versus emitter-base voltage) which is equal to 2/3 of the above t_{bb} value. It has also been shown recently [15] that for modeling base current versus base-emitter voltage

TABLE 9.1
Delay times in ps for the devices used in Fig. 9.11. The base width is given in microns; $f_{t\,\text{const}}$ is the f_t that would be obtained due to the four delays; $f_{t\,\text{max}}$ is the overall maximum attainable value of f_t. f_t values in MHz.

Device	W_b	t_{bb}	t_{em}	t_{scl}	t_{RC}	$f_{t\,\text{const}}$	$f_{t\,\text{max}}$
4	0.12	15	8	2.1	0.12	6300	5000
5	0.57	83	36	2.7	7.6	1200	470
8	12.0	19 000	4700	560	70	6.7	5.0

frequency-dependence, only about 90 percent of the above t_{em} value due to emitter charge contributes to the diffusion capacitance. Given the complexity of ascertaining the exact value of emitter charge and even of determining the exact impurity profile of the emitter region, this is not too serious an approximation in practice.

Table 9.1 gives the constant delay times for the devices used for Fig. 9.11. Note particularly the fact that the base transit time is often significantly less than the total delay time and that the sum of t_{em} and t_{scl} (the emitter and collector-base space-charge delay times) is seldom negligible. The t_{RC} term, on the other hand, may usually be neglected. It is also interesting to note that the value of f_t based on the four "current-independent" delay times, $f_{t\,\text{const}}$, is greater than the actual peak value of f_t (versus I_C) that is attainable. This is because of the fact that the $(C_{je} + C_{jc})/g_m$ term is still significant at the collector current for which the Kirk effect causes a rapid increase in base transit time t_{bb}. This current-dependence will be discussed in the following chapter.

9.6 MAXIMUM OSCILLATION FREQUENCY AND THE "OPTIMUM TRANSISTOR"

In Chap. 7 we introduced the concept of maximum oscillation frequency [Eq. (7.28)]

$$f_{m\,\text{osc}} = \sqrt{\frac{f_t}{8\pi r_{bb} C_{jc}}}$$

We are now in a position to discuss this in more detail. Neglecting the (small) time constant $R_c C_{jc}$, the expression for f_t may be written as

$$\frac{1}{2\pi f_t} = t_{bb} + t_r$$

where t_r is the sum of the delay times t_{scl}, t_{em}, and t_{qbe} and the junction capacitance term $(C_{je} + C_{jc})/g_m$. We now substitute for base transit time delay

$$t_{bb} = \frac{F_b W_b^2}{2D_n}$$

for collector-base junction capacitance

$$C_{jc} = \frac{F_c \, \varepsilon BL}{W_{scl}}$$

where F_c is an area factor to allow for the fact that the base-collector junction area is in general greater (typically by a factor of 4) than the emitter area $A = BL$, and for base resistance [Eq. (7.17)]

$$r_{bb} = \frac{L}{12\sigma_{be} \, W_b \, B}$$

where σ_{be} is the pinched base region conductivity. These substitutions yield

$$f_{m\,osc} = \frac{1}{2\pi L} \sqrt{\frac{3\sigma_{be} \, W_b \, w_{scl}}{\varepsilon F_c (F_b \, W_b^2/2D_n + t_r)}} \qquad (9.83)$$

Apart from the fact that the maximum oscillation frequency is inversely proportional to the emitter stripe width L (which explains the basic reason for ever decreasing values of L in microwave and high speed devices), it may also be seen from the above expression that an optimum condition exists for $f_{m\,osc}$. If a device is designed solely for very high values of f_t, small values of base width W_b are indicated; the numerator inside the square root term of Eq. (9.83) thus decreases (base resistance increases) and $f_{m\,osc}$ is degraded. If, on the other hand, a large base width W_b is used, the base transit time term in the denominator of Eq. (9.83) increases (f_t decreases) and eventually the maximum oscillation frequency is once again degraded.

Clearly a design condition exists which will optimize $f_{m\,osc}$. By inspection, it is evident that this condition is given by

$$\frac{F_b W_b^2}{2D_n} = t_r \qquad (9.84)$$

This is of particular significance in the choice of vertical impurity profile parameters for a transistor to be used for maximum small signal or large signal power amplification at high frequencies. In the case where the device is biased at a collector current giving maximum or near-maximum values of f_t, the dominant component of t_r will normally be the base-collector transit time delay t_{scl}. A near-optimum design would therefore satisfy the condition

$$\frac{W_b^2}{2D_n} \sim \frac{w_{scl}}{2v_d} \qquad (9.85)$$

Figure 9.20 illustrates this optimization for a typical transistor [16]. The device is an RF power transistor with $L = 4$ μm, $B = 1$ cm (a multistripe inter-digitated layout is used), epitaxial layer resistivity and thickness are 3 Ω cm and 6 μm, respectively. The base diffusion sheet resistance is held constant in the computer study at 200 Ω/sq to approximate a realistic constraint.

Notice the variation of pinched base sheet resistance in Table 9.2 (from 23.7 to 0.95 kΩ/sq). Figure 9.20 shows that for the lowest value of junction depth X_{jbc} equal to 0.9 μm, the base transit time is almost negligible. The peak f_t value of

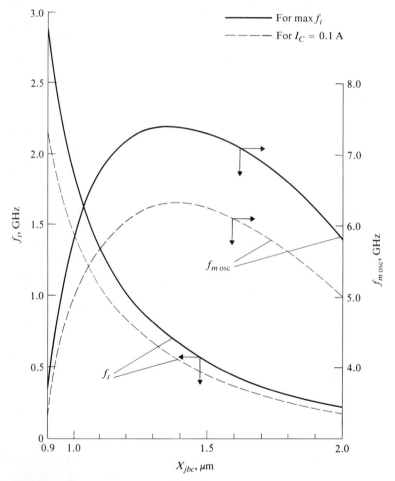

FIGURE 9.20

Computed curves of f_t and $f_{m\,osc}$ versus base-collector junction depth X_{jbc} assuming gaussian impurity profiles for emitter and base diffusions. $X_{jeb} = 0.6$ micron and the total base diffusion sheet resistance is held constant at 200 Ω/sq. The solid curves are obtained by choosing the collector current at which peak f_t occurs; the broken curves are obtained for a collector current of 0.1 A. Additional data is given in Table 9.2. (*After Roulston and Hébert* [16]. *Reprinted with permission from Solid State Electronics, Pergamon Press PLC.*)

TABLE 9.2

Variations of pinched base sheet resistance, neutral base width, and base transit time as X_{jbc} is increased (0.9–1.7 μm) maintaining a fixed base sheet resistance of 200 Ω/sq, adjusting the surface base concentration N_B

X_{jbc}, μm	0.9	0.95	1.0	1.1	1.3	1.7
N_B, cm^{-3}	2.0	1.86	1.71	1.55	1.12	0.81
R_{BE}, kΩ/sq	23.7	14.4	10.3	5.2	2.4	0.95
W_b, μm	0.09	0.13	0.16	0.25	0.41	0.75
t_{bb}, ps	1.0	1.7	5.5	16	50	70

2.8 GHz therefore indicates a corresponding delay time of 57 ps. The optimum value of t_{bb} should therefore be 57 ps. From the table, we see that this implies an X_{jbc} value of about 1.3 μm. The computed curves confirm that $f_{m\,osc}$ peaks for approximately this value, at 7.46 GHz.

The conclusions resulting from the above discussion on maximum oscillation frequency are important. Designing a transistor solely for a high value of f_t is not that difficult technologically. The peak value of f_t depends mainly on vertical delay times $(t_{bb}, t_{scl}, \text{ and } t_{em})$ and these can be reduced by using shallower junctions—this does not involve any photolithographic constraints on mask dimensions. The additional terms involved in $f_{m\,osc}$, however, invoke both a horizontal (mask) dimension L, and a constraint on the vertical base width W_b. We have seen that a well-defined optimum design criterion exists for obtaining the best value of $f_{m\,osc}$. It is worth noting that this is also likely to give a good design for switching applications in digital circuits. A high speed digital inverter requires low capacitances (junction plus diffusion) and hence a high f_t value; increasing f_t by reducing vertical base width W_b will yield a high base resistance r_{bb}. This will ultimately introduce an unacceptably large time constant at the input to the logic inverter (r_{bb} in series with the input junction and diffusion capacitances). The device with optimized $f_{m\,osc}$ will therefore be a "good" device for high speed digital applications (although a well-defined "optimum" is not obvious and would involve detailed consideration of the circuit drive and load conditions).

9.7 CONCLUSIONS

In Chaps. 8 and 9 we have provided a detailed study of the transistor with "real" impurity profiles. The importance of both accelerating and retarding field regions in the base and emitter is now evident, in determination of both static and high frequency characteristics. We have seen that while the dc collector current depends on the integeral of the base region doping, the major component of delay—the base transit time—depends on the *shape* of the base doping profile. This implies that for inverse (upward) operation, apart from changes in the depletion-layer boundaries due to the interchanging of V_{BE} and V_{BC}, and neglecting contributions from outside the area under the emitter, the I_C versus V_{BE} law will remain the same, whereas the inverse f_t will be somewhat different. Upward operation is explored further in Chap. 13.

The results of this chapter provide all the information required for understanding the low and medium current operation of the device, including charge components contributing to f_t. In the next chapter we shall examine the phenomena which create falloff of current gain and of f_t at high collector currents. We shall then be in a position to completely characterize the performance of the bipolar transistor for a wide range of applications.

PROBLEMS

All questions in this chapter concern a transistor with the impurity profile given in Prob. 8.1. The emitter geometry is $4~\mu m$ $(L) \times 100~\mu m$ (B) and the collector measures $16 \times 105~\mu m$.

9.1. Draw the doping and effective (due to BGN) doping profiles on a linear-linear graph. Use graphical (or numerical) integration from the emitter-base junction $[(\varepsilon/q)\,dE/dx = N(x)]$ to determine the electric field for several thicknesses of depletion-layer edge on the base side of the junction. By observing the depletion-layer edge d_p for which the peak field attains a value of about 5×10^5 V/cm, note the breakdown condition. Observe whether or not it is possible to represent the profile on either side of the e-b junction by a straight line for this breakdown situation.

9.2. Calculate the junction capacitance per unit area from the gradient determined in Prob. 8.2 and compare with the value obtained from Fig. 9.4(a). Calculate the side-wall and plane components of C_{je} at zero bias using Eqs. (9.5) to (9.8) and Fig. 9.4, and observe the importance of the sidewall component.

9.3. From the gradient used in Prob. 9.2, calculate from Eq. (3.126) the breakdown voltage and compare with the value given by Fig. 9.2.

9.4. From the impurity profile and effective impurity profile graph, determine graphically (or numerically by integration) the integral of the total base dope and the total effective base dope. Hence calculate the coefficient in the $I_C = f(V_{BE})$ law [Eq. (9.46)].

9.5. Using the result of Prob. 9.4 and assuming an effective emitter-base space-charge layer lifetime $\tau_{DE} = 10$ ns, calculate and plot the low current gain β versus collector current I_C (you may replace the product $W_b N_{Ab}$ in Eq. (9.10) by the base doping integral deduced in Prob. 9.4). If the surface recombination parameter $X_{fs} = 10~\mu m$, by how much will the gain be reduced?

9.6. For the base-collector junction, calculate (assuming the one-sided abrupt junction approximation) the depletion-layer thickness d_n on the collector side for $V_{CB} = 10$ V. Using graphical or numerical charge balance, deduce from the impurity profile graph of Prob. 9.4 the position of the depletion-layer edge on the base side of the junction. From the linear gradient approximation, determine the position of the depletion-layer edge on the base side of the emitter-base junction for a total voltage of 0.1 V (see Prob. 8.2).

From your impurity profile graph, estimate the values of base width regions W_{b1}, W_{b2} (of Fig. 9.8) and of the profile field parameters r_1 [Eq. (9.16)] and η [Eq. (9.22)]. Hence estimate F_{bb} in Eq. (9.31) and t_{bb} in Eq. (9.33). Compare the result for Eq. (9.31) with that obtained using the actual and effective base doping integrals (the Gummel numbers) and compare t_{bb} with $W_b^2/2D_{n\,av}$.

9.7. From the base doping integral, calculate the pinched base sheet resistance R_{be}/sq in Eq. (9.47). Using the solutions to Probs. 8.3 and 8.4, calculate the probable value of maximum current gain β.

9.8. Calculate the Early voltage V_A for $V_{CB} = 10$ and 30 V. Hence find the value of small signal (incremental) output resistance for $I_C = 1$ mA. Note that the values of Q_{be} and Q_{bo} in Eq. (9.56) may be found from your impurity profile graph.

9.9. Calculate the value of f_t for $V_{CB} = 10$ V and for $I_C = 0.1$, 10, and 50 mA.

9.10. Calculate the maximum oscillation frequency under the conditions used in Prob. 9.9. Could the values be improved by altering the depth of the emitter diffusion? If so, suggest a reasonable "best" value to aim for during thermal processing.

REFERENCES

1. D. J. Roulston and R. C. Kumar, "Peripheral emitter-base junction capacitance in bipolar transistors," *IEEE Trans. Electron Devices*, vol. ED-26, pp. 810–811 (May 1979).
2. C. T. Sah, R. N. Noyce, and W. Shockley, "Carrier generation and recombination in p-n junctions and p-n junction characteristics," *Proc. IRE*, vol. 45, pp. 1228–1243 (September 1957).
3. D. J. Roulston, N. D. Arora, and S. G. Chamberlain, "Modeling and measurement of minority carrier lifetime in heavy doped N diffused silicon diodes," *IEEE Trans. Electron Devices*, vol. ED-29, pp. 284–291 (February 1982).
4. N. G. Chamberlain and D. J. Roulston, "Modeling of emitter-base bulk and peripheral space-charge layer recombination currents in bipolar transistors," *IEEE Trans. Electron Devices*, vol. ED-23, pp. 1345–1346 (December 1976).
5. D. J. Roulston and A. A. Eltoukhy, "Modeling bulk and surface recombination in the sidewall space-charge layer of an emitter-base junction," *IEE Proceedings*, vol. 132, pt. I, pp. 205–209 (October 1985).
6. J. Lindmayer and C. Y. Wrigley, *Fundamentals of Semiconductor Devices*, Van Nostrand-Reinhold, Princeton (1965).
7. U. Zugelder and D. J. Roulston, "Analytic results for the base region of bipolar transistors based on computer simulations," *Solid State Electronics*, vol. 30, pp. 895–900 (September 1987).
8. H. K. Gummel, "Measurement of the number of impurities in the base layer of a transistor," *Proc. IRE*, vol. 49, p. 834 (April 1961).
9. J. M. Early, "Effects of space-charge layer widening in junction transistors," *Proc. IRE*, vol. 42, p. 1761 (1954).
10. D. Scott and D. J. Roulston, "I_C-V_{CE} characteristics of double diffused transistors under low level injection," *Solid State Electronics*, vol. 23, pp. 201–207 (March 1980).
11. F. Hébert and D. J. Roulston, "Modeling of narrow-base bipolar transistors including variable base-charge and avalanche effects," *IEEE Trans. Electron Devices*, vol. ED-34, pp. 2323–2328 (November 1987).
12. R. D. Thornton, D. de Witt, E. R. Chenette, and P. E. Gray, "Characteristics and limitations of transistors," *SEEC*, vol. 4, Wiley, New York (1966).
13. K. Negus and D. J. Roulston, "Simplified modeling of delays in the emitter-base junction," *Solid State Electronics*, vol. 31, pp. 1464–1466 (September 1988).
14. R. Beaufoy and J. J. Sparks, "The junction transistor as a charge controlled device," *Automat. Teleph. Elect. Comm. Jnl. (London)*, vol. 13, p. 310 (1957).

15. G. A. M. Hurkx, "A new approach to ac characterization of bipolar transistors," *Solid State Electronics*, vol. 31, pp. 1269–1275 (August 1988).
16. D. J. Roulston and F. Hébert, "Optimization of maximum oscillation frequency of a bipolar transistor," *Solid State Electronics*, vol. 30, pp. 281–282 (March 1987).
17. D. J. Roulston, "Low current base-collector boundary conditions in GHz frequency transistors," *Solid State Electronics*, vol. 18, pp. 845–847 (1975).
18. J. J. H. Van Den Biessen, "A simple regional analysis of transit times in bipolar transistors," *Solid State Electronics*, vol. 29, pp. 529–534 (May 1986).

CHAPTER 10

HIGH CURRENT EFFECTS IN BIPOLAR TRANSISTORS

10.1 INTRODUCTION

In this chapter, we will discuss various high current phenomena which affect the performance of the transistor, specifically the current gain and the transition frequency f_t and the manner in which they fall off at high collector current bias. The first effect to be considered is due to the lateral voltage drop in the intrinsic base resistance of the transistor; this is usually referred to as emitter current crowding. It does not in itself affect the gain or the transition frequency but is a contributing factor. The second effect we shall examine is high level injection in the base region; this gives rise to falloff of gain at high currents but does not significantly alter f_t. Then we will examine the very important phenomenon of base stretching or Kirk effect; this produces a rapid falloff of both gain and transition frequency. It is also accompanied by some sideway spreading of the collector current in the base region; this will also be analyzed. The last high current effect to be examined in this chapter is called *quasi-saturation*. This occurs when the terminal collector base bias is such that the transistor appears to be operating in the normal active region, whereas in fact, due to the internal voltage drop in the collector layer, the actual collector base junction is forward biased. The above effects will be examined theoretically as if each were occurring on its own. Obviously, under many operating conditions several of the high current effects will occur simultaneously.

It is only by treating them separately, however, that a clear understanding can be obtained. Very complex models (or complete numerical simulation) would be required to account for the simultaneous occurrence of the effects together.

10.2   EMITTER CURRENT CROWDING: BASE-RESISTANCE VOLTAGE DROP

As we saw in Sec. 7.4.2 there is a significant horizontal resistance through which the base current has to flow in order to inject carriers into the emitter. Referring to Fig. 10.1 and using the theory developed in Sec. 7.4.2, the total resistance of the base region may be written for a two-base contact discrete transistor as

$$R_b = \frac{L}{4B\sigma_b W_b} \tag{10.1}$$

where L is the emitter stripe width, B the total emitter stripe length, and W_b the neutral base width. The conductivity σ is given by

$$\sigma_b = q\mu_p N_A \tag{10.2}$$

where N_A is the average base doping level, μ_p the corresponding hole mobility, and q the electronic charge. Because of the voltage drop due to the base current flowing in this resistance, it is clear that the expression for collector current density at the center of the base region will be given by

$$J_{nc} = J_{cs} \exp\left(\frac{V_{BE} - I_b R_b'}{V_t}\right) \tag{10.3}$$

where R_b' is some effective resistance related to R_b and where J_{cs} is the equivalent reverse saturation collector current density, I_b is the dc base current, and V_{BE} is the terminal base emitter voltage. From this equation it is clear that as long as the base voltage drop remains less than V_t, the current density at the center of the active region will be comparable to the current density at the edge. However,

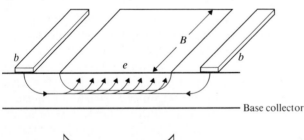

FIGURE 10.1
Emitter current crowding. (a) Cross section of base region; (b) collector current density versus distance.

when the voltage drop $I_b R_b'$ attains a value of 25 mV, it is apparent that the collector current density at the center of the device will be $1/e = 0.37$ times the current density at the perimeter of the active region, $J_{c\,edge}$. As the voltage V_{BE} is further increased this voltage drop will exceed considerably the value 25 mV and the current density at the center of the device will rapidly become much less than the current density at the perimeter. This phenomenon is known as *emitter current crowding*. The result of this effect is that at high forward bias only a small fraction of the total emitter area is actually carrying most of the current. Hauser [1] showed that in the presence of emitter current crowding the effective emitter width may be written in the form

$$\frac{L_{eff}}{L} = \frac{\sin Z \cos Z}{Z} \tag{10.4a}$$

where the total collector current $I_C = J_{c\,edge} L_{eff} B$ and where $Z \tan Z = I_b R_b / 2V_t$. A more convenient form, valid at low and high currents, but approximating the above result at currents where $I_b R_b / 2V_t$ is of order unity is

$$\frac{L_{eff}}{L} = \frac{1}{1 + I_b R_b' / V_t} \tag{10.4b}$$

We may deduce that the resistance R_b' introduced in Eq. (10.3) is in fact equal to $R_b/2$. This dc resistance may be compared to the (low current) small signal ac resistance r_{bb} which we have already shown to be equal to $R_b/3$ [see Eq. 7.17)].

Hauser also showed that the small signal resistance varies with current according to

$$\frac{r_{bb}(I)}{R_b} = \frac{\tan Z - Z}{Z \tan^2 Z} \tag{10.5a}$$

This may also be expressed in a more convenient approximate form thus:

$$\frac{r_{bb}(I)}{r_{bb}} = \frac{1}{1 + I_b r_{bb} / 2V_t} \tag{10.5b}$$

where r_{bb} is taken to be the low current (constant) value of $r_{bb}(I)$. It is clear that as the current crowding phenomenon increases in magnitude so the base current has a shorter distance to travel, thus resulting in a decreased base resistance and, more importantly, a decreased effective area. This means that high current densities will be experienced at a lower V_{BE} and lower I_C bias than would be indicated by a simple calculation based on uniform current density and emitter area.

10.3 HIGH LEVEL INJECTION IN THE NEUTRAL BASE

When the emitter base junction of the transistor is sufficiently forward biased, the carrier concentration injected into the base, $n(0)$, rises above the impurity concentration N_{Ab}. The situation is similar to that discussed for the P^+N narrow base diode in Sec. 3.5.1 [2]. Here we saw that as the minority carrier concentration rose in value above the background doping level, space-charge neutrality forced the hole and electron concentrations to become approximately equal. In this case the electric field due to the hole concentration gradient

$$E(x) = -V_t \frac{1}{n} \frac{dn}{dx} \tag{10.6}$$

creates a drift current which (as in the case of the narrow base diode) is equal in magnitude to the normal diffusion current. This results in an effective doubling of the electron diffusion constant and an exp $(V_{BE}/2V_t)$ dependence of the collector current as in Eq. (3.65). Since the emitter is normally much more heavily doped than the base region it is a fair assumption that the base enters high level injection before the emitter. Figure 10.2(a) shows a typical impurity profile for the emitter and base regions. It is difficult to convey the information on a linear scale, but the values make it clear that most of the base can quite easily enter high level injection, $n(0) \gg N_{Ab}$, while only a small part of the emitter neutral region closest to the space-charge layer enters high level injection. Figure 10.2(b) shows a simplified diagram, on a linear scale, for typical high level injection conditions in the base region. In this case, applying the Boltzmann relations across

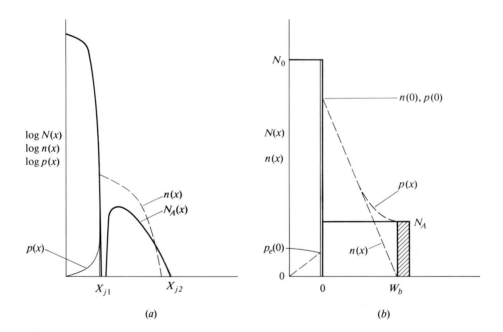

(a)

(b)

FIGURE 10.2
Impurity profile and carrier concentration distribution for emitter and base for high level injection. (a) Actual profile on log scale; (b) simplified impurity profile on linear scale.

the emitter base space-charge layer we can write:

$$\frac{n_b(0)}{p_e(0)} = \frac{N_0}{n_b(0)} \tag{10.7a}$$

where N_0 is the doping level at the space-charge layer boundary in the emitter. The base current injected into the emitter is given by

$$I_b = K p_e(0) \tag{10.7b}$$

where K contains all the emitter terms in Eq. (8.13) or Eq. (8.22). In fact, K will vary slightly as the emitter region nearest the space-charge region enters high level injection, but we will neglect this effect. The collector current is given by the expression developed for the narrow base diode under high level injection [Eq. (3.65)]

$$I_C = \frac{2qAD_n n(0)}{W_b} \tag{10.8a}$$

or

$$I_C = I_{COH} \exp\left(\frac{V_{BE}}{2V_t}\right) \tag{10.8b}$$

where $I_{COH} = 2qAD_n n_i/W_b$ as per Eq. (3.70). We can therefore write the expression for current gain using Eqs. (7.9a) and (10.7a) in the form

$$\beta(I) = \frac{\beta_0 N_{Ab}}{n_b(0)} \tag{10.9}$$

where β_0 is the (constant) gain given by Eq. (7.8) etc. It is apparent that the current gain varies inversely with injected electron concentration into the base region, i.e., gain varies inversely with collector current. Following the procedure in Sec. 3.3.5.1, the collector current at which the base enters high level injection may be written as

$$I_{\text{HLI}} = \frac{q\mu_P V_t N_{Ab}}{W_b} \tag{10.10}$$

We may therefore also write the gain as a function of current for $I_C < I_{\text{HLI}}$ in the form

$$\beta(I) = \beta_0 \left(\frac{I_C}{I_{\text{HLI}}}\right)^{-1} \tag{10.11}$$

It should be noted that when the gain becomes current-dependent, as is the case at both very high and very low currents, the small signal or ac value will be different from the dc value discussed in this section. Eventually, of course, the emitter region further away from the space-charge layer boundary will enter high level injection also; the variation of gain with current will then be different from the above result.

10.4 TRANSITION FREQUENCY UNDER HIGH LEVEL INJECTION

We have the expression for collector current under high level injection conditions [Eq. (10.8a)]. It is a simple matter to write the value of the electron charge in the base region, $qAn(0)W_b/2$. The ratio of charge to current enables us to define the base delay time as [2]

$$t_{bb} = \frac{W_b^2}{4D_n} \qquad (10.12)$$

This differs only from the low level injection value given by Eq. (7.16) by the factor 2. Thus, in the case of a uniformly doped base transistor the transition frequency, or rather the part of it which depends on base transit time, may improve by a factor of 2. However, in the case of a base region which has a significant accelerating field, it is possible that a slight reduction in transition frequency value will occur as may be seen by comparing the above result with Eq. (9.33) at high values of field factor and no retarding field region (that is, $W_{b1} = 0$). The reduction could attain a factor $2/\eta$ which would be about $1/3$ in the case of a high accelerating field. Figure 10.3 shows the manner in which current gain and f_t would vary with collector current if only high level injection effects existed in the base region. β falls off as $1/I_C$ whereas f_t can either rise slightly as in case (i) for "uniform" base doping, or decrease slightly as in case (ii) for the ideal drift transistor profile with a high η value.

10.5 HIGH CURRENT BASE STRETCHING: KIRK EFFECT

We will now study a very important cause of high current falloff of gain and f_t. The effect was originally discovered by Kirk [3] and the following theory is

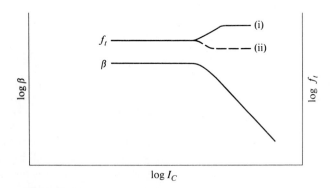

FIGURE 10.3
Variation of gain and f_t when only high level injection effects exist. For the f_t curve, (i) is for a uniformly doped base, (ii) is for a high accelerating drift field.

based on both [3] and subsequent treatments [4]. The actual mechanism consists in a vertical widening of the effective neutral base region by an amount w_k beyond the low current value W_b.

In Chap. 7 we saw that in order for the collector electron current to pass through the collector base space-charge layer, it was necessary to maintain a finite electron concentration n_c within this layer. The value of this concentration was given by Eq. (7.2), repeated here for convenience in terms of collector current density

$$n_c = \frac{J_{nc}}{qv_d} \tag{10.13}$$

We will assume that the electric field is sufficiently high for the constant drift velocity conditions to be maintained. Poisson's law in the collector epitaxial collector layer region now becomes modified as follows: for $x > X_{jbc}$ and assuming the abrupt junction approximation, we can write

$$\frac{dE}{dx} = \frac{q}{\varepsilon}(N_{epi} - n_c) \tag{10.14}$$

Integrating this gives the electric field

$$E(x) = \frac{q}{\varepsilon}[(N_{epi} - n_c)x + c] \tag{10.15}$$

Figure 10.4 shows the electric field diagrams for various values of current density, i.e., for various values of electron concentration n_c. It is clear from the preceding equation that as n_c increases toward the value N_{epi} the gradient of the electric field gradually decreases, until for $n = N_{epi}$ the gradient of the electric field in the epitaxial layer is constant. For this condition the whole of the epitaxial layer will correspond to the space-charge region. In the figure the diagrams have been drawn for an approximately constant base collector voltage. As n_c increases above the value N_{epi} the gradient of electric field reverses in sign. From this condition onward it is easier to visualize the behavior starting from the right-hand side where the N$^+$ region forces the field rapidly down to zero. For a given value of current density, hence a given value of electron concentration n_c, it is clear that the only mathematical solution with the reversed electric field gradient is such that the electric field becomes zero somewhere in the middle of the epitaxial region. Using Eq. (10.15), the result of integrating Poisson's equation for the condition $n_c > N_{epi}$ and redefining for convenience $x = 0$ to be at the NN$^+$ interface, gives

$$E(x) = -\frac{q}{\varepsilon}(n_c - N_{epi})(x - w_{scl}) \tag{10.16}$$

where

$$w_{scl} = W_{epi} - w_K \tag{10.17}$$

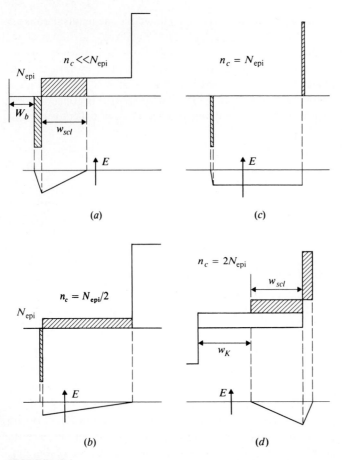

FIGURE 10.4
Electric field in the collector for various current densities. (a) $n_c \ll N_{epi}$; (b) $n_c = N_{epi}/2$; (c) $n_c = N_{epi}$; (d) $n_c = 2N_{epi}$.

Integrating this gives the total voltage

$$V(x) = \frac{q}{\varepsilon} (n_c - N_{epi})\left(\frac{x^2}{2} - w_{scl}\, x\right) \tag{10.18}$$

Here the NN^+ junction is taken as the zero voltage reference point. The total voltage is obtained when $x = w_{scl}$

$$V_{CB} = \frac{q}{2\varepsilon} (n_c - N_{epi})w_{scl}^2 \tag{10.19}$$

or

$$V_{CB} = \frac{q}{2\varepsilon} (n_c - N_{epi})(W_{epi} - w_k)^2 \tag{10.20}$$

We can thus find the space-charge layer width (recall that $n_c > N_{epi}$):

$$w_{scl} = \sqrt{\frac{2\varepsilon V_{CB}}{q(n_c - N_{epi})}} \tag{10.21}$$

At this point is is convenient to introduce once again the epitaxial layer reach-through voltage $V_{pt\ epi}$. This is the voltage which will be required under low current conditions to make the space charge extend from the base collector junction up to the NN^+ junction. It is given by the solution to Poisson's law

$$V_{pt\ epi} = \frac{qN_{epi}}{2\varepsilon} W_{epi}^2 \tag{10.22}$$

We thus obtain

$$w_K = W_{epi}\left[1 - \sqrt{\frac{V_{CB}}{V_{pt\ epi}} \frac{1}{(n_c/N_{epi}) - 1)}} \right] \tag{10.23}$$

From this we see that the value of n for which w_K is zero is given by

$$n_K = N_{epi}\left(1 + \frac{V_{CB}}{V_{pt\ epi}} \right) \tag{10.24}$$

The corresponding value of collector current density is

$$J_K = qv_d N_{epi}\left(1 + \frac{V_{CB}}{V_{pt\ epi}} \right) \tag{10.25}$$

Equation (10.23) can now be written in terms of the collector current density J_c:

$$\frac{w_K}{W_{epi}} = 1 - \frac{1}{\sqrt{[(J_c/J_K) - 1](V_{pt\ epi}/V_{CB}) + J_c/J_K}} \tag{10.26}$$

Let us consider some typical values of the parameters. For $N_{epi} = 10^{15}$ cm^{-3} and $W_{epi} = 10$ microns, we have $V_{pt\ epi} = 100$ V. For $V_{CB} = 100$ V we have J_K equal to about 3000 A/cm^2 and

$$w_K = W_{epi}\left[1 - \frac{1}{\sqrt{2J_c/J_K - 1}} \right] \tag{10.27}$$

or, for $J_c = 1.1J_K$,

$$w_K = 0.1W_{epi} \sim 1\ \mu m \tag{10.28}$$

The base width will have increased from W_b to $W_b + w_k$, (neglecting the narrow original depletion layer in the base) by 1 micron for an increase of 10 percent in the current density above the value J_K. For a transistor with an original base width of 1 micron it is obvious that the current gain will drop quite drastically as J_c increases. Since the base transit time varies as the square of the base width, it is clear that this base stretching phenomenon, or Kirk effect [3, 4], will have an even more important effect on the base transit time. In the above case, base transit time will be increased by a factor of 4 with a corresponding

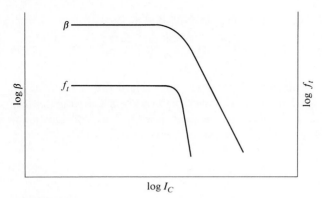

FIGURE 10.5
Variation of β and f_t versus I_C for the case where only Kirk effect exists.

reduction in f_t, while the gain will be decreased by a factor 1/2. Figure 10.5 illustrates this behavior.

10.6 LATERAL BASE WIDENING

Referring to Fig. 10.6 it is clear that at the edge of the emitter the carrier concentration cannot drop abruptly to zero, and some lateral injection of electrons will take place. Under normal low current conditions the amount of current and charge associated with this sidewall injection is usually negligible. However, at high currents, particularly at currents exceeding the Kirk current limit, this lateral injection can be very significant. Van der Ziel [5] studied this situation on the assumption that in the active base region under the emitter the current density could not increase above the Kirk current limit. Figure 10.7 illustrates the

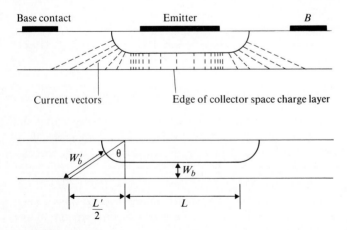

FIGURE 10.6
Lateral base widening.

variation of carrier concentration versus distance and collector current density versus base emitter voltage from below to above the Kirk current limit. It is easy to visualize the behavior if we take collector current density J_c as a reference. As J_c is increased above the value J_K, the base width increases by an amount w_k. Each successive increase in J_c corresponds to an increased carrier concentration gradient, and it is therefore clear from Fig. 10.7(a) that the value of injected carrier concentration $n(0)$ has to increase considerably faster than J_c. Since $n(0)$ is governed by the voltage V_{BE}, the characteristic of Fig. 10.7(b) results. If the base enters high level injection with $n(0)$ greater than the background doping level, the flattening of the log (J_c) versus V_{BE} curve is further enhanced.

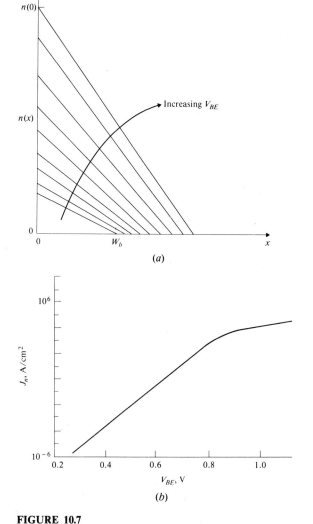

FIGURE 10.7

(a) Carrier concentration versus distance as current density is increased; (b) collector current density versus emitter-base voltage.

Van der Ziel used this fact to state that to a first approximation the collector current may not increase in the active region and that terminal increase in collector current occurs mainly due to lateral spreading from the sidewall regions. Referring to Fig. 10.6 and defining an equivalent increase in emitter stripe width L', it is apparent that the transit time associated with the increased base width W_b' at some angle θ is

$$t_{bb}' = \frac{W_b'^2}{2D_n} = \frac{W_b^2 + \frac{1}{4}(L' - L)^2}{2D_n} \tag{10.29}$$

Assuming that the collector current I_C is given by

$$I_C = B(L + L')J_K \tag{10.30}$$

we may write the modified base transit time as

$$t_{bb}' = t_{bb}\left[1 + \left(\frac{L}{2W_b}\right)^2\left(\frac{I_C}{LBJ_K} - 1\right)^2 \right] \tag{10.31}$$

This is the component of base transit time associated with the angle θ on Fig. 10.4. It is apparent that for increasing collector current this transit time can rise substantially above the low current value. This is the origin of one SPICE CAD model for f_t falloff [6].

Note that the total f_t value is due to the active base region charge $Q_n = qAn(0)(W_b + w_k)/2$ together with the additional charge associated with the integral over θ. The manner in which f_t falls off with increasing current is therefore quite complicated. The main conclusion, however, is that falloff will start when the collector current reaches a value $I_k = qAv_dN_{epi}$ and will be quite rapid, whether the Van der Ziel or Kirk effect dominates the behavior. Thus gain and f_t will still fall off roughly as in Fig. 10.5.

10.7 QUASI-SATURATION

When the ohmic drop $I_C R_{epi}$ across the collector epitaxial layer attains a value equal to the terminal voltage V_{bct}, the internal base-collector junction becomes forward biased:

$$V_{CB} = V_{bct} - I_C R_{epi} \tag{10.32}$$

Figure 10.8(a) shows the transistor in the common emitter configuration with the internal collector resistance R_{epi}. It is clear that the resistance has the same effect as a load resistance and that for large collector currents (created by increasing the base current drive), the voltage drop $I_C R_{epi}$ will eventually exceed the terminal collector-base (reverse) bias voltage, thus forward biasing the collector-base junction. The charge injected into both the base and the lightly doped collector when the foward bias V_{bc} exceeds about 0.5 V has the effect of decreasing the current gain (implicitly taken into account in the Ebers-Moll

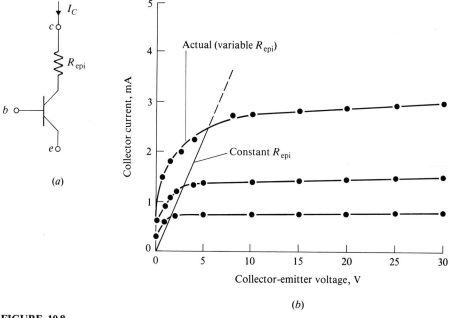

FIGURE 10.8

(a) Transistor in common emitter configuration with internal collector resistance; (b) I_C-V_{CE} characteristics showing the saturation region. (*After Hébert and Roulston* [8].)

[Eq. (7.39)] or Gummel-Poon model [Eq. (9.46)], provided the collector resistance R_{epi} is included). Figure 10.8(b) shows the resulting I_C-V_{CE} characteristics.

The collector current I_{CQS} at which the gain (and f_t) start to decrease due to this effect is given by rearranging Eq. (10.32)

$$I_{CQS} = \frac{V_{bct} + 0.5}{R_{epi}} \qquad (10.33)$$

It is instructive to relate this to the Kirk current $I_K = qAv_d N_{epi}$. Substitution for $R_{epi} = W_{epi}/(Aq\mu_n N_{epi})$ and replacing the mobility $\mu_n = v_d/E_c$ where E_c is of order 10^4 V/cm, we obtain

$$I_{CQS} = I_K \frac{V_{bct} + 0.5}{W_{epi} E_c} \qquad (10.34)$$

For a nominally optimized choice of collector thickness, W_{epi} will be such that avalanche breakdown will occur when the depletion layer reaches the N^+ junction. In this case, from Eq. (7.47) the breakdown voltage V_{cbr} may be written as $E_{br} W_{epi}/2$. The expression for the onset of nominal quasi-saturation is thus

$$I_{CQS} = I_K \left[\frac{(V_{CB} + 0.5)}{V_{cbr}} \right] \left(\frac{E_{br}}{2E_c} \right) \qquad (10.35)$$

Since the breakdown value of field discussed in Sec. 3.8 is of order 3×10^5 V/cm, it is clear that I_{CQS} will be greater than I_K unless the collector-base voltage is reduced to a value of order 1/20 the nominal breakdown value. Thus quasi-saturation will determine performance degradation only at low values of collector-base bias voltage. Otherwise, Kirk effect will be the dominant mechanism.

10.7.1 Conductivity Modulation of Collector Region in Heavy Saturation

When the hole concentration injected into the collector rises above the value N_{epi}, the collector resistance decreases in value below R_{epi}. This is why the quasi-saturation effect is not as deleterious as might otherwise be expected. Figure 10.8(b) shows the I_C-V_{CE} characteristics and the $1/R_{epi}$ locus. The characteristics extend well to the left of this line, thus reducing the saturation voltage. Clearly the collector resistance becomes nonlinear [7]. Figure 10.9 illustrates the carrier concentration versus depth in the collector in this situation. The carrier concentration gradient $dn/dx = dp/dx$ is proportional to the collector current in the conductivity modulated region. The major part of the voltage drop occurs in the nonconductivity modulated region close to the N^+ layer. This voltage drop is given by

$$V_{IR} = \frac{I_C \rho_{epi} w_R}{A} \tag{10.36}$$

where $w_R = W_b + W_{epi} - x_0$ as shown in Fig. 10.9, and where ρ_{epi} is the resistivity of the epitaxial layer.

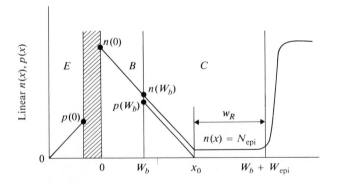

FIGURE 10.9
Carrier concentration versus depth under quasi-saturation. (*After Hébert and Roulston* [8].)

The value of internal base-collector voltage due to injection of the hole concentration $p(W_b)$ is given under high level injection by

$$V_{BC} = 2V_t \ln \left[\frac{p(W_b)}{n_i} \right]$$

Assuming that the base has also entered high level injection, the base-emitter voltage is given from Eq. (3.65) as

$$V_{BE} = 2V_t \ln \left[\frac{n(0)}{n_i} \right]$$

and the total value of collector emitter saturation voltage is thus:

$$V_{CE \, sat} = V_{BE} - V_{BC} + \frac{I_C \rho_{epi} w_R}{A} \tag{10.37}$$

The collector current and current gain may also be evaluated from Fig. 10.9. The collector current is given by

$$I_C = \frac{K_c \, q A D_n \, n(0)}{x_0} \tag{10.38}$$

where $1 < K_c < 2$ depending on whether the base region is in high or low level injection, although normally the high level injection condition will be reached rapidly once this heavy saturation condition occurs. If the base is in low level injection, then the current gain

$$\beta \propto \frac{1}{x_0}$$

If on the other hand, $n(0)$ is greater than the background base doping level, the emitter injection efficiency is degraded and, using Eq. (10.9), we see that

$$\beta \propto \frac{1}{x_0} \frac{1}{n(0)}$$

A constant value of $V_{CE \, sat}$ implies to a good approximation that the product $I_C w_R$ is constant. Under normal switch or inverter conditions, as the base drive increases, β decreases (the forced gain condition). From Eq. (10.38), $n(0)$ and x_0 will increase (to keep I_C constant), w_R decreases, and hence the saturation voltage will decrease.

The above saturation effects are important in understanding the relationship between saturation voltage, gain, and current. It is also important that they be included in CAD models for accurate representation in heavy saturation. In some SPICE CAD models this is not done.

10.8 OVERALL VARIATION
OF IMPORTANT PARAMETERS
WITH BIAS CURRENT

We have already seen (Chap. 9) that both β and f_t fall off at low collector currents due to effects associated with the emitter-base space-charge layer. We further saw that there are several delay times contributing to the f_t of the transistor (the neutral base and neutral emitter, the base-collector transit time). Figure 10.10 shows computed results for the various delay times versus collector current density for a typical VLSI transistor. Although the base transit time is normally the single most dominant term, the other terms cannot be neglected. The complex variations at high currents are due to the fact that as the base widens (in the vertical direction due to Kirk effect), the base transit time increases rapidly and then tends to settle, as the base-collector space-charge layer transit time increases initially, then decreases (when the space-charge layer is "compressed" against the N$^+$ layer) as seen in Fig. 10.4(d).

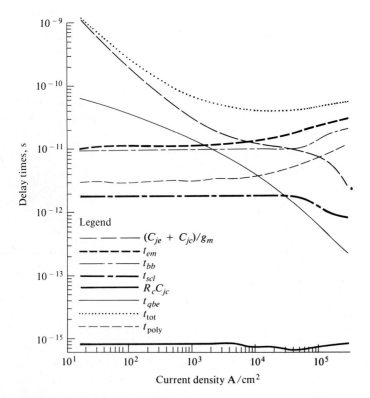

FIGURE 10.10
Variation of delay times versus collector current density for a VLSI polysilicon emitter transistor with e-b and b-c junctions of 0.18 μm and 0.35 μm, and a total epitaxial layer thickness of 1.2 μm doped 1.6×10^{16} cm^{-3}, computed with the BIPOLE program.

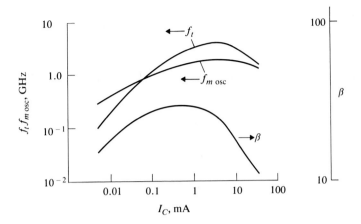

FIGURE 10.11
Computed f_t, $f_{m\,osc}$, and gain versus collector current for the transistor of Fig. 10.10 with a 5 μm $\times$ 5 μm emitter using oxide isolation.

When considering the variation of f_t versus collector current (as opposed to current density) the situation becomes even more involved. Not only do we have the above current density-dependence, but the current density varies from the center of the base region to the edge, due to the emitter current crowding phenomenon discussed in Sec. 10.2. Unfortunately, there is no simple way of estimating the amount of current crowding; this is because the part of the base region nearest the edge of the emitter will invariably be conductivity modulated (high level injection gives a majority carrier concentration, and hence a conductivity, proportional to the injected electron current density and hence to collector current density); the base will also have a decreased resistance per unit length because of the Kirk effect. Both these effects will tend to *decrease* the amount of current crowding [decrease of r_{bb} in Eq. (10.5)]. However, the current gain β decreases under these conditions and this has the effect of *increasing* the amount of current crowding [I_b in Eq. (10.5) rises faster than I_C].

The net result of the above is a very complex dependence of r_{bb}, β, and f_t on collector current. Figure 10.11 shows the f_t, $f_{m\,osc}$, and β curves for a typical transistor, versus collector current. Note the rather flat response of the maximum oscillation frequency. This is because of the square root dependence on f_t given in Eq. (7.29) and the compensating effect of r_{bb} decreasing at the same time as f_t decreases.

10.9 CONCLUSIONS

In this chapter we have examined a number of high current effects. Each of them must be taken into account when considering the design of a new structure or changes to be made in an existing process in order to improve performance. It is obvious that when more than one of the effects occurs simultaneously, accurate prediction of the gain and transition frequency falloff can only be carried out using numerical simulation. It is important to realize, however, that for design optimization purposes, it is usually only necessary to predict accurately the *bias current* at which falloff *starts to occur*. This is a much simpler task, for which the equations presented in this chapter are usually adequate. Furthermore, it is worth noting that, even if a numerical simulation tool is available, unless the user has a good understanding of the dominant physical effects discussed here, it will be difficult to predict in an efficient manner how impurity profile or mask data must be altered to effect an improvement in performance.

We shall return to a discussion of high current effects in the final chapter on CAD modeling for circuit simulation purposes.

PROBLEMS

10.1. This question concerns current crowding in a high voltage power transistor. The emitter stripe width $L = 150$ μm and the total stripe length $B = 1$ cm (this is made up of many stripes in parallel, but can be represented for base resistance calculation purposes as a long single stripe with two base contacts). The base doping level is 0.5×10^{17} cm^{-3} and the neutral base width $W_b = 10$ μm. The maximum current gain is $\beta_0 = 20$. Calculate the pinched base sheet resistance and the value of r_{bb}. Hence calculate the effective emitter stripe width L_{eff} and the effective base resistance for collector currents of 100 mA, 1 A, and 10 A (neglect all other high current effects).

10.2. This question involves β falloff in the presence of current crowding and Kirk effect. For the transistor of Prob. 10.1, the collector is doped 3×10^{14} cm^{-3}. Calculate the values of J_{HLI} and J_K. Using the results of Prob. 10.1, calculate the collector current at which the gain starts to fall off. Note that this current is less than would appear from consideration of only the area and J_{HLI} or J_K values due to emitter current crowding. Sketch the β versus I_C curve.

10.3. This question concerns the Kirk effect for a small low power transistor with an emitter 4 μm (L) $\times$ 10 μm (B). The base width is 0.5 μm, doped $N_{Ab} = 6 \times 10^{17}$ cm^{-6}, the maximum current gain is 50, the residual epitaxial layer thickness is 2.6 μm, and this layer is doped 10^{16} cm^{-3}. The value of V_{CB} is 10 V. Calculate the value of Kirk and high level injection current densities and currents (neglect current crowding but justify this). Calculate the effective base width for J_c varying from 15 000 to 40 000 A/cm^2, and plot base transit time and β versus J_C.

10.4. For the transistor considered in Prob. 10.3, plot f_t versus I_C and log (I_C) versus V_{BE}, for I_C from 0.1 to 160 mA. Assume a total junction capacitance of 2 pF.

10.5. The Van der Ziel effect is illustrated in this question. For the transistor used in Prob. 10.3, calculate the value of the effective emitter stripe width L' for I_C varying from 25 to 200 mA, assuming that the current density J_c does not exceed J_K. Hence calculate the value of t'_{bb} for current injected at the furthest angle necessary for each I_C value. Plot t'_{bb} versus I_C and compare the plot with that due to the Kirk effect in Prob. 10.3. For $I_C = 200$ mA, estimate the overall effective value of base delay time, from charge considerations.

10.6. The remaining questions concern quasi-saturation behavior. For the same transistor as that described in Probs. 10.3 and 10.4:

(i) Calculate the value of V_{CB} for which $I_{CQS} = I_K$;

(ii) Calculate the value of I_{CQS} for which $V_{CB} = 0$, that is, $V_{CE\,sat} = V_{BE}$;

(iii) Draw carefully, with values, Fig. 10.9 for this transistor for $I_C = 25$ mA, for $w_R = 0$ and estimate the values of V_{BC}, V_{BE}, $V_{CE\,sat}$, and β.

10.7. For the same transistor, use $w_R = 1$ μm, and $I_C = 25$ mA. Draw Fig. 10.9 for this condition, including doping levels and calculate approximately: (i) $V_{CE\,sat}$; (ii) the (forced) current gain.

10.8. Noting that a constant value of I_C implies (to a good approximation) a constant gradient dn/dx, for $I_C = 25$ mA, calculate and plot β versus $V_{CE\,sat}$ for the transistor used above.

10.9. Show that, for a constant current gain in the presence of base region high level injection and quasi-saturation,

$$\left[1 + \frac{n(0)}{N_{Ab}}\right]\left[1 + \frac{(W_{epi} - w_R)}{W_b}\right]$$

must be constant. Hence, starting with $w_R = 0$, calculate and plot for the transistor of Prob. 10.3. $V_{CE\,sat}$ versus I_C for a forced gain of 5.

REFERENCES

1. J. R. Hauser, "The effects of distributed base potential on emitter current injection density and effective base resistance of stripe transistor geometries," *IEEE Trans. Electron Devices*, vol. ED-11, pp. 238–242 (May 1964).

2. J. Lindmayer and C. Y. Wrigley, *Fundamentals of Semiconductor Devices*, Van Nostrand-Reinhold, Princeton (1965).

3. C. T. Kirk, "A theory of transistor cut-off frequency (f_t) falloff at high current densities," *IRE Trans. Electron Devices*, vol. ED-9, pp. 164–174 (March 1962).

4. R. J. Whittier and D. A. Tremere, "Current gain and cut-off frequency falloff at high currents," *IEEE Trans. Electron Devices*, vol. ED-16, pp. 39–57 (January 1969).

5. A. Van der Ziel and J. Agouridis, "The cut-off frequency falloff in VHF transistors at high currents," *Proc. IEEE*, vol. 54, pp. 411–412 (March 1976).

6. I. Getreu, *Modeling the Bipolar Transistor*, Elsevier, New York (1978).

7. P. L. Hower, "Optimum design of power transistor switches," *IEEE Trans. Electron Devices*, vol. ED-20, pp. 426–437 (April 1973).

8. F. Hébert and D. J. Roulston, "Unified model for bipolar transistors including the voltage and current dependence of the base and collector resistances as well as the breakdown limits," 18th European Solid State Device Research Conference, *Journal de Physique*, C4, suppl. to no. 9, vol. 49, pp. 371–374 (September 1988).

CHAPTER
11

TRANSIENT
BEHAVIOR
OF
TRANSISTORS

11.1 INTRODUCTION

A study of the "exact" way in which charge is distributed in the three regions of a transistor, and the corresponding "exact" evaluation of terminal currents and voltages, during transient turn-on or turn-off, would require very complicated analyses. Such analyses have only been performed by numerical simulation [1, 2], under certain very specific drive conditions. Fortunately, from a device design engineer's point of view, the overall switching speed of a transistor is usually determined by a combination of circuit and device parameters and is normally far slower than the transit times of charge through the base and emitter regions (and frequently also slower than the transit time through the collector epitaxial region). Also, because in a digital circuit there are several transistors interacting through circuit resistors, the precise determination of transient behavior of any single device is not, in this situation, of great importance; while one logic inverter is turning on, another is always being turned off. It is an estimate of the net result (propagation delay time) which is important, rather than a detailed knowledge of how the carriers propagate through the base, emitter, and collector regions of the device.

These factors lead us to conclude that a very crude transient analysis of the device is sufficient in nearly all practical cases for good engineering predictions of performance. It is important to understand qualitatively the mechanisms which dominate the switching process, so as to be able (if the flexibility exists) to adjust the design accordingly. It is also necessary to have a fairly simple physical model of the charging process. Such a model is in fact the Ebers-Moll or Gummel-Poon equivalent circuit nonlinear model with junction and diffusion capacitance terms incorporated [3]. This forms the industry standard for computation of propagation delay times in VLSI logic circuits. Indeed, for a very large range of design problems, orders of magnitude simplifications are made through the use of timing simulators, in which all detailed modeling of the device dynamics has been removed, but the basic delay of a logic gate retained.

An area where some more detailed information is required for the device design engineer is that of high voltage power switching. Not only does the high voltage necessitate wide (thick) lightly doped collector regions, but the device is switched under high current conditions and a number of interacting phenomena occur. We shall focus on one particular aspect of such high voltage switching—reverse base drive inductive load breakdown—later in this chapter.

The chapter starts with an introduction to the transient build-up of charge during turn-on, relating the results to widely quoted charge control equations. Simplified equations for prediction of turn-on and turn-off times are explained. We then study the high voltage switching transistor from two points of view: reverse I_B inductive load turn-off, and a brief introduction to thermal transient behavior. The study of these high voltage devices uses essentially the simple quasi-static charge control model developed at the start of the chapter, in the form of CAD models, and couples this to the other phenomena necessary for prediction of transient breakdown.

11.2 TRANSIENT BEHAVIOR OF CHARGE DISTRIBUTIONS

11.2.1 Base Voltage Drive

The circuit used for switching a bipolar transistor is shown in Fig. 11.1. Let us start by considering an extreme case where the base resistance (including the r_{bb} of the device) is zero. If a voltage step is applied, ideally this fixes the injected carrier concentrations in the emitter and base $p(0)$, $n(0)$. The charge will hence build up in a manner similar to that for a diode (refer to Figs. 4.1 and 4.2). This behavior is illustrated in Fig. 11.2. As in the diode, there will be a sudden increase in base current. The base current will decrease as charge diffuses away from the junction; simultaneously the collector current will start to rise.

It is clear from our study of diodes that if we neglect the effects of built-in electric field in the base region (whose net effect we have shown in Chap. 9 to be very small), and of the transit time in the base-collector space-charge layer, the steady state charge distribution will be reached in a time close to $t_{bb} = W_b^2/2D_n$.

Behavior in the emitter is rather more complex. The final steady state distribution is, as we have already seen in Sec. 8.2, determined largely by the strong retarding field created by the (diffused or implanted) impurity profile.

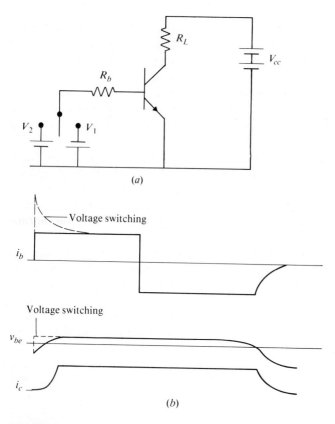

FIGURE 11.1
(a) Circuit for switching a bipolar transistor; (b) current and voltage waveforms.

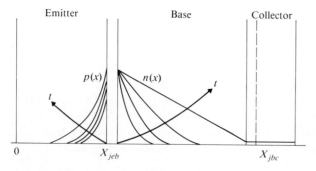

FIGURE 11.2
Build-up of charge during voltage turn-on of a transistor.

The dominant time constant will be that related to the emitter component of the diffusion capacitance. The corresponding charge is [Eq. (8.9)]

$$Q_{p\text{-}em} = F_e \, qp(0)W_e \tag{11.1}$$

and the time constant was shown [Eq. (9.73)] to be approximately

$$t_{em} = \frac{W_e W_b}{2D_n} \tag{11.2}$$

Without solving the rather complicated transient problem of the charge diffusing into the emitter region, it is fairly safe to conclude that the transient will be complete in a time comparable to t_{em}. The value of current injected from the base into the emitter at the end of the turn-on transient, i.e., when the collector current has just reached the value I_C, is I_C/β_0 where β_0 is the normal (unsaturated) dc current gain of the transistor.

It will be convenient for later use to define a factor relating the base and emitter minority carrier charges

$$F_{eq} = \frac{Q_{p\text{-}em}}{Q_{n\text{-}ba}}$$

From the results of Chap. 9 repeated above, this is given approximately by

$$F_{eq} = \frac{t_{em}}{t_{bb}} \sim \frac{W_e}{W_b} \tag{11.3}$$

From the discussion in Chap. 9 on f_t components of diffusion capacitance, we see that typically the emitter charging will be complete somewhat before the collector current has reached its final value I_C for this voltage switching condition.

11.2.2 Base Current Switching

Figure 11.3 illustrates the carrier concentration distribution versus time for the case where the resistance in series with the base and the generator voltage are both large enough that a constant base current is forced during the turn-on tran-

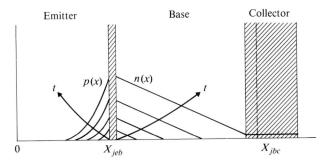

FIGURE 11.3
Transient charge distribution for current turn-on.

sient. This is the standard test condition used to evaluate the switching performance of the device. The base current initially has to charge the base-emitter depletion-layer capacitance to within 0.1 V of the normal V_{BE} bias value, typically about 0.6 V, before any substantial base charge builds up in the neutral base region, i.e., before the collector current starts to rise. If the source voltage has been initially at a (negative) value V_1 (Fig. 11.1) and is switched to a value V_2, and if the junction capacitance has an "average" effective value $C_{j\,\text{eff}}$, it is a simple matter to show that the time taken for the base-emitter voltage to rise from the initial value V_1 to a final value V_{on} is

$$t_{on1} = t_{RC} \ln\left[\frac{V_2 - V_1}{V_2 - V_{on}}\right] \tag{11.4}$$

where $t_{RC} = R_b C_{j\,\text{eff}}$ and we assume $R_b \gg r_{bb}$. For the frequent case of large forward base drive (in order to obtain a short turn-on time), and large reverse base drive (to obtain a fast turn-off, see below), this simplifies to

$$t_{on1} = t_{RC} \ln\left[1 + \left|\frac{V_1}{V_2}\right|\right] \tag{11.5}$$

For the case where $V_1 = 0$ and $V_2 \gg V_{on}$ the result becomes

$$t_{on1} = t_{RC} \ln\left[1 + \left(\frac{V_{on}}{V_2}\right)\right] \tag{11.6}$$

The value of effective capacitance may be estimated from the $C(V)$ law for a linearly graded junction where $C \propto V^{-1/3}$. Assuming an offset voltage of about 0.8 V, the capacitance at $V_{on} \sim 0.7$ V approaches twice the zero bias value. For an initial voltage V_1 equal to about -6 V, the capacitance is approximately one-half the zero bias value. Thus we see that the effective value of $C_{j\,\text{eff}}$ will lie typically somewhere between the zero bias value and about 50 percent higher, depending on the initial base-emitter bias V_1.

Once the voltage has risen to a value V_{on} which under steady state conditions would give a collector current equal to about one-tenth of its actual final value, the neutral base charging time comes into play. We cannot (contrary to the treatment used in many texts) use the charge continuity equation directly in its customary form; this is due to the fact that initially the base current provides the free carrier charge in the base, whereas toward the end of the transient, it is used to provide the charge in the emitter and collector. We can, however, very simply determine the time t_{on2} for the base charge to attain its nonsaturated steady state value, which is the time taken for the collector current to reach its steady state value, for the most common case of large base drive. If the base current ($\sim V_2/R_b$) is constant at a value I_{b1}, then this time will be determined by the base current supplying charge to the neutral emitter and the neutral base (for simplicity, bearing in mind the approximate nature of this treatment, we will again neglect the charge required within the width w_{scl} of the base-collector space-charge layer):

$$t_{on2} = Q_{n\text{-}ba}\frac{1 + F_{eq}}{I_{b1}} \tag{11.7}$$

where F_{eq} was defined in the previous section when considering voltage-switching. The expression for this part of the turn-on transient is therefore

$$t_{on2} = t_{bb}(1 + F_{eq})\beta_{forced} \tag{11.8}$$

where β_{forced} is the forced gain I_c/I_{b1}. This is basically the same as the result obtained by use of the charge control equation because this time is short compared to the final time for steady state conditions to have built up.

From a circuit point of view, the switching-on transient is now complete. The charge will continue to build up, however, in the emitter, base, and collector regions because of the fact that the base drive I_{b1} is much greater than the value required to sustain the collector current I_C. This is equivalent to saying that the forced gain β_{forced} is much less than the actual (nonsaturated) value of β. Figure 11.4 shows the continued build-up of charge.

For most modern transistor structures, it is the charge in the relatively thick epitaxial layer, thickness W_{epi}, which dominates in saturation. Continuing our assumption of a large base drive I_{b1} corresponding to a small forced gain, the magnitude of the final epitaxial layer charge will be given to a very good approximation by

$$Q_{epi} = \left(I_{b1} - \frac{I_C}{\beta_0}\right)\tau_{epi} \tag{11.9}$$

This may usually be approximated (for high base drive) by

$$Q_{epi} = I_{b1}\tau_{epi} \tag{11.10}$$

Since this charge is also given by

$$Q_{epi} = qA_b W_{epi}\, p(0) \tag{11.11}$$

where A_b is the area of the base diffusion and $p(0)$ is the value of hole concentration injected into the epitaxial collector, it follows that the value of forward

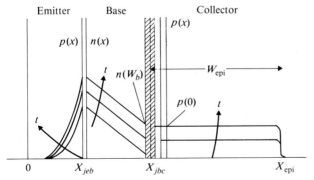

FIGURE 11.4
Charge build-up from t_{on2} to full saturation conditions.

biased base-collector voltage is given by Eq. (2.71)

$$V_{bc\,sat} = V_t \ln\left[\frac{p(0)N_{epi}}{n_i^2}\right] \qquad (11.12)$$

Substituting from Eq. (11.9), Eq. (11.11) gives

$$V_{bc\,sat} = V_t \ln\left[\frac{I_{b1} N_{epi}\,\tau_{epi}}{qn_i^2 A_b\, W_{epi}}\right] \qquad (11.13)$$

In other words, the base-collector saturation voltage increases as the log of the base drive current (assumed much greater than I_C/β_0).

We have implicitly neglected base current due to recombination in the neutral base region and base current flowing upward from the (now) forward biased base-collector junction to the surface of the extrinsic base region (under the oxide and under the base contact) shown in Fig. 11.5 These current components are discussed in more detail in Chap. 13 in connection with IIL and lateral PNP transistors. A simple measurement of inverse gain would not enable separation of the components and would simply indicate a lower than actual value of τ_{epi}. Some charge control approaches to determination of the switching times in the saturated inverter involve the inverse current gain. It is perhaps therefore worth noting at this point that in terms of the present discussion, the base current I_{b1} as approximated above would be approximately the normal base current in the inverse active region. The upward collector current is given by

$$I_c \sim \frac{qA_e D_n n(W_b)}{W_b}$$

and the upward gain $\beta_i = I_C/I_b$ is therefore given by

$$\beta_i = \frac{A_e}{A_b}\frac{L_p^2}{W_b W_{epi}}\frac{D_n}{D_p}\frac{n(W_b)}{p(0)} \qquad (11.14)$$

where $p(0)$ is the injected hole concentration into the epitaxial layer used in Eq. (11.11) and $n(W_b)$ is the electron concentration injected into the base under inverse operation (V_{BE} forward bias, V_{BC} reverse bias).

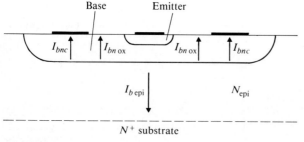

FIGURE 11.5
Additional components of base current due to injection to the surface under the oxide and under the base contact.

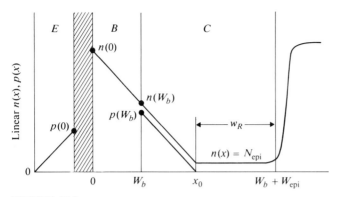

FIGURE 11.6
Steady state carrier distribution in heavy quasi-saturation.

There is no direct simple relation between this inverse gain, the saturation voltage, and the turn-off delay time. Many text book treatments are based on the fact that the excess base current accounts only for base region recombination of the excess saturation base charge; this is far from reality in most modern transistors with epitaxial collectors.

The above treatment assumes that low level injection conditions prevail in the collector. This may be the case in VLSI logic inverters. It will in general not be true for high voltage power switches. For high level injection, the conditions are as shown in Fig. 11.6 and the saturation voltage is

$$V_{bc\,sat} = 2V_t \ln \left[\frac{p(W_b)}{n_i} \right] - V_{IR} \tag{11.15}$$

The nonlinear ohmic drop V_{IR}, associated with the collector epitaxial layer, was considered in Sec. 10.7.1, see Eq. (10.36). Because of the fact that the free electron concentration extends well into the epitaxial layer in high voltage devices under large base drive, there will be an additional delay corresponding to the increased effective base width $W_{b\,add} = x_0 - W_b$ shown in Fig. 11.6 (refer to Chap. 10) [4, 11]

$$t_{bb\,add} = \frac{W_{b\,add}^2}{2D_n} \tag{11.16}$$

Since the increased width $W_{b\,add}$ depends on both collector current and voltage, the additional delay cannot be described by a unique time constant; also, the forced gain is now a function of x_0 as given by Eqs. (10.39) and (10.40).

11.2.3 Turn-off Time

Figure 11.7 illustrates, for the moderate saturation case, the evacuation of charge by a base current I_{b2}. Initially, virtually all of the excess hole charge in the collector region must be removed before the base-collector voltage falls from its saturated value down to zero volts (refer to the treatment for reverse recovery of diodes in Chap. 4). Assuming that this collector region charge is the only excess

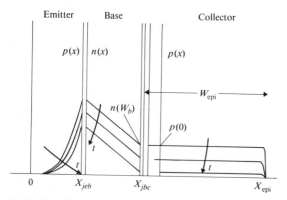

FIGURE 11.7
Evacuation of charge during turn-off for case of moderate saturation.

charge in saturation, the conventional published expression for storage delay time (the time taken for V_{bc} to fall to zero, before which the collector current remains constant) is valid, provided the lifetime in the epitaxial layer is used [5]

$$t_s = \tau_{\text{epi}} \ln \left[\frac{I_{b2} - I_{b1}}{I_{b2} - I_{b\,\text{sat}}} \right] \tag{11.17}$$

where $I_{b\,\text{sat}} = I_C/\beta_0$ and is the base current required to keep the transistor on the edge of saturation (i.e., between normal active and saturation regions).

A common test switching condition is $I_{b2} = -I_{b1}$ and in this case, for high forward base drive ($I_{b1} \gg I_{b\,\text{sat}}$) the storage delay time $= 0.7\tau_{\text{epi}}$.

Since τ_{epi} is typically over 1 μs, this is the dominant delay in the saturated inverter. The advantage of using a Schottky diode to clamp the base collector junction at a small forward bias is obvious. The properties of the Schottky diode (barrier height and area) must be chosen such that only a small amount of charge is injected into the collector in saturation. An alternate technique used in the past was to gold dope the structure to lower the lifetime. This is a difficult operation to control because too large a concentration of Au atoms will decrease the normal (active region) gain and degrade the active region performance; this will also increase the voltage drop in the collector and therefore degrade the performance in saturation.

The tail of the decay for collector current is determined by a combination of extraction of the remaining base charge (function of t_{bb} and I_{b2}) and charging of the junction capacitances. The time required to evacuate the neutral region base charge at the end of the time t_s may be simply estimated as

$$t_{\text{decay}} = -\frac{Q_b}{I_{b2}} = -t_{bb} \frac{I_C}{I_{b2}} \tag{11.18}$$

If, as is normally the case, a load capacitance exists from collector to ground, this can dominate the tail of the turn-off transient.

For the high voltage transistor in heavy saturation, the initial turn-off phase will be determined by the evacuation of the charge in the collector conductivity modulated region. The above result [Eq. (11.17)] may still be used to give the constant current turn-off delay time, if an appropriate value of τ_{epi} is used. This will be a combination of base and collector region lifetimes and $t_{bb\,\text{add}}$ given by Eq. (11.16), under high level injection conditions. As suggested by Hower [4], and implemented in some CAD models, a combination of coupled charge continuity solutions can be used to model this complex behavior.

11.3   HIGH VOLTAGE TRANSISTOR SWITCHING UNDER INDUCTIVE LOAD

One of the major problems confronting the manufacturer of high voltage power transistors lies in the fact that the load is often inductive. This gives rise to a complex behavior under reverse base drive conditions. This has been studied by several workers [2, 6, 7] and has been shown to be a source of second breakdown (catastrophic breakdown, resulting in destruction of the device). It is quite distinct from second breakdown associated with forward base current. Figure 11.8

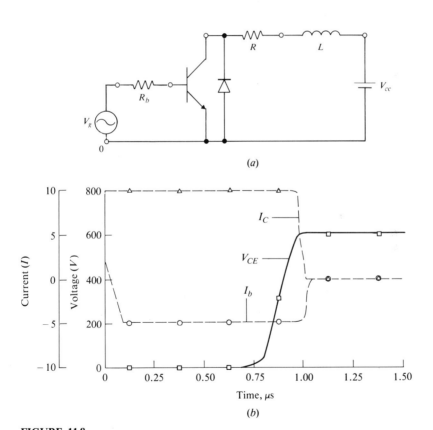

(a)

(b)

FIGURE 11.8
(a) Inductive load switching circuit; (b) computed collector current and voltage waveforms. (*After Roulston and Quoirin* [7]. *Reproduced with permission © 1988 IEE.*)

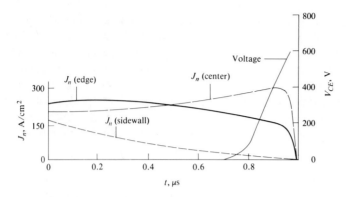

FIGURE 11.9
Computed collector current density versus time after applying the reverse base drive. The transistor is a 600 V 10 A power switch of which the details are given in Fig. 11.11.

shows a typical circuit with the corresponding collector waveforms. Because of the energy storage properties of the inductive load, the initial (small) decrease in collector current due to the reverse base drive creates a large $L \, di/dt$ voltage. The transistor thus has for part of the turn-off transient a simultaneous large I_C and large V_{CE}. Note that this does not occur in the normal resistive load circuit discussed above; in that case, the voltage only starts to rise as the collector current falls. The problem is compounded by the fact that the reverse base current tends to "focus" the collector current toward the center of the emitter stripe—the opposite to normal emitter current crowding discussed in Sec. 10.2. The current density near the center region of a stripe can exceed that at the edge, before the total current decreases significantly. This, plus the sidewall component collected outside the area under the emitter, is shown in Fig. 11.9 for a 600 V power switching transistor.

Figure 11.10 shows the current density at the center of the emitter stripe

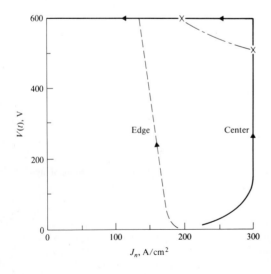

FIGURE 11.10
Computed instantaneous collector voltage versus collector current density at the center and the edge of an emitter stripe during the turn-off transient.
(*After Roulston and Quoirin* [7].
Reproduced with permission © 1988 IEE.)

and at the edge of the stripe as a function of instantaneous collector voltage, during the transient. The current density may well be above the steady state Kirk current density limit. If this occurs, a very important phenomenon can lead to runaway conditions. Let us examine the electric field distribution for currents below and above the Kirk limit as shown in Fig. 11.11 (the reader is referred to the origin of Kirk effect discussed in Chap. 10). It is clear that the peak value of field can be significantly higher if the current density exceeds the Kirk limit. This leads to lower breakdown voltages and to a phenomenon known in this switch-

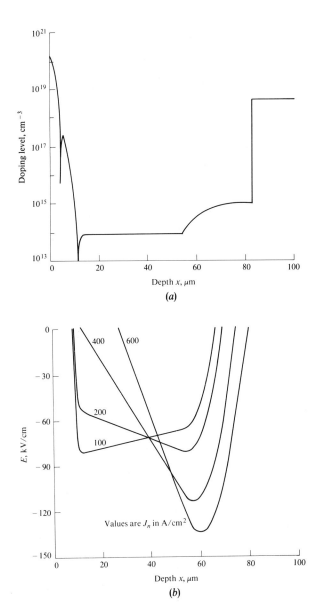

FIGURE 11.11
(a) Impurity profile of high voltage transistor used as an example; (b) electric field distribution computed at high voltage for current densities below and above the Kirk limit, $J_K = 160$ A/cm^2.

ing situation as *avalanche injection* [2, 10]. In Sec. 7.6 we discussed the runaway condition which exists in normal static operation when the base current is zero—the corresponding voltage limit is BV_{CEO}. It is clear that under inductive load turn-off, a condition can easily be encountered in which the base current due to high current avalanche multiplication cancels the normal base current. This is very similar to what happens in the static BV_{CEO} case except that the currents are then of order reverse leakage currents, whereas in the case under discussion the currents are several amperes in value. Prediction of the runaway condition therefore necessitates solution of the ionization integral at high current densities. Figure 11.12 shows a solution for the transistor switching situation under discussion. If the worst case transient V-J_n curve in Fig. 11.10 intersects the high current "BV_{CEO}" curve of Fig. 11.12, a potential runaway condition occurs which can result in destruction of the transistor.

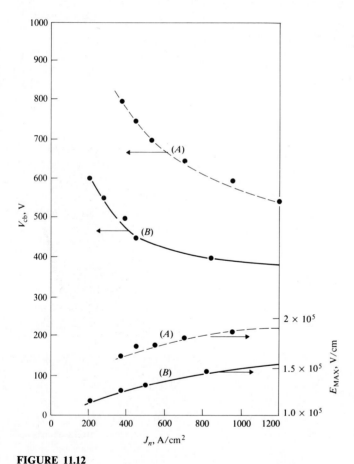

FIGURE 11.12
Computed "BV_{CEO}" versus collector current density and corresponding value of maximum electric field. Curves A use the ionization coefficients of Lee, Sze, et al.; curves B use the data of Van Overstraeten et al.—see Sec. 3.8. (*After Roulston and Quoirin* [7]. *Reproduced with permission* © *1988 IEE.*)

This is clearly a feature of primary concern in this type of switching application. Although a few computer simulation studies have been presented, there are difficulties in pinning down the exact conditions for destruction of the transistor, one such difficulty being the lack of precise values for the ionization coefficients (Fig. 11.12). Qualitatively, it is beyond doubt that the conditions which lead to failure are: inductive load, high reverse base current, and high "on" value of collector current.

It may be noted that precise modeling of these high voltage transistors during the turn-off phase (from saturation) is hindered by the fact that the collector regions are very wide (thick), with W_{epi} values of 50 μm or more. The time for the carriers to diffuse across this region is determined by $W_{epi}^2/2D$ which is of order one microsecond. This is quite comparable to the duration of the transient in some cases of large reverse base drive (the carrier lifetime is typically 20 μs). The charge control approach, and use of the charge continuity equation for determination of transient behavior (as is done in all Ebers-Moll and Gummel-Poon type CAD models), can give rise to significant errors. Furthermore, the complicated nonlinear behavior of the conductivity modulated collector region (discussed in Chap. 10) is not included in the standard CAD models.

11.4 THERMAL TRANSIENT BEHAVIOR OF TRANSISTORS

Destruction of a transistor is inevitably due to temperature rise, often very locally, in the chip. A certain time is required for local melting to occur. Although a complete treatment of thermal transients is beyond the scope of this work, in the following we present some results showing under what typical conditions second breakdown can occur.

In Chap. 7, we discussed the effect of avalanche multiplication on the breakdown conditions for a transistor. The well-known BV_{CEO} voltage was derived by setting the base current to zero. Figure 11.13 shows a complete I_C versus V_{CE} characteristic including the high current low voltage second breakdown region. The double-valued characteristic can be explained by two distinct mechanisms: (a) the avalanche multiplication effect creating a high current, low resistance condition; (b) power dissipation within the device, creating a temperature rise, with a consequent increase in current. Note that if, under steady state or transient operation, a point on the curve at which $dI_C/V_{CE} = \infty$ is reached, then the device is in an unstable situation and the current is uncontrolled. A rise in current produces a rise in temperature, leading to a potential thermal runaway situation and destruction. The collector current rise with temperature can be seen as follows. From the normal active region, Eq. (7.39) gives

$$I_C = I_{ES}(1 - \alpha_F) \exp\left(\frac{qV_{BE}}{kT}\right)$$

and

$$I_{ES} \propto \exp\left(-\frac{E_g}{kT}\right)$$

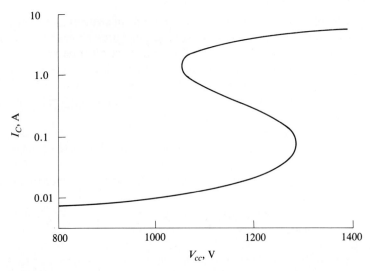

FIGURE ‡1.13
General I_C-V_{CE} characteristic including second breakdown region.

This gives

$$I_C \propto \exp\left(\frac{qV_{BE} - E_g}{kT}\right) \tag{11.19}$$

The term $qV_{BE} - E_g$ is typically -0.4 eV and, for a given V_{BE}, I_C will double for approximately 15 degrees rise in temperature.

The conditions for electrical or thermal runaway are clearly complex. It can be purely electrical in origin (as in the case of the avalanche multiplication triggered process) although final destruction will be thermal [8, 9]; it can be explained in some cases by one-dimensional thermal effects; other situations require the consideration of two- or three-dimensional heat flow to adequately model the phenomena. Let us examine one method of studying the general thermal transient problem.

Figure 11.14 shows the transistor with part of the header. It is safe to assume that most of the heat generated within the collector-base space-charge layer (where most of the voltage drop occurs with the collector current flowing)

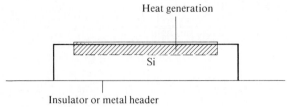

Heat generation

FIGURE 11.14
Transistor cross section with header.

Insulator or metal header

flows down through the silicon to the header material and to the heat sink. For low frequency applications, the collector may be connected directly to the case and hence to the heat sink. For RF devices, the collector must often be insulated from the case (otherwise the large capacitance of the metal case and sink will degrade the high frequency performance); BeO is a common insulator used which is also a good conductor of heat.

The heat flow equation can be expressed as

$$\text{div} (K \text{ grad } T) = -\rho c \frac{dT}{dt} \tag{11.20}$$

where K is the thermal conductivity, T the absolute temperature, ρ the density, c the specific heat, and t the time. If the total silicon thickness is represented by a single thermal resistance R_T (°C/W) and thermal capacitance C_T (J/°C), the temperature at the top of the (thick) silicon chip is given by

$$T(t) = T_0 + P_T R_T - R_T C_T \frac{dT(t)}{dt} \tag{11.21}$$

This equation tells us that for any situation in which power, P_T, is being dissipated, with a consequent rise in temperature with time, the larger the thermal capacitance, the lower will be the temperature rise at any given time. For the specific case of dc operation, the equation defines the thermal resistance of the transistor.

$$R_T = \frac{T_j - T_0}{P_T} \tag{11.22}$$

where T_j is the steady state junction temperature. A very close analog can be made with an electrical RC network in which voltage replaces temperature, current replaces power, and electrical resistance and capacitance replace thermal resistance and capacitance, respectively.

Figure 11.15 shows such an electrical analog representation for the silicon chip with current sources into the thermal grid representing power supplied at the silicon surface (the collector space-charge layer is normally much closer to the surface than to the substrate and header) [9]. Voltage at nodes in the thermal grid corresponds to temperature. The thermal properties of the header and heat sink are represented in this model by a single resistance capacitance pair. Figure 11.16 shows how the combined electrical and thermal network may be used to represent the interaction occurring in the transistor. For this purpose, a relatively simple Ebers-Moll representation is adequate for the characteristics of the transistor, with the temperature, T, available as a parameter for each cell.

This scheme has been used successfully for both dc and transient simulation of power transistors with a nonlinear network analysis program, WATAND adapted specially [9]. It should be noted that the thermal resistance of silicon is temperature-dependent, and hence is nonlinear in the electrical analog representation, although approximate results may be obtained using an average constant value.

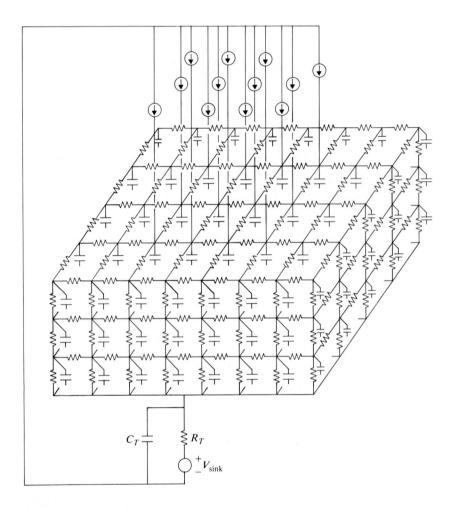

FIGURE 11.15
Electrical analog representation of heat flow. The current sources correspond to electrical power dissipated. (*After Latif and Bryant* [9]. *Reproduced with permission* © *1982 IEEE.*)

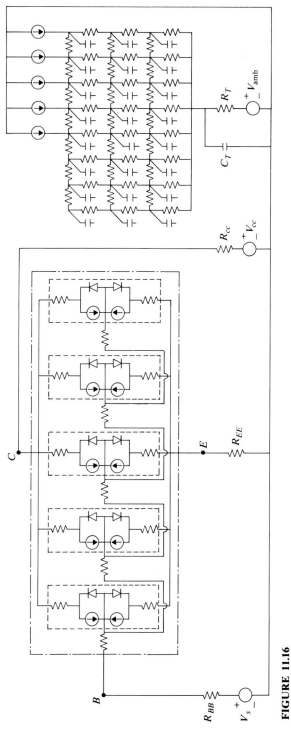

FIGURE 11.16
Combined electrical and thermal equivalent circuit. (*After Latif and Bryant* [9]. *Reproduced with permission* © *1982 IEEE.*)

The reader should note that for many cases, useful results may be obtained using the one-dimensional one-section representation, in which case a simple Fortran or similar program can be written. In the case of a large power transistor, however, two- and three-dimensional thermal effects become important, particularly for low frequency interdigitated structures with wide emitter stripes; in this situation considerable temperature differences can occur over the chip. Figure 11.17 shows computed results for dc operation for such a power transistor.

Figure 11.18 shows computed curves for collector current versus time for a high voltage transistor. It is seen that for this specific case, the modeling of the thermal resistance of the silicon can make quite a difference. The curves for variable thermal conductivity indicate a runaway condition. Note the time scale—for this power device the runaway occurs about 50 ms after the application of the V_{CE} step.

Figure 11.19 shows the results of multiple computer solutions for runaway conditions (both electrical at fixed temperature, and thermal). The manufacturer's Safe Operating Area curve is also shown. A 45 degree line would correspond to constant power dissipation—region "b." The rest of the static limit is due to second breakdown phenomena and interaction of avalanche multiplication with temperature rise. The transient results show that, as is to be expected, for short pulses of dissipated power, it is possible to apply much higher peak currents and voltages than under static conditions.

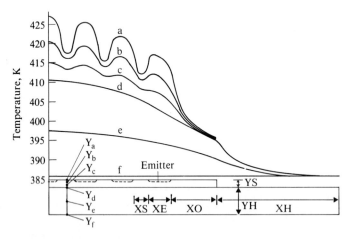

FIGURE 11.17
Computed static temperature distribution in a high voltage interdigitated power transistor. The emitter stripe width XE = 270 μm, XS = 216 μm, XH = 4000 μm, YS = 200 μm (the silicon thickness), YH = 1400 μm. The respective depths for the curves are: a, 0 μm; b, 66.6 μm; c, 133.2 μm; d, 200 μm; e, 900 μm; f, 1600 μm. (*After Latif and Bryant* [9]. *Reproduced with permission* © *1982 IEEE.*)

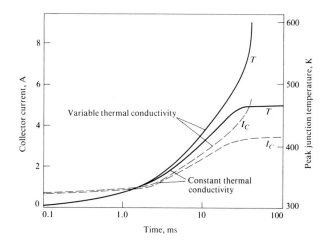

FIGURE 11.18
Computed collector current and temperature versus time for a high voltage power transistor using a two-dimensional electrical-thermal model. The conditions used are: $V_{BE} = 0.68$ V, $V_{CE} = 41.5$ V, step at $t = 0$. (*After Latif and Bryant* [9]. *Reproduced with permission © 1982 IEEE.*)

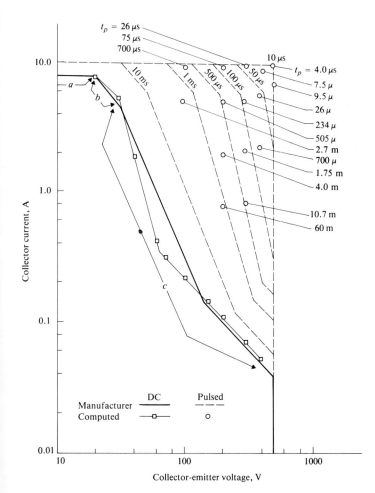

FIGURE 11.19
Computed and manufacturer's safe operating area for the high voltage transistor used in Fig. 11.13. The times indicated are for rectangular pulses of dissipated power. (*After Latif and Bryant* [9]. *Reproduced with permission © 1982 IEEE.*)

11.5 CONCLUSIONS

In this chapter, we have given an overview of some of the more important theoretical aspects of transistor transient operation. The charge control approach, with suitable modifications for modern epitaxial collector devices, is adequate for most logic circuit inverter applications. It may also give approximate results for high voltage devices in some cases. However, only significant extensions to the conventional CAD model equations, or complete numerical simulation, can cope with the detail of high voltage transistor switching—both for inductive and resistive loads. Thermal transients can be studied using network analysis techniques as shown for a typical situation by an example.

PROBLEMS

11.1. A transistor has an emitter whose dimensions are $L = 4$ μm, $B = 40$ μm, with a 16 μm $\times$ 50 μm collector. The base width is $W_b = 0.5$ μm and the residual collector epitaxial layer width is 2.6 μm. The base and collector doping levels are 5×10^{17} and 10^{16} cm^{-3}, respectively. The carrier lifetime in the lightly doped collector is 0.05 μs. The peak gain is 100. The transistor is used in a conventional (common emitter) logic inverter with a supply voltage $V_{cc} = 10$ V, a load resistance between collector and V_{cc} equal to 10 kΩ. The base is driven through a 100 kΩ resistance by a source which starts at $V_1 = -5$ V, rises to $V_2 = 5$ V at time $t = 0$, and after a time $t_{fs} = 1$ μs, falls back to the value $V_1 = -5$ V. Assume the emitter-base junction capacitance is 0.3 pF at zero bias, and use the linearly graded junction capacitance law. Calculate all significant time constants and the saturation voltage for the above conditions. Sketch versus time, base and collector currents, and base-emitter and collector-emitter voltages. Repeat the sketch if the forward drive only lasts for a duration $t_{fs} = 40$ ns.

11.2. Repeat the essential calculations and the sketches for Prob. 11.1 if the source starts and ends at a value $V_1 = 0$ V.

11.3. Repeat the essential calculations of Prob. 11.1 with an epitaxial layer lifetime $\tau_{epi} = 0.01$ μs, and sketch the carrier distribution in the base and collector regions for the saturation conditions, for $\tau_{epi} = 0.05$ and 0.01 μs.

11.4. Equation (11.11) gives the charge injected into the epitaxial layer in saturation. Write a similar expression for the excess (saturation) charge in the base region and derive an expression for the ratio of saturation recombination current in the base to that in the collector layer. Your expression should include the ratios τ_{epi}/τ_b, N_{epi}/N_{Ab}, W_b/W_{epi}, where τ_b is the carrier lifetime in the base region and N_{Ab} is the base doping level.

11.5. In order to decrease the constant current storage delay time, one possibility is to reduce τ_{epi} as in Prob. 11.3. The penalty is an increased $V_{CE\,sat}$. Use Eq. (11.13) to calculate $V_{CE\,sat}$ versus τ_{epi} and determine the upper and lower limits of validity of Eq. (11.13), for the conditions used in Prob. 11.1.

11.6. For a high value of τ_{epi} and a high base drive, Eqs. (11.9) and (11.11) indicate that the collector enters high level injection. The nonlinear quasi-saturation operation discussed in Sec. 10.7.1 then applies. For the forward drive conditions of Prob. 11.1, but with collector and base resistances decreased by a factor of 10, make any necessary calculations and sketch the carrier concentration in the base and collector for the case of high τ_{epi}. Discuss briefly the major differences in transient behavior, including storage time t_s.

11.7. Refer to the inductive load turn-off shown in Fig. 11.8 for a high voltage transistor. If the load inductance is 0.1 mH, calculate the incremental fall in collector current associated with the V_{CE} change in the range 0.8 to 0.9 μs. Is this ΔI_C compatible with the data on the figure? How low would the inductance have to be in order for the current I_C to start to fall significantly as V_{CE} rises?

11.8. During the reverse base drive transient, the collector current density in Fig. 11.9 is seen to increase at the center by a factor of about 2 compared to the value at the edge. If this is due to a "reverse current crowding" effect [see Eq. (10.3)], estimate the voltage increase between the edge and the center of the base region. How do you explain that this does not appear to be compatible with the known value of base resistance for this transistor, which is 0.5 Ω? Estimate the average carrier concentration in the base.

11.9. Refer to the thermal derating curves for the power transistor shown in Fig. 11.19. On the assumption that the maximum junction temperature is limited to 150°C, calculate the thermal resistance of this device. Thermal transients are not simple exponentials. However, if we use a law of the form: $\Delta T(t) = \Delta T_{max} [1 - \exp - (t/\tau)]$, we can determine the "time constant" τ from the derating curves for different pulse durations. From the results in the figure, determine the time constant for various peak pulse powers (various pulse durations) and show that it is not constant.

REFERENCES

1. O. Manck and W. L. Engl, "Two-dimensional computer simulation for switching a bipolar transistor out of saturation," *IEEE Trans. Electron Devices*, vol. ED-22, pp. 339–347 (June 1975).
2. K. Hwang, D. H. Navon, T-W Tang, and P. L. Hower, "Second breakdown prediction by two-dimensional numerical analysis of BJT turn-off," *IEEE Trans. Electron Devices*, vol. ED-33, pp. 1067–1072 (July 1986).
3. I. Getreu, *Modeling the Bipolar Transistor*, Elsevier, New York (1978).
4. P. L. Hower, "Application of a charge-control model to high-voltage power transistors," *IEEE Trans. Electron Devices*, vol. ED-23, pp. 863–869 (August 1976).
5. M. S. Ghausi, *Electronic Devices and Circuits: Discrete and Integrated*, Holt, Rinehart, and Winston, New York (1985).
6. L. Turgeon and D. H. Navon, "Two-dimensional nonisothermal carrier flow in a transistor structure under reactive circuit conditions," *IEEE Trans. Electron Devices*, vol. ED-25, pp. 837–843 (July 1978).
7. D. J. Roulston and J.-B. Quoirin, "Simulation of high voltage power switching transistors under forced gain and inductive load turn-off conditions," *IEE Proc. Part I*, vol. 137, pp. 7–12 (February 1988).
8. M. Latif and P. R. Bryant, "Multiple equilibrium points and their significance in the second breakdown of bipolar transistors," *IEEE Jnl. Solid State Circuits*, vol. SC-16, pp. 8–14 (February 1981).

9. M. Latif and P. R. Bryant, "Network analysis approach to multidimensional modeling of transistors including thermal effects," *IEEE Trans. on Computer-Aided Design of Integrated Circuits and Systems*, vol. CAD-1, pp. 94–101 (April 1982).

10. P. L. Hower and G. K. Reddi, "Avalanche injection and second breakdown in transistors," *IEEE Trans. Electron Devices*, vol. ED-17, pp. 320–327 (April 1970).

11. F. Hébert and D. J. Roulston, "Unified model for bipolar transistors including the voltage and current dependence of the base and collector resistance as well as the breakdown limits," *Journal de Physique*, C4 Suppl. to no. 9, vol. 49, pp. 371–374 (September 1988).

CHAPTER
12

DISCRETE BIPOLAR TRANSISTOR STRUCTURES

12.1 INTRODUCTION

In this chapter, we study various "real" discrete bipolar transistor structures. While the theory already presented applies to all structures, there are characteristics relating to specific applications which make certain aspects of the theory more applicable than others, or which require development of additional theory in order to determine the electrical terminal characteristics.

We start this chapter with two discrete devices: the microwave or RF power transistor, followed by the high voltage power switching transistor. Both these devices are made in such a way (usually with interdigitated techniques) that they follow very closely the theory of the previous chapters. However, the RF device is usually designed for high power gain and operates at high power levels, so we shall discuss briefly the concept of emitter ballasting to equalize the temperature distribution over the chip. The high voltage low frequency switching device is usually characterized by its saturation voltage $V_{CE\,sat}$ under forced gain conditions. A discussion of this device and the Darlington pair is followed by an overview of thyristor (SCR) operation.

12.2 THE MICROWAVE POWER TRANSISTOR

Figure 12.1(a) is a typical impurity profile for an RF power transistor with phosphorus-diffused emitter and boron-diffused base. This structure has an f_t

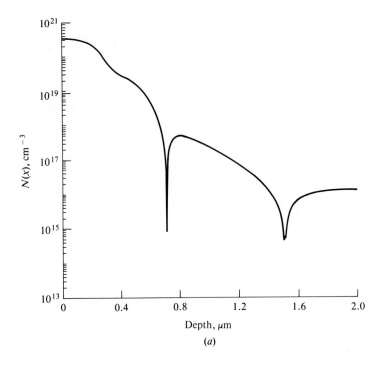

(a)

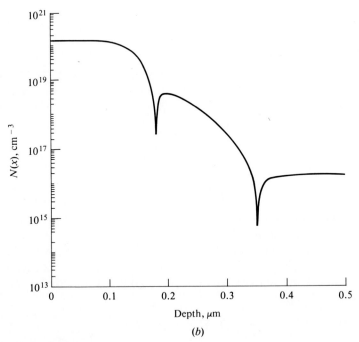

(b)

FIGURE 12.1
Impurity profiles for RF and microwave power transistors. (a) Phosphorus emitter; (b) arsenic emitter.

peaking at 2 GHz and a 50 V collector-base breakdown. Emitter-base junction depths less than 0.5 μm give rise to spiking problems from the aluminum if the metal is deposited directly. Polysilicon contacted emitters are now common and provide improved performance as explained in Chap. 8. For shallow junction emitters, arsenic is the preferred donor dopant because of the rather "square" impurity profile obtained. This is illustrated in Fig. 12.1(b). Note that as well as a strong (beneficial) retarding field in the emitter, giving high gain and low stored minority carrier charge, the (detrimental) retarding field region in the base is reduced in extent compared to that of the phosphorus-doped emitter. This transistor gives a peak f_t of 5 GHz with a polysilicon contacted emitter.

Figure 12.2 illustrates a typical discrete transistor in which the required emitter area for high current operation is obtained by interleaving a number of emitter "stripes" or "fingers" with base contacts. The reason for using such a structure is as follows: for a given high frequency performance, a low value of emitter diffusion width (base contact to base contact direction) is necessary, according to Eq. (9.83), in order to achieve a high value of maximum oscillation frequency. A large value of emitter stripe length B (in the "z" direction) is therefore required for high collector currents; because of the finite resistance of the (aluminum) metallization, there is a voltage drop in the "z" direction; it is a simple matter to calculate the value of this drop for a given collector (i.e., emitter) current; if it exceeds $kT/q \sim 25$ mV there will be a significant amount of "crowding" similar to that created by the base resistance voltage drop discussed in Sec. 10.2. To reduce this voltage drop and associated current crowding, the total required emitter "z" direction length B must be reduced to a value B_s and n interdigitated stripes used, where $n = B/B_s$.

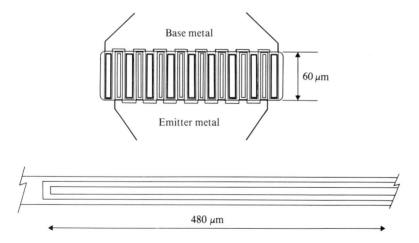

FIGURE 12.2
Interdigitated discrete RF power amplifier transistor and equivalent single stripe structure. Each emitter stripe is 4 μm wide. (*After Bryant and Roulston* [2]. *Reproduced with permission* © *1988 IEEE.*)

The relation between total maximum collector current, I_{ctm}, B/L ratio, metal sheet resistance, and allowed voltage drop is found as follows:

The emitter current per stripe is

$$I_s = \frac{I_{ctm}}{n} \tag{12.1}$$

The total voltage drop along one stripe is

$$V_{st} = I_s \left(\frac{B_s}{2L}\right) \rho_{sq} \tag{12.2}$$

where the factor 2 comes from the fact that the emitter current along a stripe decreases linearly with distance from the contact end, and where ρ_{sq} is the sheet resistance of the aluminum. Substituting Eq. (12.1) gives

$$V_{st} = \left[\frac{I_{ctm}}{n}\right]\left[\frac{B}{n}\right]\frac{\rho_{sq}}{2L}$$

$$= I_{ctm}\left(\frac{B}{2L}\right)\frac{\rho_{sq}}{n^2}$$

The number of stripes, n, required for a given voltage drop is therefore

$$n = \sqrt{I_{ctm}\left(\frac{B}{2L}\right)\frac{\rho_{sq}}{V_{st}}} \tag{12.3}$$

For example, the resistivity of aluminum is 2.7×10^{-6} $\Omega \cdot$cm. For a one-micron layer of metallization, the sheet resistance is thus about 0.03 Ω/sq. For a design requiring a collector (emitter) current of 1 A, using 2 micron wide emitter stripes, the value of B is determined by the maximum collector current density; this will normally be the Kirk limit $J_K = qv_d N_{epi}$. For an epitaxial layer doping level of somewhat less than 10^{16} cm^{-3}, $J_K \sim 10^4$ A/cm^2. Since $I_{ctm} = BLJ_K$, we have $B = 0.5$ cm for the 1 A required. Assuming that the metal voltage drop V_{st} must remain at most equal to 25 mV gives, from Eq. (12.3), the number of stripes $n = 39$; each stripe will have a length $B/39 = 128$ μm.

It is worth noting that the design of such a transistor proceeds in quite a logical manner. For a given operating voltage, the breakdown voltage and hence the epitaxial layer doping and thickness are determined. This fixes the Kirk current density; for a given collector current and emitter stripe width, the number of stripes and length of each stripe are thus determined as just described.

12.2.1   Emitter Ballast Resistance

Because of the high power density encountered in such devices (simultaneous high voltage and high current for sinusoidal output signal) there can be a significant amount of heat generated. The temperature will in general be greatest near the center of the silicon chip. This gives rise to higher current densities in this region, and thermal runaway is a possibility. To reduce the likelihood of thermal runaway and to ensure a more uniform current distribution, small "ballast" resistances can be inserted in series with each emitter finger (distributed ballast is also a possibility) [1]. This can be done, for example, by using nichrome resistors. It is particularly easy to add the ballast if polysilicon contacted emitters are being used, since the resistance can in this case be implemented by a small section of doped polysilicon. Such an arrangement is shown in Fig. 12.3. The action of a ballast resistance R_e can be readily seen by examining the relation for collector (emitter) current

$$I_e = I_{c0} \exp \left[\frac{q(V_{BE} - I_e R_e)}{kT} \right]$$

where

$$I_{c0} \propto n_i^2 \propto \exp \left(-\frac{E_g}{kT} \right)$$

This gives the relation:

$$I_e \propto \exp \left[-\frac{(E_g - qV_{BE})}{kT} \right] \exp \left[-\frac{qI_e R_e}{kT} \right] \tag{12.4}$$

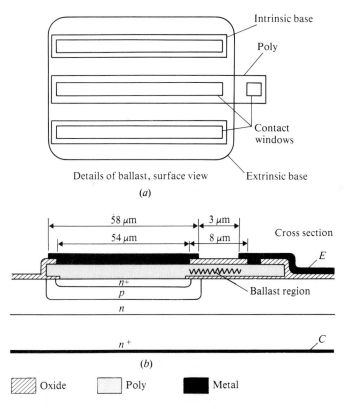

Details of ballast, surface view

(a)

Cross section

(b)

FIGURE 12.3
Implementation of emitter ballast resistors using polysilicon. (a) Plane view; (b) cross section.

The argument of the first exponential is of order $-(1.1 - 0.8)/0.025 \sim -10$ and varies inversely as the absolute temperature. Clearly, the effect of the second exponential, with an argument proportional to $-I_e$, is to compensate by limiting the rise in current. The determination of the value of ballast resistance is not simple. There are a number of conflicting requirements. If a very large value of resistance is used, the thermal stability is assured. However, the electrical performance of the transistor is degraded, as can be seen from a circuit point of view, where the series resistance acts as a negative feedback term in the common emitter configuration, and reduces the voltage gain.

Figure 12.4 shows computer generated results of temperature distribution for an RF power transistor, with no ballast resistances. Measured results obtained using thermochromic LCD material are superimposed [2]. Table 12.1 shows the computed temperature difference (between maximum and minimum active chip temperatures) without and with ballast resistances.

It is clear (i) that emitter ballasting can be effective in creating a more uniform temperature (and hence current) distribution over the silicon surface, and (ii) that inevitable reduction in electrical performance is the logical consequence. At the present time, there does not appear to be any simple way of predicting

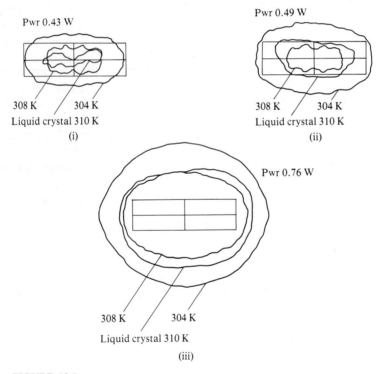

FIGURE 12.4
Computed and measured temperature contours for the RF power transistor of Fig. 12.2: (i) 0.43 W, (ii) 0.49 W, (iii) 0.76 W. (*After Bryant and Roulston* [2]. *Reproduced with permission © 1988 IEEE.*)

TABLE 12.1

Computed values of maximum temperature difference over active chip area for two power levels and the corresponding values of power gain, G_p, at 900 MHz [2]

R (ballast total), Ω	0	0.33	3.9
T_{diff} ($T_{\text{max}} = 100$)	24	17	15
T_{diff} ($T_{\text{max}} = 150$)	59	41	24
G_p at 900 MHz, dB	16.2	15.1	10.4

such trade-offs; only numerical simulation or expensive experimental iterations can give the required information to the device design engineer.

12.2.2 High Voltage Power Switching Transistors

Since the goal here is invariably to switch high currents, multiemitter stripe structures are again used. The major differences compared to the RF power device are (i) wide emitter stripes, and (ii) lightly doped thick collectors for the high breakdown voltages. The main performance criteria are: breakdown voltage, saturation voltage $V_{ce\,\text{sat}}$ for specified forced gain and collector current, and switching speed. In many applications (such as speed control of motors) switching is performed with inductive loads; an important characteristic is therefore the sustainable current/voltage for a given reverse base drive current, discussed in Chap. 11.

A typical impurity profile for a 600 V transistor is shown in Fig. 12.5. Note the deep emitter-base and base-collector junctions to enable a large radius of curvature to be obtained at the sidewalls, hence maintaining adequate breakdown voltages. For the drive conditions typical of these switching applications, the emitter-base breakdown is at least 10 V. The two-level doping profile in the collector is widely used in these devices and represents a compromise in high-voltage high-current switching performance. Double epitaxial growth is a common fabrication technique.

The emitter dopant is normally phosphorus. For the base, boron is now fairly standard although gallium has found frequent use because of its considerably faster diffusion rate. Aluminum is also used. The time taken to diffuse beyond a depth of about 10 micron is not a negligible part of the processing cost of these devices. Use of gallium, however, precludes planar processing because SiO_2 does not act as a barrier to Ga atoms; a blanket wafer diffusion for the base is required, with suitable etch and passivation to maintain the required high breakdown voltage.

For currents of tens of amperes, a large interdigitated or matrix layout is obligatory. Contrary to the RF device, quite wide emitter stripes are used in

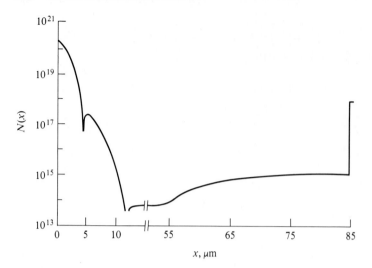

FIGURE 12.5
Impurity profile for 600 V power switching transistor. (*After Roulston and Quoirin* [3]. *Reproduced with permission © 1988 IEE.*)

order to simplify the mask constraints, achieve high current operation and good reliability (very large area devices with small design features lead to repro- ducibility and reliability problems.) Figure 12.6 shows a typical cross section for a 600 V 15 A transistor. In the high voltage switching application, the saturation voltage $V_{CE\,sat}$ for a low forced gain (typically 5 or 10) is a crucial design param- eter. As discussed in Chap. 9, only part of the collector epitaxial layer remains non-conductivity modulated at high currents for $V_{CE\,sat}$ values of order 1 to 2 V.

Figure 12.7 is a curve of current gain versus collector current for this tran- sistor. The device enters quasi-saturation at quite a low current, as can be seen by applying the formula of Chap. 10. Equation (10.35) gives a ratio of approximately 1/5 for I_{CQS}/I_K at the 5 V bias used for the plots of Fig. 12.7. The value of Kirk current density J_K for the low doped collector is 160 A/cm². For the active

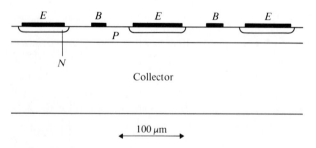

FIGURE 12.6
Cross section for 15 A, 600 V power switching transistor.

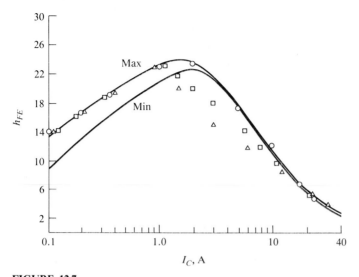

FIGURE 12.7

Gain versus collector current for the transistor under discussion. Solid curves measured, points computed using the BIPOLE program for various conditions. (*After Roulston and Quoirin* [3]. *Reproduced with permission © 1988 IEE.*)

(emitter) surface area of 80 μm × 10 cm, the value of quasi-saturation current, I_{CQS}, is approximately 2.6 A. This is seen to correspond well with the start of gain falloff in the figure.

Figure 12.8 shows the important $V_{CE\,sat}$ versus I_C curve for the same transistor. The transistor is in saturation throughout this plot, where the forced gain $\beta = 5$. The reader is referred to the theoretical discussion of Chap. 10, and it is instructive to calculate the extent of the nonconductivity modulated region of the collector for a given collector current.

Figure 12.9 shows the electron concentration in the base and collector regions of the transistor corresponding to the impurity profile of Fig. 12.5. The computer simulations correspond to a fixed V_{CB} of 2 V. As the current density increases, the electron distribution penetrates further into the collector epitaxial layer, in order to keep the $I_C R_{epi}$ voltage drop at an approximately constant value (lateral collector current-spreading alters the apparent behavior somewhat and the values are not strictly current "densities"). Figure 12.10 shows the same results superimposed on the impurity profile using a logarithmic scale. This transistor is designed for use at a forced gain of 5 and it is seen that for the 2 volt V_{CB} (corresponding approximately to $V_{CE} = 1.1$ V) the base charge is constrained roughly to the left of the increased collector doping level. This charge and the corresponding carrier lifetime determine the time taken to turn off the transistor. An estimate of this time may be obtained using the charge control theory presented in Chap. 11 but because of the very nonlinear operating conditions, with part of the collector and base operating under high level injection conditions, the

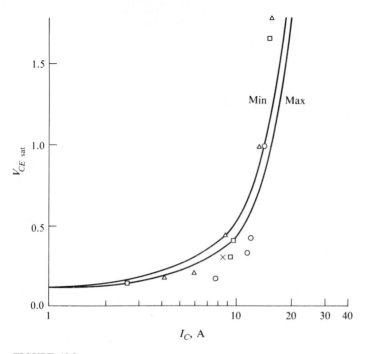

FIGURE 12.8

$V_{CE\,sat}$ versus I_C for a forced gain of 5 for the device under discussion. Solid curves measured, points computed using the BIPOLE program for various conditions. (*After Roulston and Quoirin* [3]. *Reproduced with permission* © *1988 IEE.*)

predictions can only be approximate. A further approximation stems from the fact that the charge being evacuated from the collector cannot strictly be represented by a quasi-static distribution. The diffusion transit time $W^2/2D$ for a 50 micron thick region is about 1 μs; this is comparable to the switching time when large reverse base drive is used.

The second breakdown inductive load turn-off phenomenon is important in a device of this type. The switching behavior is discussed in Chap. 11 and the example used there is the transistor described above. Because the failure mechanism for inductive load turn-off is due to a concentration during turn-off of high current density at the center of an emitter stripe, a proposed remedy involves omitting the emitter diffusion at the center as shown in the sectional diagram of Fig. 12.11. By suitable choice of the space L_s between the two diffused regions, some improvement in performance is possible, for a given large emitter stripe width L. Because of emitter current crowding under high current "on" conditions, the value of L_s can be a significant fraction of L before degradation in the "on" characteristics are observed (increased $V_{CE\,sat}$ for a given forced gain and collector current).

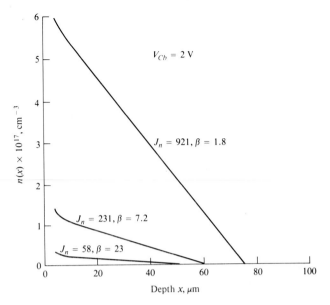

FIGURE 12.9
Electron concentrations in the base of the transistor corresponding to the profile of Fig. 12.5, computed using the BIPOLE program. Nominal current densities are indicated, together with the corresponding current gains.

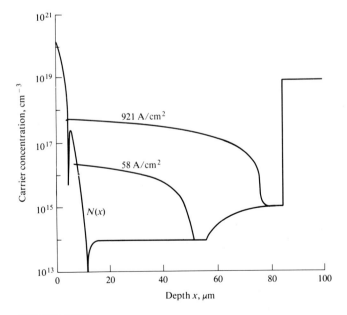

FIGURE 12.10
The information of Fig. 12.9 superimposed on the impurity profile of Fig. 12.5 using a logarithmic vertical scale.

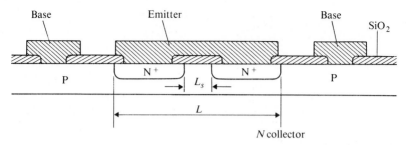

FIGURE 12.11
Cross section of the structure with center part of the emitter diffusion removed. See, for example, [12].

12.2.3   The Darlington Pair

For some applications the current gain achievable from a single transistor is insufficient. This is frequently the case for a high voltage, high current transistor switch; as we have seen, the device operates in heavy saturation with a large base current or low forced gain. The higher the base current, the more charge in the lightly doped collector for a given collector current, and hence the lower the $V_{CE\,sat}$ voltage. Supplying adequate base drive current in this case can be a problem. A common solution is to incorporate two transistors with common collectors in a Darlington configuration as shown in Fig. 12.12. Since the collectors are common, the two devices can be fabricated directly on a common substrate—hence this is not really an "integrated circuit" structure.

The performance characterized by the total collector current, the $V_{CE\,sat}$ and the total forced gain (the product of the individual gains) is clearly superior to that of the output transistor on its own. For a given collector current and saturation voltage, the input current is reduced by the (forced) gain of the first transistor. Since both devices are operating in heavy saturation, it is not a trivial matter to determine the optimum split between the areas of the driver and output transistors; a typical area split is in the range 3:1 or 10:1 [4].

The resistances R_1 and R_2 in Fig. 12.12 are necessary in order to provide a current path for evacuation of the charge during turn-off. Without any resistances, the driver transistor would normally complete its turn-off transient first, leaving a high impedance path in the base of the output device. The base-to-emitter resistance must be small in order to provide a low impedance path during turn-off, but not so small that the forced gain is reduced. This compromise is very easy to satisfy in most cases and the performance is relatively insensitive to precise resistance values. Furthermore, the resistances may readily be formed from the P base diffusion using suitable masks. An additional low gain transistor is often included to accelerate removal of charge.

Choice of Design Parameters
for High Voltage Switching Transistors

It is clear from the above discussion on the single and Darlington high voltage transistors that several fundamental design trade-offs exists. First of all, the collector lightly doped layer is chosen to satisfy the collector-base breakdown requirements V_{cbr} or BV_{CBO} (Fig. 7.12 applies). In many cases it is the collector-emitter voltage limit BV_{CEO} which is of significance. For the discrete transistor, this is given by Eq. (7.54) as $V_{cbr}/\beta_0^{1/n}$. Since for a given total gain, the individual gains of the Darlington pair can be substantially lower than for the single transistor, it follows that the Darlington pair has an advantage.

For a given collector doping level and thickness (i.e., for a given breakdown voltage), a particular value of forced gain determines the collector current density under quasi-saturation conditions for a given value of saturation voltage, $V_{CE\,sat}$. The lower the current density, the lower will be $V_{CE\,sat}$. However, the area of the device may always be increased to increase the value of collector current I_C for a required forced gain and $V_{CE\,sat}$. Conversely, the area may be increased (current density lowered) to obtain a lower $V_{CE\,sat}$ at a given forced gain. When comparing the Darlington pair to the single transistor, since the former operates at a lower forced gain (for the same overall gain as the single device), it follows that the Darlington pair will have a lower $V_{CE\,sat}$ for identical collector doping and thickness.

In terms of switching speed, it is mainly the charge under saturation which is the determining factor. This was discussed in Chap. 11 and it was seen that the higher the breakdown voltage, the slower the switching speed.

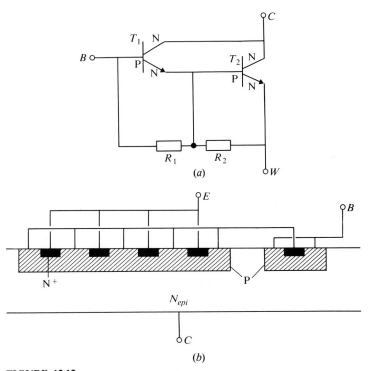

FIGURE 12.12
(a) Circuit configuration for a Darlington pair; (b) cross section view, input (driver) transistor on right.

12.3   THE SILICON CONTROLLED RECTIFIER

In this section we give an overview of the basic operation of the thyristor or SCR. This is used mainly for very high voltage applications at lower switching speeds than the bipolar transistor switch. Most of the operation can be explained on the basis of two overlapping bipolar transistors for which the impurity profile is shown in Fig. 12.13(a), and in simplified form in Fig. 12.13(b). The collector of the NPN device formed by the lightly doped background material X_{j2} to X'_{j1} forms the base of the PNP device. Injection of minority carriers into the N and P emitters, regions W_1 and W_4, and across the P base, region W_2, are determined as in a normal bipolar transistor by the voltages applied across the junction. We will neglect neutral region recombination in the following treatment, for sim

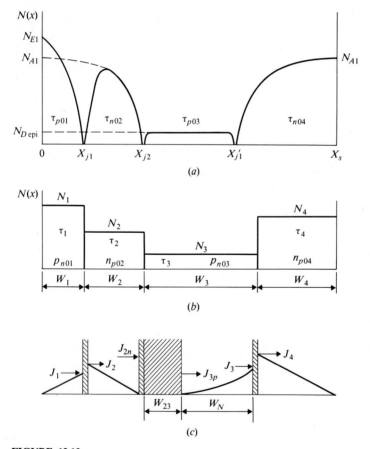

FIGURE 12.13

Real (a) and simplified (b) impurity profile of SCR; (c) current densities. (*After Roulston and Nakhla* [5]. *Reproduced with permission* © *1979 IEEE.*)

plicity. The basic operation will be deduced in terms of the two-terminal thyristor or Shockley diode.

The current densities in the two emitters may thus be written in simplified form as

$$J_1 = J_{01}(X_1 - 1) \tag{12.5}$$

$$J_4 = J_{04}(X_3 - 1) \tag{12.6}$$

where $J_{01} = qD_n n_i^2/N_1 W_1$, $J_{04} = qD_n n_i^2/N_4 W_4$ are the extrapolated reverse values determined by the effective doping profiles (and recombination in the real case) in the two regions W_1 and W_4 and where $X_1 = \exp(V_{12}/V_t)$, $X_3 = \exp(V_{34}/V_t)$ and V_{12}, V_{34}, are the applied voltages across X_{j1} and X'_{j1} shown in the figure.

The electron current in the diffused NPN base is given by

$$J_2 = J_{02}(X_1 - X_2) \tag{12.7}$$

where $J_{02} = qD_p n_i^2/N_2 W_2$ is determined by the base doping profile, as in a normal base region of a bipolar transistor, and where $X_2 = \exp(V_{23}/V_t)$.

The operation of the thyristor is controlled by the remaining lightly doped region—typically of order 100 micron thick for high voltage operation Since this region is lightly doped, the depletion-layer penetration is significant and determines the neutral base width W_N of the PNP device, see Fig. 12.13(c). The hole current in this region is given by

$$J_3 = J_{03}(X_3 - X_2) \tag{12.8}$$

where J_{03} is not a constant, but depends on W_N:

$$J_{03} = \frac{qD_p n_i^2}{N_3 W_N} \tag{12.9}$$

$$W_N = (X'_{j1} - X_{j2}) - w_{23} \tag{12.10}$$

and where w_{23} is the depletion-layer thickness determined by the sum of the applied voltage V_{23} and the barrier potential V_{b23}

$$w_{23} = \sqrt{\frac{2\varepsilon}{qN_3}(V_{23} + V_{b23})} \tag{12.11}$$

This point is crucial to understanding the *I-V* characteristic of the thyristor.

The operation of the device may be readily explained if we represent the effect of the space-charge recombination currents J_{r12}, J_{r34} [both of which are proportional to $\exp(V/mkT)$, $m \sim 2$, as in Eq. (3.106)] by variables R_{12} and R_{34}

$$R_{12} = \frac{J_1}{J_1 + J_{r12}}$$

$$R_{34} = \frac{J_4}{J_4 + J_{r34}} \tag{12.12}$$

We can thus write

$$\frac{J_1}{R_{12}} = \frac{J_{01}(X_1 - 1)}{R_{12}} = J_{3p} = J_{03}(X_3 - X_2) \tag{12.13}$$

$$\frac{J_4}{R_{34}} = \frac{J_{04}(X_3 - 1)}{R_{34}} = J_{2n} = J_{02}(X_1 - X_2) \tag{12.14}$$

$$\frac{J_{03}}{J_{01}} = A_1 = \frac{N_1 W_1}{N_3 W_N} \tag{12.15}$$

$$\frac{J_{02}}{J_{04}} = A_4 = \frac{N_4 W_4}{N_2 W_2} \tag{12.16}$$

We can solve Eqs. (12.13) and (12.14) to obtain

$$(1 - X_2) = \frac{1 - A_1 A_4 R_{12} R_{34}}{A_4 R_{34}(A_1 R_{12} + 1)} (X_3 - 1) \tag{12.17}$$

$$(X_1 - 1) = \frac{A_1 R_{12}}{A_4 R_{34}} \frac{1 + A_4 R_{34}}{1 + A_1 R_{12}} (X_3 - 1) \tag{12.18}$$

Note that the product $A_1 A_4$ is the same as the product of the gains $\beta_n \beta_p$ of the two transistors, in the absence of all recombination currents, and that A_1 depends on the voltage across the center junction via W_N. R_{12} and R_{34} are current-dependent, since for increasing current J_1 increases faster than J_{r12}; therefore, R_{12} and R_{34} increase, tending toward unity, as can be seen from Eq. (12.12). It is worth noting that the condition $A_1 A_4 R_{12} R_{34} = 1$ corresponds to the classical turn-on condition [5], $\alpha_n + \alpha_p = 1$, or $\beta_n \beta_p = 1$, as can readily be deduced from Eqs. (12.12), (12.15), and (12.16). Equations (12.17) and (12.18) provide the fundamental relations between the voltages V_1 (via X_1), V_2 (via X_2), and V_3 (via X_3) and the current ratios, or gains, via R_{12}, R_{34}, A_1, and A_4. Let us consider the different regions shown in Fig. 12.14.

When the device is in the "off" or high voltage region, W_N is large but $A_1 A_4$ is greater than 1 (except for devices with poor injection efficiency) and $R_{12} R_{34} < 1$ which keeps $A_1 A_4 R_{12} R_{34} < 1$ and, from Eq. (12.17), X_2 will be less than 1, that is, V_2 is reverse-biased and V_1 and V_3 can be calculated from Eqs. (12.17) and (12.18). In this regin $A_1 A_4 R_{12} R_{34}$ keeps increasing toward unity as V_2 increases in the reverse-bias direction.

In the negative resistance part of the characteristic $A_1 A_4$ decreases as V_2 decreases in magnitude, but $R_{12} R_{34}$ increases toward unity so that $A_1 A_4 R_{12} R_{34} < 1$ and V_2 is still negative.

In the "on" low voltage region $A_1 A_4 R_{12} R_{34}$ will be slightly greater than unity and the value of $R_{12} R_{34}$ will be practically equal to unity.

In the case where space-charge recombination and avalanche multiplication current are both negligible, $R_{12} R_{34} = 1$ and two possibilities exist: (a) $A_1 A_4 > 1$

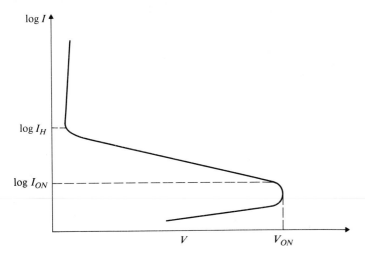

FIGURE 12.14
Static I-V characteristics of the thyristor.

for a small total voltage (the case of a high-gain device); (b) $A_1 A_4 < 1$ even for moderate values of V_2 (reverse bias). In case (a) the SCR will be in the "on" state as soon as junctions 1 and 3 are forward-biased. Equation (12.17) indicates that V_2 will also be a forward-bias voltage. This case is represented in Fig. 12.15 curve (a). In case (b), the device will be "off" at low and moderate reverse-bias V_2 values. However, as the depletion layer widths increase, the neutral regions will tend toward zero widths and $A_1 A_4$ will increase toward unity. When $A_1 A_4$ approaches unity, V_3 will start to increase rapidly to satisfy Eq. (12.17). V_1 will also increase [Eq. (12.18)] and the current will rise. This case is illustrated by Fig. 12.15 curve (c).

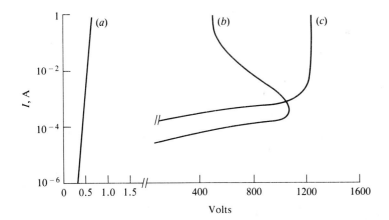

FIGURE 12.15
I-V characteristics for extreme conditions.

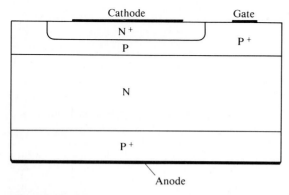

FIGURE 12.16
Complete SCR structure.

Note that avalanche multiplication will not, on its own, allow the normal "on" condition to be reached since it will have a negligible effect when V_2 is close to zero. Space-charge recombination current appears, therefore, to be the only internal mechanism which can explain the complete characteristics. It is, of course, evident that a gate-to-cathode resistance path (internal or external) will have the same general effect as space-charge recombination since the resistive current is approximately constant $= V_1/R$ and the gain increases with increasing V_1. This is equivalent to setting J_{r12} equal to a constant value, V_1/R, and is a widely used technique for controlling the characteristics.

The above discussion explains the *I-V* characteristics. In use, operation is switched from the low current "off" state to the high current "on" state by a control gate, a base terminal. The structure is shown in Fig. 12.16.

A particular problem with SCR operation is their property of widthstanding a voltage ramp dV/dt without turning on. Figure 12.17 illustrates a typical

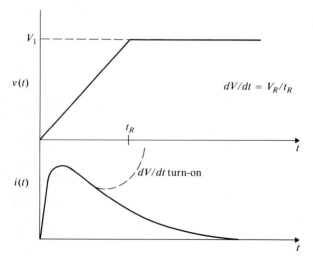

FIGURE 12.17
Voltage ramp creating switching.

test situation. If the value of dV/dt is high enough, and of long enough duration, the capacitive current in the depletion layer $i = d(cv)/dt$ can place the device in the negative resistance region of the I-V characteristic, eventually triggering it to the low resistance "on" state. This is a critical aspect of SCR characterization.

Note the shape of the current versus time curve. After the initial rise (determined by the source resistance and circuit), the current decreases slowly with time, since the depletion-layer capacitance decreases in value as V increases in the reverse bias direction. The critical value of dV/dt is that for which the instantaneous values of (i, v) are above the negative slope on the static I-V characteristic. This latter can be determined by solving Eqs. (12.17) and (12.18) for a given device (given the dependence of A_1 on W_N and hence on the large reverse bias across the lightly doped layer). It should be noted that the instantaneous values of (i, v) must be significantly above the static I-V curve before the end of the ramp dV/dt, so that the complete steady state charge distributions can be obtained.

The critical dV/dt value can be improved by use of a gate to cathode resistive path, sometimes referred to as a "short"; this can be implemented conveniently at the mask design stage by suitable positioning of contact "holes" or windows [12]. The effect is analogous to decreasing the carrier lifetime in the forward biased junction—the current gain is reduced at low bias in a controlled manner.

Another important rating of a thyristor is the maximum rate of applying current—the dI/dt limit. As the device is turned on, the active region nearest the gate will initiate conduction and heat will be generated in this area before the current spreads uniformly over the whole active cathode area. A high degree of interdigitation will clearly lead to an improvement in this characteristic.

A particular class of thyristor has the ability to be turned to the "off" state by application of a suitable negative bias to the gate. These devices are known as *gate turn off* or GTO thyristors. The turn-off is achieved by extracting excess carriers (holes) from the P base region. There are two major aspects in the design of GTO structures which facilitate this operation. First, the P base region is more heavily doped than in a normal SCR in order to keep the "base" resistance as low as possible and avoid concentration of high current densities at the centers of the emitter stripes. Second, the spacing between the gate (base) and emitter edge must be kept constant. The evacuation of charge is a two-dimensional process and the structure will tend to turn off at the emitter edges before the center is debiased enough to stop conduction. This is a very similar situation to the high voltage transistor turn-off discussed in Sec. 11.3.

Although the emitter stripe interdigitated geometry can be used to distribute the evacuation of charge as uniformly as possible, it does not lend itself to high current thyristors which utilize the whole (circular) wafer area. An alternate geometry is the involute structure shown in Fig. 12.18 [7]. The control gate is the center element connected to all the narrow (base contact) stripes. The emitters are within the wider stripes. This structure distributes the turn-off process very evenly over the whole area and gives excellent turn-off and turn-on times with a large improvement in dI/dt capability.

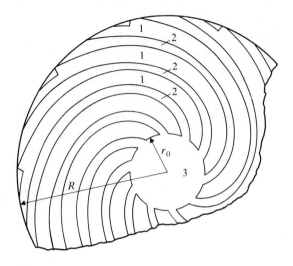

FIGURE 12.18
Involute mask layout used in high current GTO thyristors. Regions emitters are within region 1, regions 2 are base (gate) contacts connected to the central gate region 3. (*After Storm and St. Clair* [7]. *Reproduced with permission* © *1974 IEEE.*)

12.4 OTHER THREE-TERMINAL HIGH VOLTAGE SWITCHING DEVICES

Apart from the single bipolar transistor, the Darlington pair and the various thyristor structures, there are some devices which combine field effect and bipolar operation in one structure. An example is the gated diode structure, also known as a *field controlled diode* (FCD) or *field controlled thyristor* (FCT), shown in Fig. 12.19 [8]. The operation of this device makes use of the junction field effect to cut off the supply of minority carriers to the lightly doped N region. This can be seen by visualizing a reverse bias applied between the P^+ gate and the N^+ cathode. As the reverse bias increases, the depletion layer in the N layer surrounding the P^+ diffused region extends sideway, thereby suppressing progressively the area in direction of vertical current flow (cathode to anode). Eventually the diode is turned off. With no bias between gate and cathode, the structure behaves as a normal PIN rectifier as discussed in Sec. 5.4.

Another class of high voltage switch, gaining rapid popularity in many applications, is a structure based on a combination of metal oxide semiconductor or MOS and bipolar technology. MOS power transistor switches have been used for a number of years and offer the advantage of very good safe operating area, relatively high speed and, of course, high input impedance. Their major draw-

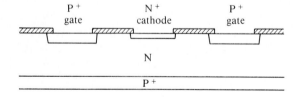

FIGURE 12.19
The field controlled diode or field controlled thyristor cross section. (*After Baliga* [8].)

backs are high "on" voltage for a given (high) breakdown voltage and given area (operation is confined to lower current densities than the equivalent bipolar structure in order to keep the "on" voltage acceptably low).

The new class of device combines MOS and bipolar technology in one device as shown in Fig. 12.20 and is referred to as the *insulated gate transistor* (IGT) [9] or the *COnductivity Modulated FET* (COMFET) [10]. This is essentially a four-layer thyristor-like structure. Operation is as follows. With no bias applied to the gate electrode, the device is normally "off" and will block current for both directions of "emitter" to "collector" bias as in the almost identical thyristor situation. If a sufficient bias is applied to the gate so that an inversion layer forms in the P base region under the gate, a path exists for electrons to flow from the N^+ emitter to the N base. The electrons flowing from the emitter through the P base to the N base cause the junctions to become forward biased and the N base becomes heavily conductivity modulated, thus enabling the structure to exhibit "on" *I-V* characteristics similar to those of a PIN diode (although with an offset voltage before forward conduction occurs). This device offers the advantages of both the separate MOS and bipolar high voltage switches.

12.5 CONCLUSIONS

We have studied various discrete devices in this chapter and seen how the significant design characteristics vary with applications. The basic theory derived in Chaps. 8 to 11 is applicable in all cases. Even basic thyristor operation does not require any fundamental new "theory," but rather the careful application of diode and transistor concepts.

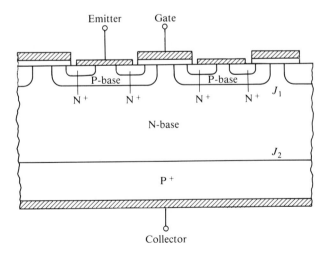

FIGURE 12.20
The insulated gate transistor cross section. (*After Baliga et al.* [9]. *Reproduced with permission © 1984 IEEE.*)

PROBLEMS

12.1. An integrated RF power transistor is built with a technology which uses 4 μm wide emitter stripes. The collector epitaxial doping level is 3×10^{15} cm^{-3}. The emitter has a junction depth of 0.5 μm and gives a gain of 50 for a pinched base sheet resistance of 5 kΩ/sq. Aluminum metallization is used with a sheet resistance of 0.03 Ω/sq.
   (a) Determine the total stripe length B for a collector current of 1 A, assuming that performance is limited by the Kirk effect ($V_{CB} \ll V_{pt}$).
   (b) Hence determine the length of each stripe and the number of stripes for a maximum acceptable voltage drop along each metal stripe of 0.01 V.
   (c) Calculate the value of ballast resistance to be placed in series with each stripe on the assumption that an additional 0.025 V drop is introduced at the maximum current of 1 A total.
   (d) Using the curves of Fig. 7.12(a), determine the maximum collector-base break-down voltage and the BV_{CEO}. Find also the reach-through value of W_{epi}.
   (e) For $V_{CB} = 40$ V, calculate the approximate value of base width W_b for maximum $f_{m\,osc}$ using the results of Sec. 9.6.
   (f) Calculate C_{je} assuming 10^{-7} F/cm^2 as per Sec. 9.2.1 and C_{jc} assuming 4 μm base contact stripes and 4 μm spacing between emitter diffusion and base contacts. Hence sketch f_t versus I_C.
   (g) Calculate the corresponding values of $r_{bb}, f_{m\,osc}$, and maximum f_t.

12.2. A Darlington pair is to be designed for a total collector current of 200 mA using a collector epitaxial layer doped 3×10^{15} cm^{-3}. Assume for the design that the maximum current is limited by Kirk effect at low voltage ($V_{CB}/V_{pt} \ll 1$) and calculate the total area required using the information of Prob. 12.1 where required. The process gives a maximum current gain of 50 for each transistor and uses 4 μm emitter stripe widths, 4 μm wide base contacts, and 4 μm spacing between emitter diffusion and base contact. Calculate a reasonable layout for the case where the combined peak gain is 2500.

12.3. Using the information of Prob. 12.2, calculate reasonable dimensions and layout for the case where the combined current gain under forced gain (saturation) conditions is 25 at I_C equal to 200 mA. Use the theory of Chap. 10 to estimate the value of $V_{CE\,sat}$ assuming high level injection conditions. Use $W_b = 0.5$ μm, $W_{epi} = 4$ μm. If a single transistor were used, of the same total area, for the same value of forced gain equal to 25, what would be the value of $V_{CE\,sat}$? Note the improvement in $V_{CE\,sat}$ obtained using the Darlington configuration. Calculate the values of resistors R_1 and R_2 which are as low as possible consistent with not reducing the overall forced gain by more than 10 percent for the Darlington pair.

12.4. A thyristor has a lightly doped layer 100 μm wide doped 10^{14} cm^{-3}. By focusing attention only on the A_1 factor given by Eq. (12.15), calculate the maximum possible value (i.e., assuming a low value of A_4) of peak voltage that can be applied before the reduced voltage increased current region (i.e., the negative resistance region) occurs. By calculating the actual maximum turnover voltage for $W_N = 10$ and 0 μm, show that the voltage value is relatively insensitive to other design parameters.

12.5. Using the nomenclature in the text, the thyristor of Prob. 12.4 has equivalent Gummel numbers given by: $G_{u1} = N_1 W_1 = 10^{15}$ cm^{-2}, $G_{u2} = N_2 W_2 = 10^{14}$ cm^{-2}, $G_{u4} = N_4 W_4 = 10^{15}$ cm^{-2}. The area is 0.1 cm^2. The recombination current of junction 3/4 is negligible, that is, $R_{34} = 1.0$. The recombination current of junction 1/2 is such that $R_{12} \sim 0.01\sqrt{I}$. If the start of the negative resistance region occurs for a current of 1 μA, calculate and sketch the complete negative resistance portion of the *I-V* characteristic (use the fact that the product $A_1 A_4 R_{12} R_{34}$ is very close to unity in this region).

12.6. The thyristor of Prob. 12.5 has a dV/dt ramp applied, ending with a voltage V_{on}. Calculate the maximum "safe" value of dV/dt (for which the transient *I-V* point coincides with the negative resistance static *I-V* curve) for different values of V_{on} and sketch the curve of limiting dV/dt versus V_{on}.

REFERENCES

1. R. P. Arnold and D. S. Zoroglu, "A quantitative study of emitter ballasting," *IEEE Trans. Electron Devices*, vol. ED-21, pp. 385–391 (July 1974).
2. P. R. Bryant and D. J. Roulston, "An experimental verification of thermal simulation by WATAND for a bipolar RF power transistor," *IEEE Intl. Symp. on Circuits and Systems Proc.*, pp. 397–400, Helsinki, Finland (June 1988).
3. D. J. Roulston and J.-B. Quoirin, "Simulation of high-voltage power switching transistors under forced gain and inductive load turn-off conditions," *IEE Proc.-I*, vol. 135, pp. 7–12 (February 1988).
4. C. F. Wheatley, Jr. and W. G. Einthoven, "On the proportioning of chip area for multistage Darlington power transistors," *IEEE Trans. Electron Devices*, vol. ED-23, pp. 870–878 (August 1976).
5. D. J. Roulston and M. Nakhla, "Efficient modeling of thyristor static characteristics from device fabrication data," *IEEE Trans. Electron Devices*, vol. ED-26, pp. 143–147 (February 1979).
6. B. J. Baliga, *Modern Power Devices*, John Wiley & Sons, New York (1987).
7. H. F. Storm and J. G. St. Clair, "An involute gate-emitter configuration for thyristors," *IEEE Trans. Electron Devices*, vol. ED-21, pp. 520–522 (August 1974).
8. B. J. Baliga, "Breakover phenomena in field-controlled thyristors," *IEEE Trans. Electron Devices*, vol. ED-29, pp. 1579–1587 (October 1982).
9. B. J. Baliga, M. S. Adler, R. P. Love, P. V. Gray, and N. D. Zommer, "The insulated gate transistor: a new three-terminal MOS-controlled bipolar power device," *IEEE Trans. Electron Devices*, vol. ED-31, pp. 821–827 (June 1984).
10. J. P. Russell, A. M. Goodman, and J. M. Neilson, "The COMFET—a new high conductance MOS-gated device," *IEEE Electron Device Letters*, vol. EDL-4, pp. 63–65 (March 1983).
11. A. Nakagawa, Y. Yamaguchi, K. Watanabe, and H. Ohashi, "Safe operating area for 1200 V nonlatchup biopolar mode MOSFET's," *IEEE Trans. Electron Devices*, vol. ED-34, pp. 351–355 (February 1987).
12. M. S. Adler, K. W. Owyang, B. J. Baliga, and R. A. Kokosa, "The evolution of power device technology," *IEEE Trans. Electron Devices*, vol. ED-31, pp. 1570–1591 (November 1984).

CHAPTER
13

INTEGRATED TRANSISTORS

13.1 INTRODUCTION

This chapter concerns integrated bipolar transistors. Firstly we discuss the conventional IC device and the important considerations of base and collector resistances. The significance of buried layers and alternate methods of obtaining a low collector resistance are discussed. The Schottky transistor is briefly described. We then develop some simple theory for the lateral PNP transistor, widely used in integrated circuits. The chapter concludes with an analysis of the *integrated injection logic* (ILL) structure, an upward operating NPN transistor and an overview of ECL and BICMOS logic structures.

13.2 THE CONVENTIONAL JUNCTION ISOLATED STRUCTURE

The conventional junction isolated integrated circuit NPN transistor is shown in Fig. 13.1 [1]. The essential difference between this and the discrete device resides in four factors: (i) it is common to use only one base contact to facilitate layouts and reduce the area; (ii) a square "minimum dimension" emitter diffusion pattern is often used; (iii) in order to have the collector contact at the top surface of the device, current must flow horizontally under the base region—to keep the value of this horizontal resistance at a low value a heavily doped buried layer is normally used; (iv) for isolation purposes, the substrate is lightly doped P type material with P^+ isolation "walls" around each transistor (or group of common

(a)

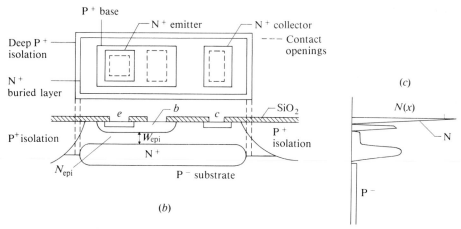

FIGURE 13.1

Conventional NPN integrated transistor. (a) Mask layout; (b) cross section view; (c) impurity profile through active area.

collector transistors); a substrate to collector capacitance therefore adds to those already present in the discrete transistor.

The first two factors lead to a significantly different value for base resistance than that presented in Sec. 7.4.2. The collector resistance and the desirability of a buried layer merits some discussion, particularly since in some technologies (e.g., BICMOS), incorporation of a buried layer is not readily achieved. Finally the existence of the additional collector resistance and the substrate capacitance leads to modification of the expression for f_t. We will study each of these points in turn.

13.2.1 Base Resistance

Figure 13.2 shows the paths for the base current from the base contact to the base region under the emitter. The sheet resistance of the extrinsic base region $R_{b\,sq}$ is typically of order 200 Ω/sq; that of the intrinsic or pinched base region (under the emitter diffusion), $R_{be\,sq}$, usually lies in the range 1 to 10 kΩ/sq for IC structures. Clearly the calculation of both extrinsic and intrinsic base resistances is a two-dimensional current flow problem. Two extreme scenarios exist, however, for which accurate values may be determined—other cases may be deduced approximately by interpolation.

High Resistance Extrinsic Current Path

The value of the resistive path from the base contact around the emitter, noting that the current is maximum on the base contact side and decreases to zero on

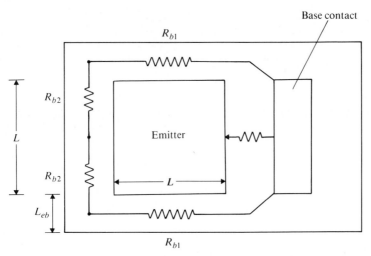

FIGURE 13.2
Base current paths in the one-base contact integrated transistor.

the far side (as it progressively enters the region under the emitter), may be estimated as:

$$R_{b\,\text{ext}} = \frac{1}{6}(R_{b1} + R_{b2}) = 0.25\,\frac{L}{L_{eb}}\,R_{b\,\text{sq}} \qquad (13.1)$$

where L_{eb} is the width of the extrinsic base region around the emitter. If the value of $R_{b\,\text{ext}}$ is very high, all the base current will enter from the base contact side. In this case the intrinsic base resistance will be given, following the approach in Sec. 7.4.2, by

$$r_{bb} = \frac{R_{be\,\text{sq}}}{3} \qquad (13.2)$$

The necessary condition for Eq. (13.2) to be valid is therefore

$$R_{b\,\text{ext}} \gg r_{bb}$$

Let us examine some typical numbers. For the extrinsic base, $R_{b\,\text{sq}} = 300$ Ω/sq; for the intrinsic base, assume a low value of $1000\ \Omega/\text{sq}$ for $R_{be\,\text{sq}}$. Equation (13.2) gives $r_{bb} = 330\ \Omega$. Assuming rather extreme values of spacing, $L = 6\ \mu\text{m}$, $L_{eb} = 1\ \mu\text{m}$, Eq. (13.1) gives $R_{b\,\text{ext}} = 450\ \Omega$. Clearly the inequality involved in using Eq. (13.2) is only approximately valid and one can at best make a crude estimate of the way in which the intrinsic and extrinsic resistances "split" in the overall equivalent circuit. In this case, for example, the actual value of r_{bb} will certainly be lower than $330\ \Omega$, since some base current enters on all four sides of the active region; likewise the actual value of $R_{b\,\text{ext}}$ will be less than the above $450\ \Omega$ value. A more likely set of mask values would be: $L = 6\ \mu\text{m}$, $L_{eb} = 2\ \mu\text{m}$, $R_{be\,\text{sq}} = 10\ \text{k}\Omega/\text{sq}$, and $R_{b\,\text{sq}} = 200\ \Omega/\text{sq}$. In this case we find: $r_{bb} = 3300\ \Omega$,

$R_{b \text{ ext}} = 150\ \Omega$, and the extrinsic resistance path is certainly much lower than r_{bb}, instead of higher as was our original assumption.

Low Resistance Extrinsic Current Path

In this common scenario, where the base current enters the active region more or less equally on all sides, the exact result for a circular structure [2] may be used to good effect. Figure 13.3 shows the ideal square and circular structures. For the latter, the value of intrinsic base resistance is given by [2]

$$r_{bb} = \frac{R_{be \text{ sq}}}{8\pi} \tag{13.3}$$

For the second set of values used above, we therefore find $r_{bb} \sim 400\ \Omega$, and the $150\ \Omega$ value for the extrinsic resistance is small enough that one can rely on this r_{bb} estimate to within about 30 percent precision. Other intermediate cases can only be calculated by two-dimensional solution of the two-dimensional current flow, although [8] gives useful empirical formulas based on exact solutions for a range of geometries.

13.2.2   Collector Resistance

Figure 13.4 shows a cross section and plane view of the region pertaining to collector current flow horizontally from the emitter area to the collector diffusion contact area. Note that a shallow N^+ collector diffusion is necessary to obtain an ohmic contact; this requires no extra processing since it can be done simultaneously with the emitter diffusion. In order to eliminate the vertical resistance component between the collector contact diffusion and the buried layer, a deep N^+ diffusion may be used instead. This is done in all high performance devices, such as high speed *emitter coupled logic* (ECL)

The buried layer is a standard process in nearly all integrated bipolar transistors. This layer is formed by implantation and/or diffusion before growing the epitaxial layer. The value of the horizontal collector resistance is determined by the buried layer sheet resistance $R_{bl \text{ sq}}$ (typically 15 to 40 Ω/sq) [1], and the number of squares between the collector contact and the emitter. This will be of order 1 for a collector placed adjacent to the emitter, and of order 2 to 3 when

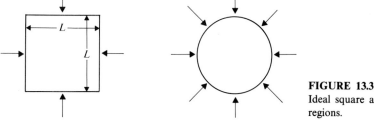

FIGURE 13.3
Ideal square and circular base regions.

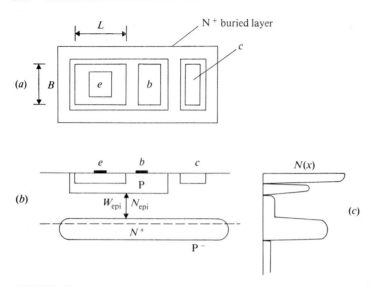

FIGURE 13.4
Collector resistance region in the integrated transistor. (*a*) Plane view; (*b*) cross section; (*c*) impurity profile.

the collector is located adjacent to the base contact. Note that the buried layer is always chosen to be maximum dimension to just fit inside the P^+ isolation diffusion, and will normally be slightly wider than the base diffusion.

It is clear that the use of a buried layer enables the horizontal component of collector resistance to be kept at a low value, limited only by the sheet resistance $R_{bl\,sq}$. For one or two squares, this gives a resistance of order 15 to 80 Ω. It is interesting to note that this value is virtually independent of design rules. Thus it presents a more severe problem for older, large design rule, higher current devices than for modern small geometry devices, because of the larger substrate capacitance and larger currents associated with the older devices.

Let us examine the situation in which no buried layer is used. The normal vertical component of collector resistance is given by

$$R_{cv} = \frac{\rho_{epi} W_{epi}}{BL} \tag{13.4}$$

(we will assume that the device is operating at low V_{CB} voltage, i.e., near saturation, where the full thickness W_{epi} is a charge neutral region). The horizontal value of collector resistance due to current flowing in the epitaxial layer is

$$R_{ch} = \frac{K_c \rho_{epi} L}{BL} \tag{13.5}$$

The factor K_c depends on the position of the collector contact and will be of order 1 to 3 as discussed above with reference to current flow in the buried layer.

The ratio of the two resistance values is thus

$$r_{chv} = \frac{R_{ch}}{R_{cv}} = K_c \left(\frac{L}{W_{epi}} \right)^2 \tag{13.6}$$

Typical values are: $L = 6\ \mu m$, $W_{epi} = 1.5\ \mu m$, $K_c = 1.0$, giving $r_{chv} = 16$. In Chap. 10 we derived the expression for the current I_{CQS} at which the transistor enters quasi-saturation for a terminal collector base V_{CB} [Eq. (10.35)]. Rearranging this to give the vertical voltage drop for a given current (which within roughly 0.5 V is the same as the quasi-saturation onset voltage V_{CQS} for a given current)

$$V_{CQS} \sim \frac{V_{cbr}}{20} \frac{I_{CQS}}{I_K} \tag{13.7}$$

This voltage was derived on the basis of the voltage drop in the vertical resistance R_{cv} and will be increased in the ratio $(1 + r_{chv})$ due to the additional horizontal voltage drop. In order to see the practical effect of this additional horizontal resistance, consider a nominal 50 volt V_{cbr} transistor operating at one-tenth of its Kirk current: $I_{CQS} = I_K/10$. The collector voltage drop for the discrete device is 0.25 V. For the integrated device with the above r_{chv} value of 16, we see that the collector voltage drop now exceeds 4 V. The actual value of r_{chv} will not change substantially with structure, since small geometry devices are invariably associated with moderate-to-low breakdown voltages, thus maintaining a ratio L/W_{epi} greater than unity. For example, a 2 micron square emitter transistor for low voltage ECL VLSI applications would have a residual epitaxial layer thickness $W_{epi} \sim 0.5\ \mu m$, thus giving the same value for r_{chv} as that used above.

The above example illustrates the absolute necessity of reducing the horizontal value of collector resistance in an integrated transistor. One technique to reduce the need for a buried layer is to run the collector contact diffusion around three sides of the emitter as shown in Fig. 13.5. This has the severe disadvantage of increasing the overall area, and consequently increasing the substrate capacitance. We will discuss this point further in Sec. 13.6 on BICMOS structures.

Collector-Substrate Capacitance

The capacitance between the collector and the P^- substrate is made up of two components: the area under the buried layer, where the depletion layer extends primarily into the lightly doped substrate under reverse bias; and the P^+ isolation diffusion component where the depletion layer extends primarily into the N epitaxial layer under reverse bias. Figure 13.6(a) shows the regions under consideration, and (b) is the modified small signal equivalent circuit including the additional collector resistance and substrate capacitance components.

It is apparent that the value of f_t (which, we recall, is defined as a function of the ac collector current flowing in a short-circuited output) will be modified. If the additional resistance and capacitance terms are lumped into an equivalent single RC section as shown in Fig. 13.6(c), it is a simple matter to show that the

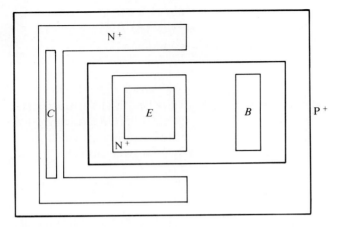

FIGURE 13.5
Collector diffusion on three sides of emitter to reduce collector resistance in the absence of a buried layer.

effect on f_t is to introduce an additional delay term $t_{\text{sub } s}$ given by

$$t_{\text{sub } s} = \frac{K_s R_{bl} C_{\text{sub } s}}{\beta} \tag{13.8a}$$

with the overall f_t value now given by

$$\frac{1}{2\pi f_{t(\text{total})}} = \frac{1}{2\pi f_{t(\text{discrete})}} + t_{\text{sub } s} \tag{13.8b}$$

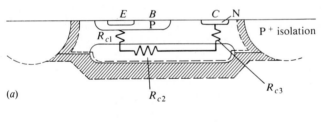

(a)

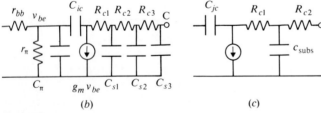

(b)   (c)

FIGURE 13.6
(a) Collector-substrate capacitance depletion layer in IC device (hatched region); (b) modified small signal equivalent circuit; (c) simplified small signal circuit for output.

where K_s is the reduction due to the distributed nature of $R_{c\,sub}\,C_{c\,sub}$ and, for typical layouts, will be of order 0.3. The value of $t_{sub\,s}$ will vary from a value of order 100 ps for 6 μm design rule transistors to 0.1 ps for 2 μm design rules (the reader can easily estimate the values from the depletion-layer widths, assuming doping levels of 10^{16} cm^{-3} for the epitaxial layer and 10^{15} cm^{-3} for the substrate). The overall contribution to f_t in Eq. (13.8) is thus normally small. Nevertheless, in a real circuit, it is seen from Fig. 13.6 that use of f_t as a measure of performance may be optimistic. For a resistive load (wide band amplifier, logic inverter) all of the substrate capacitance will appear across the load and can significantly degrade the performance, whence the necessity of its inclusion in the CAD model.

13.2.3 The Schottky Clamp

In a large number of saturated logic inverters, a Scottky diode is used to clamp the forward bias base-collector voltage at a value well below that at which significant injection of charge occurs into the collector. It is outside the scope of this text to enter into the theory of the Schottky barrier but a treatment of the bipolar integrated circuit transistor would not be complete without mentioning this widely used means of improving the switching speed of some bipolar logic families.

The Schottky barrier is formed by a metal-semiconductor junction [3]. A contact between aluminum and heavily doped P material is ohmic and may be thought of as similar to two heavily doped P type materials in contact. A contact between aluminum and a heavily doped N type region (as in the case of emitter and collector contacts) may be thought of as being similar to a PN junction heavily doped on both sides. In the limit, the junction will have exceeded its Zener breakdown voltage, due to the built-in barrier and resulting electric field, and behave like a tunnel diode for both forward and reverse currents; in other words it also exhibits the properties of an ohmic contact in the vicinity of zero bias. A Schottky barrier, on the other hand, is formed when a suitable metal is in contact with moderately doped N type material, in particular the epitaxial layer. The resulting junction has nonlinear I-V characteristics which are due to a combination of thermionic emission, diffusion, and tunneling. The thermionic emission theory is generally considered the most applicable to Schottky clamps; the I-V characteristic is given by

$$I = A_S J_S\left[\exp\left(\frac{qV}{kT}\right) - 1\right]$$

where A_S is the area of the junction, and J_S is a function of temperature, Richardson's constant, and the barrier height.

By choice of a suitable metal and for an acceptably small area, it is possible to fabricate a Schottky diode which conducts significantly more current than the base-collector junction; in other words, it is possible to incorporate a Schottky diode which will "cut-in" before the base-collector junction becomes sufficiently

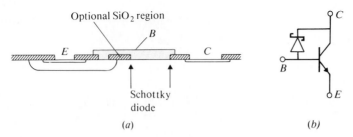

FIGURE 13.7
The Schottky clamped transistor. (*a*) Cross section; (*b*) circuit representation.

forward biased to inject a large amount of charge into the epitaxial layer (the major cause of storage delay time t_s during the turn-off phase as discussed in Chap. 11). Since the Schottky clamp is essentially a majority carrier device, the only additional delay is that associated with the depletion-layer capacitance; this may be calculated in the same manner as for a P^+N junction, where the N type epitaxial layer determines the depletion-layer properties. Figure 13.7 shows the basic Schottky clamped transistor. In order to gain maximum area for the Schottky clamp, the metal is normally run directly across the P^+N junction.

13.3 THE LATERAL PNP INTEGRATED TRANSISTOR

In many circuit applications it is required to incorporate PNP devices. The lateral PNP transistor is widely used. Although its performance (in terms of gain and f_t) is mediocre, it can be fabricated in a standard vertical NPN process with no additional processing steps. Of course, for discrete structures the vertical PNP is always possible, but for integrated circuits, the additional processing steps necessary to incorporate a vertical PNP device would normally be prohibitive. Figure 13.8 shows the cross section diagram of a lateral PNP made using an NPN base diffusion as the PNP emitter and collector. Also shown are two possible mask layouts, parallel geometry, and square geometry. The latter arrangement has the advantage of providing self-isolation from parasitic PNP effects (see below).

We will now proceed to estimate the current gain and f_t of the lateral PNP transistor. Figure 13.9 indicates the various minority carrier charge and current components which enter into the analysis. A long double collector parallel geometry is assumed.

Collector Current

This may be easily estimated by recognizing that the effective area from which holes are injected from the emitter is mainly the sidewall of the emitter diffusion.

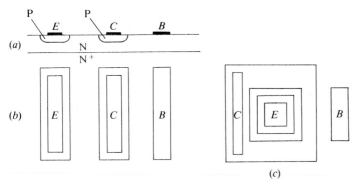

FIGURE 13.8
(a) Cross section diagram of lateral PNP transistor; (b) parallel geometry lateral PNP; (c) square geometry lateral PNP structure.

The neutral base width W_b is determined by the diffusion spacing less the depletion-layer widths (function of bias). We can write the collector current as

$$I_C = \frac{2q(F_1 X_j)BD_p p(0)}{F_2 W_b} \qquad (13.9)$$

where $p(0)$ is the injected hole concentration $= (n_i^2/N_{epi}) \exp(V_{BE}/V_t)$. The factor $F_1 > 1$ represents the fact that some current is injected from the bottom part of the emitter junction, thus increasing somewhat the effective area; the factor $F_2 > 1$ represents the fact that the average path length for carriers crossing the base is greater than W_b because current is also collected in the bottom part of the collector junction and, as already mentioned, injected from the bottom part of the emitter junction. To a first approximation, these two factors may be expected to cancel; we assume this cancellation in the rest of the analysis.

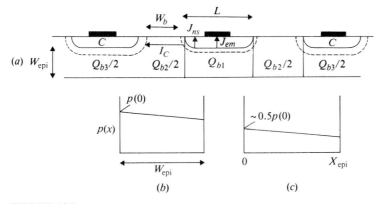

FIGURE 13.9
(a) Simplified cross section of lateral PNP device for gain and f_t analysis (the z direction has a length B perpendicular to the diagram); (b) charge distribution Q_{b1} in vertical direction; (c) charge distribution Q_{b2} in vertical direction midway between emitter and collector.

Base Region Minority Carrier Charge

We have shown three charge components in the diagram. Q_{b1} is a near rectangular charge distribution due to carriers injected vertically down into the epitaxial layer. At the buried layer, these carriers encounter a retarding field. The carrier concentration remains therefore almost constant (the diffusion length L_p being typically of order 50 μm and the residual epitaxial layer thickness being a couple of microns—see the P^+NN^+ discussion in Chap. 3 and in particular Fig. 3.4) over the vertical distance W_{epi} and then drops rapidly toward zero as shown in the diagram. The expression for this important charge component is therefore

$$Q_{b1} = qp(0)BLW_{epi} \tag{13.10}$$

where L and B are the emitter dimensions shown in Fig. 13.9. Figure 13.10 shows results from a two-dimensional computer solution [4] of charge distribution. Following a similar reasoning in the region between emitter and collector and recognizing that at the surface the carrier concentration varies linearly from a value $p(0)$ at the emitter to essentially zero at W_b, the base-collector depletion-layer boundary as can be seen from Fig. 13.10, we may write

$$Q_{b2} = F_{b2}\,qp(0)BW_b\,W_{epi} \tag{13.11}$$

where $F_{b2} \sim 0.5$. The third base region charge is that under the collector

$$Q_{b3} = F_{b3}\,Q_{b1} \tag{13.12}$$

From the numerical results of Fig. 13.10, it is clear that this charge is almost negligible; a reasonable value is to use $F_{b3} = 0.1$. The total base charge is

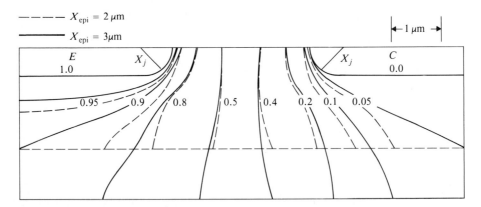

FIGURE 13.10
Computer-simulated results for base charge in a lateral PNP transistor. The numbers represent hole concentration normalized with respect to $p(0)$. Two values of epitaxial layer thickness were used and the concentration falls abruptly to zero below the bottom (buried layer) interface where an abrupt NN^+ junction is assumed in the computer analysis. (*After Eltoukhy and Roulston* [4]. *Reprinted from Solid State Electronics, with permission © 1984 Pergamon Press PLC.*)

given by

$$Q_{bt} = Q_{b1} + Q_{b2} + Q_{b3} \sim 1.4Q_{b1} \tag{13.13}$$

where we have assumed a base width W_b approximately one-half the emitter diffusion width L. It is important to note that two factors contribute to reduction of the neutral base width: the reverse biased collector depletion-layer thickness and the simultaneous lateral diffusion of the emitter and base regions during processing. If both these factors are not included carefully at the design stage, a base punch-through situation can easily occur ($W_b = 0$). For diffused junctions, the junction due to lateral diffusion is approximately 80 percent of the vertical junction.

Base Current

The base current is made up of three components: recombination of the above epitaxial layer charge with a lifetime τ_{epi}; injection of electrons into the P diffused emitter under the metal contact; injection of electrons into the emitter under the oxide on either side of the metal.

The first case can be represented by a vertical current density

$$J_{pv} = \frac{qp(0)W_{epi}}{\tau_{epi}} = \frac{Q_{b1}}{BL\tau_{epi}} \tag{13.14a}$$

Adding the above discussed components of minority carrier charge gives a base current

$$I_{b1} = J_{pv}BL(1.4) \tag{13.14b}$$

Because the emitter is normally an NPN base diffusion step, the surface boron concentration is of order 10^{19} cm^{-3}. Bandgap narrowing and Auger recombination have therefore only a small influence here (compare to a standard emitter diffusion where the surface concentration is of order 10^{21} cm^{-3} and both these high doping effects dominate the behavior of the injected carriers). A significant fraction of the upward injected electron current reaches the surface and the nature of the Si-SiO$_2$ interface is important in determining the total current. It is customary to characterize the effect of the surface on minority carriers by its "effective surface recombination velocity" S_{ox}

$$J_{ns} = qn_s S_{ox} \tag{13.15}$$

where J_{ns}, n_s, are the values of the electron current density and electron concentration at the surface. The value of S_{ox} is very process-dependent, and values ranging from 10 to 10^5 cm/s have been reported in the literature [5]. Recombination in the bulk of the diffused layer is another quantity which is somewhat process-dependent. The analysis of Chap. 8 may be used, in particular Eqs. (8.7) and (8.10) with appropriate changes in nomenclature. We may simplify the situation, however, by recognizing that the PNP transistor currents which we wish to calculate are only part of the total and by admitting that, in view of the assumptions already made concerning the factors introduced to account for various two-dimensional current flow quantities, it would not really be worthwhile to pursue a more exact analysis.

Proceeding then, with some substantial simplifications, we can write the component of base current injected into the emitter under the metal as

$$J_{em} = \frac{qn(0)D_n}{F_{em} X_j} \tag{13.16}$$

where F_{em} is the retarding field factor given by Eq. (3.40) and lies typically in the range 10 to 100. $n(0)$ is the injected electron concentration and this may be taken, to a first approximation, as comparable to $p(0)$ on the base side of the emitter-base junction [in fact, this junction is forward biased, and therefore $n(0)$ will be very close to $p(0)$; however, the forward bias results in a small retarding field region on the base (epitaxial layer) side before the N_{epi} doping level is reached and a reduction factor of about 2 should really be included in Eq. (13.14) for a more accurate derivation].

This may be related to the current density J_{pv} as follows:

$$\frac{J_{em}}{J_{pv}} = \frac{1}{F_{em}} \frac{D_n \tau_{epi}}{X_j W_{epi}}$$

$$= \frac{1}{F_{em}} \frac{D_n}{D_p} \frac{L_p^2}{X_j W_{epi}} \tag{13.17}$$

where we have used $L_p^2 = D_p \tau_{epi}$. This ratio clearly depends on the structure details; for the purpose of evaluating the gain below, we will use a typical ratio of 5.

The current density under the oxide may be written as:

$$J_{eox} = J_{em} r_{oxm} \tag{13.18}$$

For typical surface recombination and bulk recombination data, the factor $r_{oxm} < 1$ is of order one-half, and we shall use this in subsequent calculations although, of course, it must be borne in mind that the factor could be greater or smaller than 0.5 depending on the diffused region properties.

The total base current, made up of epitaxial layer recombination, injection under the metal, injection under the oxide, substituting typical values for a 1 μm junction depth, 2 μm W_{epi} structure, is thus:

$$I_b = J_{pv} BL(1.4) + J_{pv} BL_m(5) + J_{pv} BL_{ox}(2.5)$$

$$= J_{pv} BL(1.5 + 1.7 + 1.7) = 5J_{pv} BL \tag{13.19}$$

We have used $L_c = \frac{1}{2}L_{ox} = \frac{1}{3}L$. While the exact values will change with different structure and process parameters, it turns out that over a wide range of conditions, the total base current lies in the range 2 to 10 times the value contributed by the charge in the epitaxial layer under the emitter [the Q_{b1} term of Eq. (13.10)].

Gain Calculation

Using the above results, and equating $p(0)$ to $n(0)$, we find the ratio of collector to base currents

$$\beta = \frac{I_C}{I_b} = \frac{2X_j BD_p \tau_{epi}}{5W_b W_{epi} BL} \tag{13.20}$$

This may be rearranged slightly to enable easier identification of the ratio

$$\beta = \frac{2}{5} \frac{X_j}{W_b} \frac{L_p}{W_{epi}} \frac{L_p}{L} \tag{13.21}$$

Using $X_j/W_b = \frac{1}{2}$, $L_p = 30$ μm, $W_{epi} = 3$ μm, $L = 3$ μm, gives $\beta \sim 20$, somewhat on the high side compared to typical measured values, but nevertheless a good indication of the performance possible with this type of structure.

f_t Derivation

The maximum value of transition frequency may be found directly from the diffusion capacitance approach, following the method of Sec. 7.4 [see Eqs. (7.14), (7.15), and (7.23)]:

$$f_t = \frac{1}{2\pi t_{diff}}$$

where

$$t_{diff} = \frac{dQ_b}{dI_C} = \frac{Q_b}{I_C} \tag{13.22}$$

Substituting the above result for base charge and collector current gives, from Eqs. (13.9), (13.10), and (13.13),

$$t_{diff} = 1.4 \frac{W_b^2}{2D_p} \frac{L}{W_b} \frac{W_{epi}}{X_j} \tag{13.23}$$

Substituting the values used above gives $t_{diff} = 17(W_b^2/2D_p)$. Note that this is substantially greater than the base transit time. For a base width of 3 μm the value of t_{diff} is 76 ns, corresponding to a peak f_t value of 2 MHz. It is clear that using conventional technology with vertical NPN process compatibility, there is not much scope for improvement of this figure, and f_t values for lateral PNP transistors seldom exceed several tens of MHz.

Layout Considerations

From a practical point of view, the square layout of Fig. 13.8(c) is preferred. It is important to note that in the parallel geometry structure there will be a parasitic PNP at the ends from the emitter to the P^+ isolation diffusion, unless a deep N^+ region is used to separately isolate the PNP emitter. This effect is shown in Fig. 13.11. Since the substrate is always reverse biased (in a normal NPN chip), there

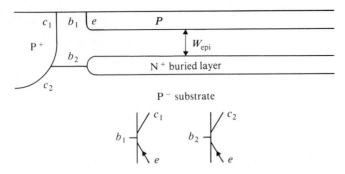

FIGURE 13.11
Parasitic PNP at end of parallel stripe structure.

will be considerable collection of carriers in the P^+ regions from the ends of the normal P emitter. In the cross section diagram of the square structure, virtually all of the base charge in the epitaxial layer is swept into the collector, leaving none to penetrate to the P^+ isolation, thus ensuring high gain. Circular structures may also be used and have the advantage of uniform base width (by eliminating the four corners of the square geometry) and better breakdown voltages. They are, however, incompatible with some mask layout programs.

Finally, it may be noted that the above-mentioned parasitic PNP exists in a normal NPN transistor, where the P diffusion in Fig. 13.11 is now the base of the vertical NPN structure. There is a small current also collected directly by the P^- substrate between the P^+ diffusion and the end of the N^+ buried layer. This can present problems if the NPN transistor operates in saturation, when the base-collector junction becomes foward biased. The net effect is to produce a lower forced gain in saturation than would otherwise be expected. The obvious solution is to increase the spacing between the isolation and base P diffusions, or to use an additional deep N^+ diffusion, which acts as a barrier to the diffusing holes (i.e., a retarding field).

13.4 INTEGRATED INJECTION LOGIC (IIL)

Closely related to the lateral PNP device is the *integrated injection logic* or IIL cell [6, 7]. This finds applications where digital functions are required on an otherwise analog chip, as well as in its own right as an efficient high packing density low power delay product bipolar logic family. Figure 13.12 shows a basic IIL cell in plan and cross-sectional view, with the electrical circuit representation.

The lateral PNP transistor acts as a (constant) current source, and replaces a conventional pull-up resistance. The PNP emitter is often in the form of a long "rail" called the injector rail, with multicollector IIL cells running off on either side. The current to the injector rail may be set off-chip using a low voltage supply and a suitable resistance. The IIL emitter is the epitaxial region, which is connected to ground. The transistor is thus operating in the upward or inverse

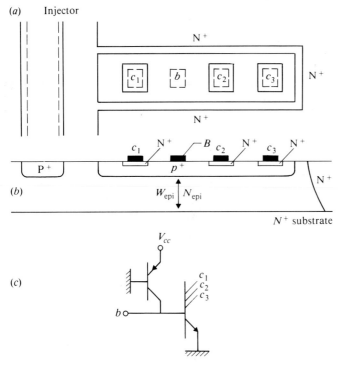

FIGURE 13.12
(a) Mask view of a typical IIL cell; (b) cross section view; (c) electrical circuit representation.

mode, and the analysis which follows may also be used to predict the upward operation of any NPN transistor. In the IIL circuit, with which we are not concerned here, each collector is wired to a base. It is the logic state of a particular collector (saturated or not) that determines whether or not the following base will receive the injector current. We have already studied the lateral PNP structure. We will now examine the upward gain of the IIL cell and its switching speed.

Upward Gain of IIL Structure

Clearly, the charges and currents in the IIL cell are closely related to those discussed above for the lateral PNP transistor. In both cases, we have a base diffusion with a forward biased base to N type epitaxial layer junction. The upward components of current density under the metal base contact and under the base surface oxide layer will be identical to those in the lateral PNP emitter. Likewise the major component of minority carrier charge (holes) in the epitaxial layer will be the same as the Q_{b1} charge in the lateral PNP transistor. We only need to examine one additional current in order to evaluate the IIL cell performance; this is the upward electron current to one collector. Figure 13.13 shows the charge and current components in the IIL cell. The upward collector current can be

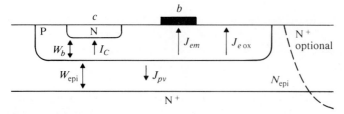

FIGURE 13.13
Charge and current components in an IIL cell.

written directly from our knowledge of the base transport properties of a normal NPN transistor examined in detail in Chap. 9. The Gummel equation, Eq. (9.46), showed that only the total base majority carrier charge integral entered into the determination of collector current for a given applied emitter-base voltage. Although this integral depends slightly on bias voltage (cf the discussion on Early voltages in Sec. 9.4.1), to a good approximation we may write the IIL collector current as

$$I_C = qA_c D_n \frac{n(0)}{W_b}$$

$$= \frac{qA_c D_n n_i^2}{W_b N_{Ab}} \exp\left(\frac{V_{BE}}{V_t}\right) \tag{13.24}$$

where N_{Ab} is an effective "constant" base doping level, or we may replace the product $W_b N_{Ab}$ by the Gummel integral.

The base current is, following the treatment for the PNP transistor,

$$I_b = J_{pv} A_B + \frac{Q_{b1}}{\tau_p} + J_{e\,ox} A_{ox} + J_{em} A_m \tag{13.25}$$

where the areas A_B, A_{ox}, and A_m are those of the base diffusion, the oxide area on the base surface, and the base contact area, respectively, and where the minority carrier charge Q_{b1} is under the whole of the base diffusion in the epitaxial layer. We assume the presence of an (optional) deep N^+ isolation diffusion around the base to prevent this charge diffusing laterally. The reciprocal of the current gain is thus, using Eqs. (13.10) through (13.18), after some rearrangement of terms, and equating $n(0)$ to $p(0)$:

$$\beta^{-1} = \frac{I_b}{I_c} = \left[\frac{W_{epi}}{L_p}\right]^2 \left[\frac{W_b}{W_{epi}}\right]\left[\frac{A_b}{A_c}\right] + \left[\frac{W_b}{X_j}\right] \frac{1}{F_{em}} \frac{(A_m + r_{oxm} A_{ox})}{A_c} \tag{13.26}$$

Although r_{oxm}, defined by Eq. (13.18), is less than unity, the area of the oxide in an IIL cell is much greater than the base contact area, so it is the current under the oxide, plus the current due to the holes in the epitaxial layer, which dominate in determining the gain of the IIL cell. Typical values for 6 micron design rules are: $A_b = 2500 \ \mu m^2$, $A_c = 400 \ \mu m^2$, $A_m = 400 \ \mu m^2$, thus giving $A_{ox} = 1300 \ \mu m^2$

for a two-collector structure. For the data used above in the PNP example, the reader can verify that this gives an upward gain slightly greater than unity, typical of IIL cells. It should be noted that there will be additional epitaxial layer charge diffusing away from the base; this will increase the total recombination current and decrease the gain. This effect may be kept to a minimum by using a deep N^+ isolation diffusion around the three sides of the cell, as shown in Figs. 13.12 and 13.13. If the deep N^+ isolation overlaps the P^+ base diffusion, the sidewall capacitance will be increased, thus degrading low current performance. This point will be examined below.

The switching speed of the IIL cell may be examined using the charge control approach based on the static charges already evaluated. For high current operation, the diffusion capacitance determined by Q_{b1} will dominate. The corresponding time constant is given by:

$$t_{\text{diff}} = \frac{Q_{b1}}{I_C} = \frac{W_b^2}{2D_n} \frac{W_{\text{epi}}}{W_b} \frac{A_b}{A_c} \tag{13.27}$$

For the values under consideration, with a base width of 0.5 μm, this gives a time equal to 4.7 ns (an upward peak f_t of 34 MHz). Clearly the average IIL structure is rather slow. At low currents (set off-chip as we recall, by the bias to the injector rail), the power delay product is a significant parameter. Consider the dominant depletion-layer capacitance of the junction at low bias. The amount of charge to be switched, ΔQ, is related to the (constant) charging current, I, and time, t_d, by

$$\Delta Q = C_j \Delta V$$

$$= I t_d \tag{13.28}$$

The power is given by

$$P = IV$$

Since the voltage swing ΔV is from about 0.1 to 0.6 V, it follows that to an excellent approximation, the power delay product is given by

$$PD = C_j \Delta V^2 \tag{13.29}$$

This is constant over a very wide range of (low current) bias conditions, since the depletion-layer capacitance is almost independent of bias current. An idea of this quantity may be obtained by using the areas and capacitances per unit area associated with a standard diffused junction. Table 13.1 gives computed zero bias values for a process with a 0.7 μm emitter base junction and a 1.5 μm base collector junction of the type used above for illustrative purposes. For the IIL emitter-base junction, the zero bias value should be approximately doubled to give the average value under 0.3 V forward bias. For the areas used above, this gives a total capacitance of approximately $0.4 + 1.3$ pF $= 1.9$ pF. For a peak voltage of 0.6 V, the power delay product is therefore approximately 0.7 pJ. Note that this figure reduces drastically for 2 micron design rules.

TABLE 13.1

Capacitance values (plane and sidewall) for conventional process with junction depths of 0.7 and 1.5 microns

	N^+ base junction	Base—N_{epi} junction
C plane F/cm^2	10^{-7}	0.5×10^{-7}
C perimeter F/cm	0.2×10^{-10}	0.5×10^{-13}

The IIL cell is seen to offer the advantage of a compact, high packing density logic, with the potential for a very low power delay product under low current operation. Furthermore the power and switching speed can be traded by altering the off-chip bias.

It is of interest to see qualitatively the effects of process modifications to IIL structures. Figure 13.14 shows the relative effect of various changes or enhancements to the basic process. The curves are simply an indication of relative trends. Note the degradation which occurs at low currents if the deep N^+ isolation intersects the P base diffusion (due to increased junction capacitance). Reduction in epitaxial layer thickness affects only the minimum switching speed [through the

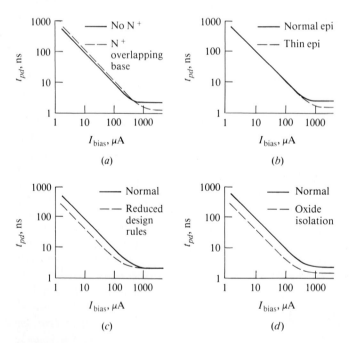

FIGURE 13.14

Improvements in IIL for various changes in process or masks. (a) Effect of deep N^+ isolation around P base; (b) effect of epitaxial layer thickness; (c) effect of design rule reduction; (d) effect of oxide isolation.

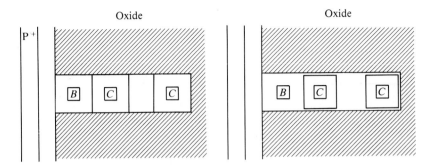

FIGURE 13.15
Oxide isolated IIL structure. (*a*) For low current operation; (*b*) for high current operation.

TABLE 13.2
Computed results for a conventional and an advanced IIL/PNP structure
The conventional structure uses 6 micron design rules with 0.7 and 1.5 micron junction depths, $W_{epi} = 1$ micron. The advanced process uses 2 micron design rules and has junction depths of 0.2 and 0.35 micron, $W_{epi} = 0.2$ micron. Current densities are normalized with respect to J_{pv} [Eq. (13.14*a*)].

	J_{eox}	J_{em}	J_c	t_{diff}	β(up)	β(PNP)	f_t(PNP)
Conventional	5.3	13.2	300	15 ns	13	33	30 MHz
Advanced	130	550	2900	4.1 ns	1.8	32	300 MHz

W_{epi} term in Eq. (13.27)]. Reduction of design rules affects only low current operation (junction capacitances). Finally, the use of oxide isolation (see Chap. 14) improves the performance at all bias, because of the considerable reduction in both junction capacitance and in minority carrier stored charge. The resulting structure is shown in Fig. 13.15. It should be noted that for high current operation, the oxide wall must not intersect the N^+ collectors and a space must be left; this is so that there will be a low base current path around each collector.

Table 13.2 shows computed results for a conventional 6 micron process with 0.7 and 1.5 micron junction depths, $W_{epi} = 1.0$ μm; and for an advanced junction isolated process using 2 micron design rules, with junction depths of 0.2 and 0.35 μm. Note the relative magnitudes of J_{eox} and J_{em}. In spite of the factor of order 1/3, it is the current injected upward under the large area of oxide which usually limits the upward gain.

13.5 EMITTER COUPLED LOGIC (ECL) TRANSISTORS

Since one of the main applications of bipolar technology today in VLSI is *emitter coupled logic*, or ECL, it is of some interest to examine the main transistor characteristics pertaining to this situation. An ECL gate uses transistors in which current is "switched" from one device to another, with the emitters being fed from a common current source as in Fig. 13.16. The transistors are operating in either the normal active region or in cut-off. One would therefore expect high performance ECL circuits to perform best using devices with very high f_t and $f_{m\,osc}$ values.

In fact, the determination of propagation delay time for an ECL gate is not a trivial matter and can only be accomplished with circuit simulation methods using programs such as SPICE or WATAND. Chor et al. [9] studied this problem by assigning weighting constants to all the delays and time constants and determining the values of the weighting constants using SPICE simulations for various transistor parameter and circuit resistance values. The interested reader is referred to [9] for complete details. The important conclusions concerning ECL operation can be summarized, however, by stating that for a wide range of load resistance values (250 Ω to 2 kΩ) and transistor parameter values (intrinsic base resistance r_{bi} from 0 to 5 kΩ, extrinsic base resistance r_{bx} from 0 to 1 kΩ, collector and emitter junction capacitances C_{jc}, C_{je} from 2 fF to 1 pF, load capacitance C_L from 0 to 1 pF, forward delay time t_f from 2 to 200 ps), the weighting factors were found to be approximately constant. Neglecting the less important time constant contributions, the results of Chor et al. can be simplified by stating that over 90 percent of the total propagation delay time can be accounted for by the following terms

$$t_{pd90} = 1.2t_f + r_{bi}[3.83C_{jc} + 0.74C_{je} + 0.97C_D]$$
$$+ r_{bx}[6.16C_{jc} + 0.81C_{je} + 0.96C_D]$$
$$+ R_L[2.54C_{jc} + 0.25C_{je} + 0.16C_{js} + 0.30C_L]$$

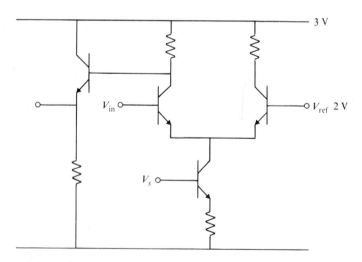

FIGURE 13.16
ECL gate as simulated by Chor et al. [9]. (*Reproduced with permission © 1988 IEEE.*)

where C_{jcx} is the part of C_{jc} outside the active region, C_{js} is the collector to substrate capacitance, and C_D is the diffusion capacitance (a function of t_f).

This equation demonstrates clearly the extreme importance of minimizing the base resistance and collector junction capacitance in high speed bipolar logic. Of course, since the diffusion capacitance is proportional to the sum of the delay times contributing to asymptotic peak f_t as given in Chap. 9 [Eqs. (9.81) and (9.82)], it is clear that if the base resistance reduction is accomplished by increasing the base width W_b, the performance will ultimately be limited by f_t. It should be noted that the study performed in [9] uses a process in which the emitter and collector resistances contribute almost negligible time constants. Since ECL is a high performance logic, mask layouts and diffusions/implants will be chosen so as to keep these values to a minimum.

Using the above method, but including a total of 24 time constants and weighting coefficients, Chor et al. [9] reported an ECL speed of 320 ps for 6 μm geometries using a conventional process and 160 ps using a polysilicon self-aligned process. A gate delay of 20.4 ps was computed for a fully optimized 0.4 μm ECL process [9].

13.6 BICMOS INTEGRATED STRUCTURES

While *complementary metal oxide semiconductor* (CMOS) technology finds widespread use in VLSI integrated circuits due to its excellent logic level characteristics (good noise margins), low drive current (essentially zero for dc), and virtually zero standby power consumption, there are increasing requirements for bipolar transistors on CMOS chips, particularly for drivers, and high performance analog stages. Combinations of CMOS and bipolar circuit arrangements have also been used for ECL random access memory applications in order to reduce the otherwise high power dissipation of pure ECL RAMs—the HI-BICMOS concept [10]. Here we will discuss briefly some of the methods used to implement bipolar transistors in CMOS technology.

Figure 13.17(*a*) shows a basic CMOS cross section diagram using a P substrate. The N well is used for the P channel MOS device. The P$^+$ and N$^+$ source and drain regions are either implanted or diffused junctions. The simplest arrangement for incorporating bipolar devices would be to use the N well for the

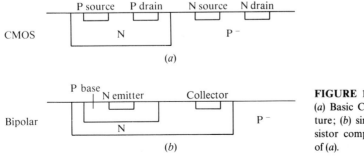

FIGURE 13.17
(*a*) Basic CMOS N well structure; (*b*) simplest bipolar transistor compatible with CMOS of (*a*).

bipolar collector and to choose junction depths for the P^+ regions to be compatible with base implants or diffusions for the bipolar device, which would then use the existing N^+ regions for emitters and collector contacts. Thus, in principle, it would be possible to have a bipolar process on a CMOS chip with no extra processing steps, as shown by the cross section diagram of Fig. 13.17(b). This assumes, however, that the emitter N^+ and base P^+ regions would be compatible with the CMOS source and drain requirements.

In general, for CMOS which is already in existence with a given set of design rules, source and drain implants/diffusions, and channel implants (for controlling threshold voltages), at least one additional step is required in order to incorporate a bipolar transistor on the same chip. Use of the N well for the collector, and the N^+ source and drain for the emitters/collectors with an extra "deep" P^+ diffusion for the bipolar base is one possibility; this is shown in Fig. 13.18(b). An alternate choice is shown in Fig. 13.18(c) where the existing P channel MOS P^+ diffusion is used for the bipolar base region, with a separate shallow N^+ implant/diffusion for the bipolar emitter.

In all of the above cases, the impurity profile of the N well can be tailored to give a compromise in collector doping between high breakdown voltage and low collector series resistance. The profile is sometimes referred to as triple diffused, as shown in Fig. 13.19. The compromise is evident from the study in Sec. 13.2.2, where the constant epitaxial layer doping can be replaced by an effective or average value of the slowly varying doping level associated with the deep N^+ well. In order to keep the horizontal component of collector resistance at an acceptably low level, the arrangement shown in Fig. 13.20 can be used as suggested by Zimmer et al. [11]. The bipolar transistor is rectangular with one emitter stripe, two base contact stripes, and two outside collector contact stripes. The reported resistance values for a 115 μm (direction collector to collector) by

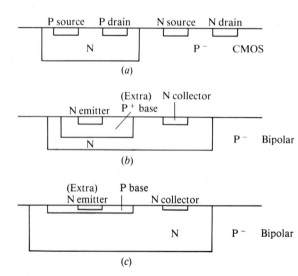

FIGURE 13.18
(a) CMOS structure; (b) compatible bipolar transistor using extra P^+ diffusion; (c) CMOS compatible bipolar transistor using extra shallow N^+ diffusion.

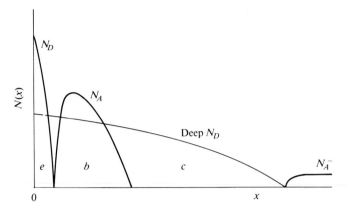

FIGURE 13.19
Triple diffused impurity profile for bipolar transistor in an N well of a CMOS process.

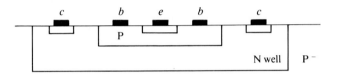

FIGURE 13.20
Double collector symmetrical layout for CMOS compatible bipolar transistor to reduce collector resistance.

200 μm transistor using a 7.5 μm deep N$^+$ well and a 2.3 μm deep base collector junction were 15 Ω for the collector and 5 Ω for the base [11].

Of course, if increased process complexity is allowed, high performance bipolar transistors can be incorporated in a CMOS chip. In [11], both N$^+$ and P$^+$ buried layers are used, with epitaxial collector regions and deep N$^+$ collector contact diffusions. The emitters are polysilicon contacted. The resulting performance is comparable to that obtained with high performance bipolar structures.

13.7   CONCLUSIONS

In this chapter we have studied aspects of bipolar transistor operation pertaining to integrated circuit structures. The base and collector resistances were evaluated for typical geometries. The lateral PNP transistor was analyzed and shown to have an inherently low f_t. The lateral PNP analysis was extended to the IIL structure and the basic characteristics of this logic family were examined. The chapter concluded with a brief overview of bipolar transistor operation in the widely used VLSI families: ECL and BICMOS.

PROBLEMS

13.1. The geometry of a double diffused integrated circuit transistor as shown in Fig. 13.1 is as follows. Emitter contact window, $6 \mu m \times 6 \mu m$; emitter diffusion mask, $18 \mu m \times 18 \mu m$; base diffusion mask, $30 \mu m \times 42 \mu m$; collector diffusion mask, $18 \mu m \times 30 \mu m$; P^+ isolation mask, $60 \mu m \times 96 \mu m$. The junction depths are 0.75 and 1.6 μm. The total epitaxial layer of 0.6 Ω cm resistivity is 9 μm thick, and the buried layer upward diffusion extends a distance of 6 μm. The total base diffusion has a sheet resistance of 300 Ω/sq. Estimate the value of total base resistance (intrinsic plus extrinsic) for two cases: (i) pinched base sheet resistance $R_{be \, sq} = 1k\Omega$/sq, (ii) $R_{be \, sq} = 10$ kΩ/sq.

13.2. For the transistor in Prob. 13.1, estimate the total collector resistance (vertical and horizontal components) if the buried layer sheet resistance is 50 Ω/sq. Repeat the calculation if the collector contact diffusion is placed on the alternate side of the base to that shown in Fig. 13.1. For the arrangement of Fig. 13.1, calculate the total collector resistance if the shallow N^+ collector diffusion is replaced by a deep diffusion. For each of these cases, determine the increase in $V_{CE \, sat}$ at a collector current of 2 mA.

13.3. For the device described by Fig. 13.1, with the data given above, estimate from the given information the vertical delay times and the maximum possible f_t of the transistor. Estimate the junction capacitance values for the normal forward base-emitter bias, and $V_{CB} = 5$ V (assume a capacitance of 10^{-7} F/cm^2 for the emitter-base junction at zero bias). Hence calculate and sketch f_t versus I_C.

13.4. In the above described transistor, if the substrate to collector bias is 5 V and the P type substrate is doped 10^{15} cm^{-3}, calculate the total substrate to collector capacitance. Hence calculate the additional pole created in the frequency response of an amplifier using the transistor, with a load resistance of 1, 10, and 50 kΩ. Compare to the pole corresponding to the β cutoff frequency for $\beta = 50$, when the transistor is biased for peak f_t.

13.5. Consider a lateral PNP transistor with the geometry of Fig. 13.9, made using the process and mask design rules given in the above questions. Calculate the maximum gain and f_t of the PNP device for three values of PNP emitter to collector mask spacing: 6, 4, and 3 μm. (*Note*. Be careful to include lateral diffusion effects.) Assume a V_{BC} bias of 5 V and an epitaxial layer lifetime $\tau_{epi} = 1$ μs. For other data, use the values given in the text examples.

13.6. Repeat the PNP calculations of Prob. 13.5 for the case where the residual epitaxial layer thickness is reduced (by suitable choice of epitaxial layer and buried layer processing) to $W_{epi} = 0.5$ μm and $W_{epi} = 0.2$ μm.

13.7. Consider a three-collector IIL cell made using the same mask design rules and process as described in the above questions. Calculate, making any reasonable approximations (i) minimum delay time, (ii) low current power delay product for a voltage swing of 0.6 V, (iii) current gain. Sketch the delay time versus power curve.

13.8. Repeat necessary parts of Prob. 13.7 for an IIL transistor with a residual epitaxial layer thickness of 0.5 and 0.2 μm.

13.9. For the IIL cell of Prob. 13.7, estimate the value of total base resistance from the injector rail side of the base to the furthest collector. Hence deduce the maximum operating current before the furthest collector becomes significantly debiased.

REFERENCES

1. A. B. Glaser and G. E. Subak-Sharpe, *Integrated Circuit Engineering*, Addison-Wesley, Reading, Massachusetts (1977).
2. A. B. Philips, *Transistor Engineering*, McGraw-Hill, New York (1962).
3. S. M. Sze, *Physics of Semiconductor Devices*, Wiley, New York (1981).
4. A. A. Eltoukhy and D. J. Roulston, "A complete analytic model for the base and collector currents in lateral PNP transistors," *Solid State Electronics*, vol. 27, pp. 69–75 (1984).
5. M. H. Elsaid and D. J. Roulston, "A method for the characterization of P^+NN^+ diodes using dc measurements," *IEEE Trans. Electron Devices*, vol. ED-15, pp. 1365–1368 (December 1978).
6. K. Hart and A. Slob, "Integrated injection logic—a new approach to LSI," *IEEE Jnl. Solid State Circuits*, vol. SC-7, pp. 346–351 (October 1972).
7. H. H. Berger and S. K. Wiedman, "Merged transistor logic—a low cost bipolar logic concept," *IEEE Jnl. Solid State Circuits*, vol. SC-7, pp. 340–346 (October 1972).
8. F. Hébert and D. J. Roulston, "Base resistance of bipolar transistors from layout details including two-dimensional effects at low currents and low frequencies," *Solid State Electronics*, vol. 31, pp. 283–290 (February 1988).
9. E.-F. Chor, A. Brunnschweiller, and P. Ashburn, "A propagation-delay expression and its application to the optimization of polysilicon emitter ECL process," *IEEE Jnl. Solid State Circuits*, vol. 23, pp. 251–259 (February 1988).
10. K. Ogiue, M. Odaka, S. Miyaoka, I. Masuda, T. Ikeda, and K. Tonomura, "13 ns, 500 mW 64 k bit ECL RAM using HI-BICMOS technology," *IEEE Jnl. Solid State Circuits*, vol. SC-21, pp. 681–685 (October 1986).
11. G. Zimmer, B. Hoeffinger, and J. Schneider, "A fully implemented NMOS, CMOS, bipolar technology for VLSI of analog-digital systems," *IEEE Trans. Electron Devices*, vol. ED-26, pp. 390–396 (April 1979).

CHAPTER
14

ADVANCED
TECHNOLOGY
DEVICES

14.1 INTRODUCTION

In this chapter we introduce the reader to devices made using "advanced technology." In some cases (polysilicon emitters) the technology is highly developed and incorporated in commercial structures; in other cases (gallium arsenide heterojunction devices) the technology is just emerging for high performance commercial applications.

We start the chapter by reviewing briefly the advanced technologies available to enhance the performance of conventional devices already discussed in the previous chapters. We then proceed to describe advanced devices including polysilicon and double polysilicon transistors, oxide walled devices, and transistors made using various etch techniques. The chapter concludes with an introduction to heterojunction transistors, specifically the GaAlAs/GaAs structure. It is not the intention to enter into extensive details of technology. The reader is referred to the appropriate comprehensive texts on this subject (e.g., [1], [2], [3]) and to the papers referenced in the following pages.

14.2 SUMMARY OF ADVANCED FABRICATION TECHNIQUES

Ion Implantation

For shallow junction devices, the conventional diffusion of impurities does not give sufficient control. Ion implantation is now widely used to introduce the

donor and acceptor atoms in carefully controlled doses, giving much more repro-
ducible profiles. The implanted profile is approximately gaussian in shape, but
with a peak concentration below the surface at a depth dependent on the high
energy implant and drive-in conditions. The implant step is always followed by
an anneal step to remove damage which occurs during implantation.

Polysilicon and Double Polysilicon Contacts

The polysilicon emitter has already been mentioned in Chap. 8, since it forms the
basis of most advanced commercially used transistors. Polysilicon is deposited
conveniently using *low pressure chemical vapor deposition* (LPCVD) techniques.
The polysilicon may be doped during deposition or subsequently. Use of P and
N type polysilicon for the same device enables double self-alignment to be
obtained, thus providing very small area devices.

Oxide and Other Isolation Methods

This has also been mentioned in Chap. 13. The technique is now standard for
high performance VLSI devices. Various isolation methods are available
(isoplanar, LOCOS, trench, selective epitaxial growth). The essential feature of
oxide isolation [1] is that instead of a deep isolation diffusion, a deep groove is
formed by wet or dry etch. This groove is then filled with SiO_2. The resulting
oxide isolation may be positioned on the final structure so that it intersects the
various diffusions, thus removing the sidewalls and their associated deleterious
junction capacitances, although this can create high leakage currents in some
cases. At the same time, a minimum active area is achieved.

Wet and Dry Etch Techniques

Use of various orientation-dependent etch techniques enables improved structures to be made. *Anisotropic reactive ion etch* is now widely used and enables
controlled vertical etches to be readily incorporated. KOH etching offers the
possibility of V-groove structures with sidewalls at a well-defined angle to the
vertical.

Rapid Thermal Anneal (RTA)

Although in itself not intrinsic to novel device structures, RTA is a technique
which is acquiring widespread use as a means of annealing shallow junction
devices. Annealing times of less than one minute are used and this requires the
use of special equipment both to enable the rapid rise and fall in temperature to
be created, and also to enable the very short duration temperature changes to be
monitored.

Molecular Beam Epitaxy (MBE) and Metal Organic Chemical Vapor Deposition (MOCVD) Systems

Controlled growth of atomic layers of crystal enable concepts to be implemented which formerly could only be imagined on paper. Molecular beam epitaxy and metal organic chemical vapor deposition are two methods now used to deposit layers of semiconductor with different characteristics (the so-called superlattice structures). In particular, it is possible to deposit layers of gallium arsenide with variable doping and bandgap (through the addition of controlled fractions of aluminum). This enables the fabrication of heterojunction bipolar structures, with different bandgaps in the emitter and base regions.

14.3 POLYSILICON AND DOUBLE POLYSILICON STRUCTURES

In Chap. 8, we looked at the theoretical performance obtained using doped polysilicon as a contact to the N^+ emitter of a bipolar transistor. Let us here review briefly the fabrication steps of this single polysilicon device. Figure 14.1 shows the basic process steps involved.

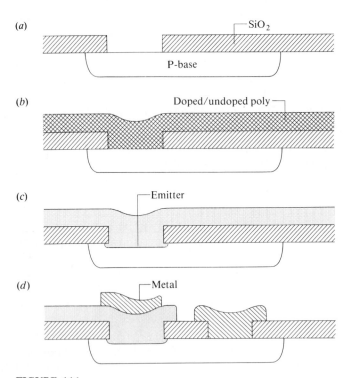

FIGURE 14.1
Basic fabrication steps for the single poly process. (a) Window opened in oxide; (b) undoped poly-silicon deposited over whole wafer; (c) dopant atoms introduced by ion implantation and diffusion; (d) unwanted polysilicon removed, contacts formed.

The process is normal up to formation of the boron doped base region. The first step in the polysilicon emitter process consists in an oxide layer being grown and a window opened up using normal photolithographic techniques. The surface is cleaned (for example, using an HF dip etch), and then an undoped polysilicon layer is deposited (for example, using low pressure chemical vapor deposition). The polysilicon is then ion implanted with donor impurities (normally arsenic, although phosphorus can also be used). The dopant atoms are then used to create the base emitter junction by diffusion to a depth of order $0.1 \, \mu m$ below the monosilicon surface. The final steps consist of removal of unwanted polysilicon and formation of the metal contact.

As mentioned in Chap. 8, a very thin oxide layer will normally exist between the polysilicon and monosilicon regions. Experiments have shown that this has an almost negligible effect when an HF dip etch is used, but when an RCA clean is used prior to polysilicon deposition, an oxide layer of about 14 angstroms is formed; as we saw in Chap. 8, this can have a very significant effect on the electrical characteristics.

Variations on the basic process, of which only the most essential steps are shown here, consist in: (i) in situ doping of the polysilicon during deposition—this technique has been shown [4] to yield layers with extremely low sheet resistance and very shallow junction depths in the monocrystalline silicon (less than $0.05 \, \mu m$); (ii) diffusing the dopant from the surface (with no implant)—this is the simplest technique but gives very poor control of the dopant profile in the monocrystalline region. Note that the final metallization pattern shown in Fig. 14.1(d) can be located in any position, not necessarily directly over the emitter—a direct consequence of the "self-alignment" feature for which the process was originally developed. It is of course obvious that if the metal contact is offset too much, an additional series resistance will appear, which will degrade the performance.

The two N type dopants, arsenic and phosphorus, currently used give essentially similar results; a minor difference [5] appears to be a reduced amount of recombination current in the monocrystalline region for phosphorus doping compared to arsenic doping when using the RCA clean. This results in a higher gain for the phosphorus-doped RCA clean device.

Although reliable high performance polysilicon structures are now fabricated routinely in VLSI circuits, there is still some uncertainty concerning the transport mechanisms in the polysilicon layer and in the nature of the grains and grain boundary and interface regions. *High resolution electron microscopy* techniques [6] can provide valuable insight into the effects of various heat treatments on the grain and interface properties, and thus assist in developing valid electrical models for current transport.

Double self-aligned polysilicon bipolar transistors are now standard in VLSI applications. In these structures, the emitter is self-aligned to the base contacts, thus giving a greatly reduced total surface for the device. Several schemes are in use. Sakai et al. [7, 8] were among the first to propose a double self-aligned structure using silicon nitride and selective etch techniques. Figure 14.2 shows the main steps in their SST-1A process [8]:

(i)

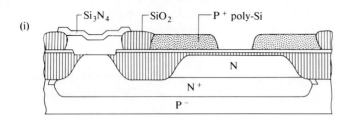

(ii)

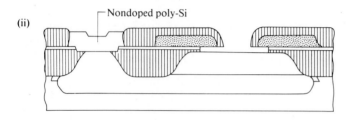

(iii)

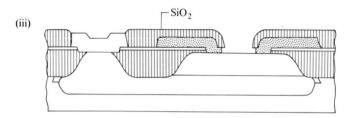

(iv)

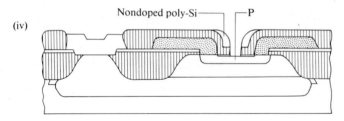

(v)

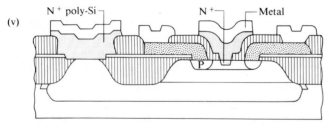

FIGURE 14.2
Essential process steps for SST-1A double self-aligned technology. (*After Sakai et al.* [8]. *Reproduced with permission* © *1983 IEE.*)

(i) SiO_2, Si_3N_4, and polysilicon are formed consecutively and the unnecessary parts of the polysilicon are then oxidized. The polysilicon is removed by etching in the area defined to be the active device region.

(ii) The polysilicon is oxidized, the Si_3N_4 is side-etched, and the SiO_2 etched away.

(iii) Polysilicon is deposited and etched away, filling up the space under the overhanging part of the first polysilicon layer.

(iv) SiO_2 is formed by thermal oxidation, and the base area is formed by B^+ ion implantation through the thin SiO_2. Then CVD SiO_2 and polysilicon are deposited consecutively. The emitter window is opened using a dry etch.

(v) The emitter is now formed by arsenic diffusion through arsenic implanted polysilicon.

The resulting structure gives the self-aligned emitter device combined with an emitter self-aligned to the base contacts and separated by the so-called "spacer oxide" of order 0.5 μm. In 1983, the performance figures quoted for a 0.5×5 μm^2 emitter were $f_t = 12.4$ GHz at $V_{CE} = 3$ V. A propagation delay time of 90 ps was reported at 0.59 mW/gate.

Tang et al. [9] and Ning et al. [10] proposed an alternate method of double self-alignment. Figure 14.3 shows the essential steps of this process.

• First, an in situ doped P^+ polysilicon layer and an SiO_2 layer are deposited as shown in diagram (a).

• Next the SiO_2 and polysilicon layers are patterned using reactive ion etching in CF_4 completely through the SiO_2 layer and partially through the polysilicon layer. Patterning of the polysilicon layer is then completed using a preferential etch; this etches the heavily doped polysilicon and not the lightly doped substrate as shown in diagram (b).

• An oxide layer of about 20 nm is now grown thermally, followed by deposition of a thicker 200 nm layer as shown in diagram (c).

• The extrinsic base is now formed by diffusion of boron from the P^+ polysilicon layer.

• The two oxide layers are now removed everywhere except on the polysilicon sidewalls; this is done using a maskless reactive ion etch in $CF_4 + H_2$, which has a very low etch rate for silicon. This opens the emitter window and leaves a spacer oxide of order 0.4 μm which will separate the emitter from the base contacts.

• The intrinsic base and emitter are finally formed by suitable steps involving ion implantation to obtain the required shallow junctions.

In present-day technology, the emitter is invariably formed using methods similar to one of the above-mentioned polysilicon techniques. Ning et al. [10] report an emitter-to-collector area ratio of only 2:1 for the case of one base contact (3:1 for two base contacts) thus giving a significant reduction in

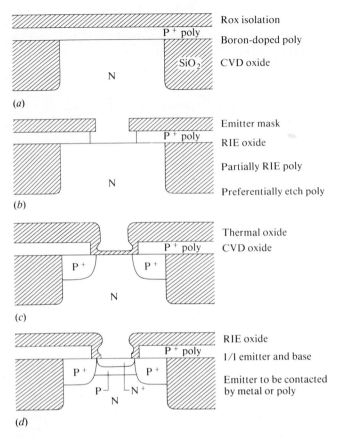

Rox isolation

Boron-doped poly

CVD oxide

(a)

Emitter mask

RIE oxide

Partially RIE poly

Preferentially etch poly

(b)

Thermal oxide
CVD oxide

(c)

RIE oxide

1/1 emitter and base

Emitter to be contacted
by metal or poly

(d)

FIGURE 14.3
Essential process steps for alternate process providing double self-aligned technology. (*After Ning et al.* [10]. *Reproduced with permission © 1981 IEEE.*)

"unwanted" parasitic collector junction capacitance for nonsaturating logic such as ECL, but also reducing considerably the excess stored charge in saturating logic such as TTL.

Other schemes have been suggested for achieving the goal of self-aligned base contacts and emitters. Cuthbertson and Ashburn [11] suggested using concentration-dependent oxidation properties as shown in Fig. 14.4. In this scheme, a light boron implant is carried out for the base region and an undoped 0.4 μm layer of polysilicon is deposited as shown in diagram (a). Arsenic is implanted into the polysilicon at low energy so that no penetration to the underlying silicon occurs. The polysilicon layer is now patterned and wet oxide grown. Due to the large differential oxidation rate between lightly doped polysilicon and the surrounding lightly doped extrinsic base, a thick oxide is selectively grown over the emitter region as shown in diagram (b). A heavy boron implant is now performed, penetrating through the very thin oxide region to form the extrinsic

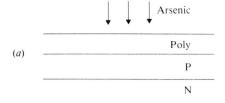

(a)

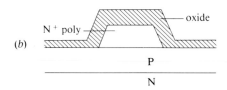

(b)

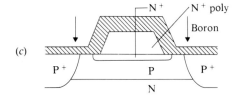

(c)

FIGURE 14.4
Details of a novel double self-aligned technology.
(*After Cuthbertson and Ashburn* [11]. *Reproduced with permission* © *1985 IEEE.*)

base. The extrinsic base is thus automatically aligned to the emitter, as shown in diagram (c).

All of the above schemes provide self-alignment of the base contact with the emitter and make use of a spacer oxide which determines the actual spacing between the emitter region and the base contact.

14.4 EXPERIMENTAL STRUCTURES USING POLYSILICON AND ETCH PROCESSES

The purpose of this section is to introduce the reader to the possibilities offered by combining some of the silicon technologies now available. The three structures chosen as examples are solely experimental in nature but give a good idea of improvements possible through area and sidewall reduction for given design rules.

14.4.1 V-Groove Bipolar Transistor

When $\langle 100 \rangle$ oriented silicon is etched in aqueous KOH, a V-groove shaped etch is formed whose sidewalls are at an angle of 55 degrees with the horizontal, as shown in Fig. 14.5. Provided the etch is stopped before it reaches the self-limiting V shape, a flat bottom is formed whose dimensions are determined by the depth of the etch and the original oxide (mask) opening. The length of the flat bottom,

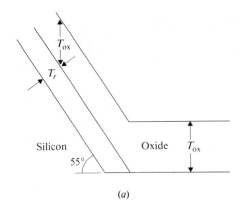

(a)

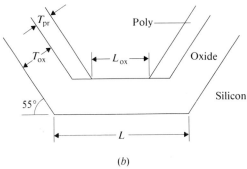

(b)

FIGURE 14.5
(a) Formation of V groove and oxide for novel bipolar technology; (b) use of sacrificial polysilicon layer for V-groove structure. (*After Solheim et al.* [35]. *Reprinted with permission from Solid State Electronics, Pergamon Press PLC.*)

L, is given by

$$L = L' - 2 \frac{X_g}{\tan 55}$$

where L' is the length of the oxide cut determined by the mask and X_g is the depth of the V groove. If a uniform thickness of oxide is deposited or thermally grown, the vertical thickness of the sidewall oxide will be greater than that of the layer on the bottom. If an unmasked *reactive ion etch* (RIE) step is now performed, some oxide will remain on the V-groove sidewalls when the bottom oxide is removed. For a completely anisotropic RIE process, 40 percent of the original oxide will be left on the walls. The remaining bottom region can now be used for the formation of the bipolar transistor with considerably reduced dimensions (from L to L') compared to those of the original mask-determined oxide opening.

An alternative approach is to use a sacrificial layer of polysilicon to reduce the dependency on the anisotropicity of the RIE process as shown in Fig. 14.5(*b*). This polysilicon layer is deposited after oxidation of the V groove. Because of the conformal nature of the polysilicon deposition, the vertical thickness on the sidewalls is greater than on the bottom of the V groove. Performing the unmasked

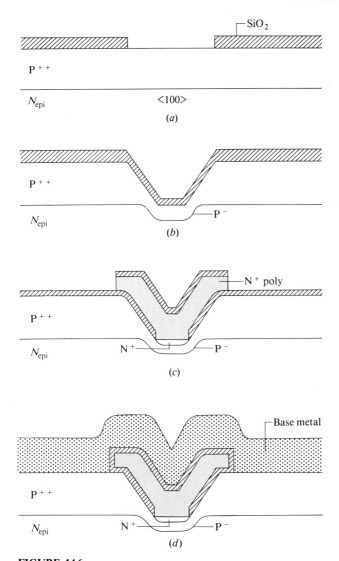

FIGURE 14.6
Process steps for fabrication of a V-groove bipolar transistor. (*After Hébert et al.* [12]. *Reprinted with permission from Solid State Electronics, Pergamon Press PLC.*)

RIE step leaves a polysilicon layer only over the sidewall oxide. This mask is then used to protect the sidewall oxide during the etch of the oxide on the V-groove bottom, resulting in 100 percent of the original thickness remaining after the etch; this gives consequent reduction in parasitic MOS capacitance. Also since the selectivity between SiO_2 and polysilicon is very large (infinite for HF and at least 10 for CF_4 and H_2) the required mask thickness is minimized, thus reducing the requirements for RIE anisotropicity. The endpoint detection of the RIE is also simplified using this approach. There is a large reduction of polysilicon area at the end of the etch of the sacrificial layer; analysis of the exhaust gases for endpoint detection is thus facilitated. Yet another advantage of the use of the polysilicon is a further reduction in minimum feature size [12] as can be deduced from Fig. 14.5(b).

Figure 14.6 shows the steps used in the fabrication of a V-groove bipolar transistor. The deep extrinsic base region is formed conventionally. The emitter V groove is then created to a depth approximately equal to the depth of the extrinsic base diffusion. The intrinsic base is then implanted or diffused and the V groove is oxidized. The previously described steps are implemented, leaving the bottom of the V groove ready for N^+ polysilicon deposition for the emitter. In order to form the self-aligned base contacts, the polysilicon is patterned using RIE and the etch continued until the extrinsic base is exposed. The polysilicon sidewalls are then oxidized and RIE again used to clear the extrinsic base region, thus providing oxide spacers on the polysilicon sidewalls. The base metal covers the emitter region and the N^+ polysilicon is contacted at the side of the device. Figure 14.7 shows plan and cross section diagrams of the completed structure. The advantages of this process include feature size reduction, overall reduction in

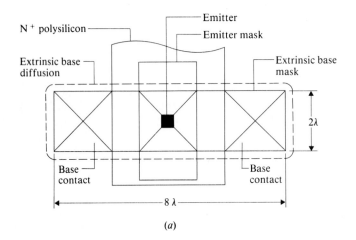

(a)

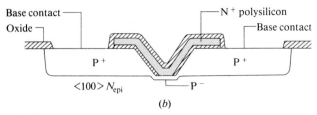

(b)

FIGURE 14.7
Completed V-groove bipolar transistor. (*After Solheim, Hébert, and Roulston* [35]. *Solid State Electronics, reprinted with permission, Pergamon Press PLC.*)

device size, reduced extrinsic base resistance, no extrinsic base encroachment, and self-aligned extrinsic base. The major disadvantage is that the process is non-planar.

14.4.2 Base Etched Self-Aligned Transistor Technology (BESTT)

This process uses a single layer of polysilicon. The emitter can be walled on all four sides, thus eliminating completely this component of capacitance. The base and emitter contacts are self-aligned, thus yielding a minimum area transistor without the need for two polysilicon layers. Figure 14.8 shows the process steps [13]. Fabrication up to formation of the base region is conventional. The base window is opened and the intrinsic base formed by implant or diffusion. A blanket etch is then carried out to remove any oxide grown during base drive-in. N^+ polysilicon is now deposited and oxidized as shown in diagram (a). The poly-silicon is patterned using RIE and the etch is continued past the penetration of

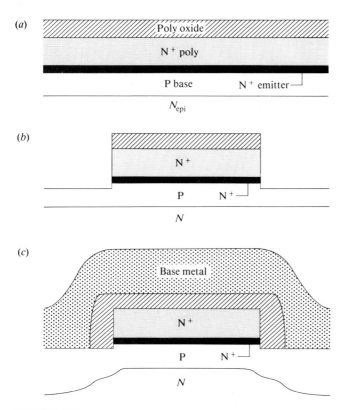

FIGURE 14.8
Process steps for the base-etched self-aligned transistor technology (BESTT) structure. (*After Solheim et al.* [13].)

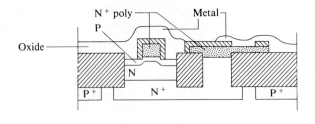

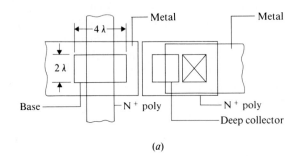

(a)

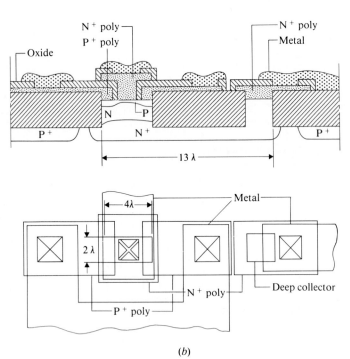

(b)

FIGURE 14.9
Comparison of BESTT and double polysilicon structures; λ is the minimum feature size. (*After Solheim et al. [13].*)

the emitter into the monocrystalline silicon as shown in diagram (*b*). The poly-silicon oxide sidewall spacers are then formed by oxidizing the wafer and per-forming an unmasked RIE step to remove the oxide on the base contact. Because of the anisotropicity of the reactive ion etch, the oxide spacer is left on the poly-silicon sidewalls. Since the RIE is stopped when the extrinsic base region is cleared, this also forms base contacts which are self-aligned to the edge of the base windows and the spacer oxide. The base link-up region is critical and can be tailored by implants before and after formation of the sidewall spacer [13]. A good link is ensured due to the lateral spread of the extrinsic base implant, even for very narrow base widths. The base metal covers the emitter electrode and the emitter polysilicon is contacted at the side of the device.

Since the emitter is defined by the polysilicon etch, the emitter extends to the edge of the base window. This gives a minimum area base window of $2\lambda \times 4\lambda$ for minimum contact dimensions of $2\lambda \times 2\lambda$, even if the base diffusion is not walled. This compares to the double polysilicon double self-aligned process for the case of a walled base diffusion.

Figure 14.9 shows a BESTT device and a double poly device for compari-son. It is seen that the collector area of the BESTT device is 62 percent of the area of the double polysilicon device. This process provides a structure which is comparable to the more commonly used single base contact double polysilicon device.

14.4.3 Polysilicon Base (Pedestal) Transistor

Here we describe a transistor using a boron-doped polysilicon layer for the intrinsic and extrinsic base regions giving excellent upward (inverse) character-istics. The process uses standard polysilicon deposition plus doping-dependent recrystallization.

Various authors [14] have reported the reduction of parasitic junction capacitance using self-alignment and by locating the extrinsic base region on top of a thin oxide layer. The latter method normally requires high temperature epi-taxial layers to be grown for the base. Here we describe a similar process but using a polysilicon rather than an epitaxial base; the main advantage is process simplification, coupled with good performance. Figure 14.10 shows the completed structure. The device is fabricated on a standard epitaxial layer. A field oxide is

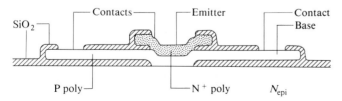

FIGURE 14.10
Pedestal emitter bipolar structure. (*After Hébert* [16].)

then grown (to approximately 0.5 μm thickness) and the intrinsic base is patterned using chemical etching of the oxide. An undoped LPCVD polysilicon layer of approximately 0.5 μm is then deposited at about 590°C [16].

Since undoped polysilicon does not easily realign epitaxially, whereas phosphorus-doped polysilicon layers do [15], a low concentration of phosphorus is introduced into the undoped polysilicon layer in order to enhance recrystallization and epitaxial realignment with the substrate. Rapid thermal anneal is used to recrystallize the polysilicon film [17].

The base layer is now boron doped to provide a low extrinsic base sheet resistance (of order 150 Ω/sq). Patterning in KOH gives a good idea of the crystalline nature of the base layer. Typical doped or undoped polysilicon films etch in about 4 min, while the base of this pedestal transistor requires 42 min to etch; this indicates that the layer morphology has been altered significantly. The combination of phosphorus doping and RTA is responsible for the changed layer morphology, since the undoped polysilicon on a test wafer annealed for 120 min at 1150°C cleared in 13 min using KOH.

The emitter window is defined using patterning of the base oxide followed by deposition of approximately 0.4 μm of in situ phosphorus-doped polysilicon. After RTA for activation of the emitter doping, the contact windows and metallization steps are carried out.

14.5 OXIDE ISOLATION

We have mentioned, when dealing with integrated devices, the advantages to be gained from using some form of oxide isolation. Figure 14.11 shows the basic integrated transistor using oxide isolation.

The basic LOCOS and ISOPLANAR oxide isolation schemes use a nitride layer as a barrier to the growth of silicon dioxide [1]. Isolation regions are etched to a depth of about 55 percent of the desired oxide thickness. Steam oxidation is then used to grow SiO_2 until the oxide is level with the surface of the wafer. One of the problems with this technique is the uncontrolled "bird's beak" shape at the transisition between the recessed oxide region and the silicon nitride mask. This is particularly severe in small geometry VLSI structures. Various forms of

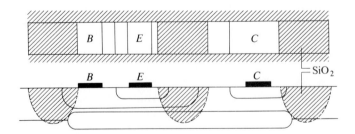

FIGURE 14.11
The oxide isolated transistor. (*a*) Plan view; (*b*) cross section view.

dielectric isolation have been proposed and are now used. Trench isolation [18] is based on the use of reactive ion etch, and the final structure consists of poly-silicon, silicon dioxide (and nitride). Other variations have been developed [19, 20]. The goal in each case is the elimination of the parasitic capacitance associated with the conventional junction isolation and a reduction of "wasted" area between individual transistors. Figure 14.12 shows cross section diagrams for two processes.

In general, two or three sides of the emitter are "walled," plus all four sides of the base; in addition, the deep P^+ junction isolation is replaced by oxide isolation. This results in a very substantial reduction in overall junction capacitance due to elimination of the various sidewall components and a large part of the otherwise "wasted" area. Reported results [21] for a polysilicon emitter oxide walled device with a $3.25 \times 3.5 \ \mu m^2$ emitter, e-b junction depth of 0.145 μm, c-b

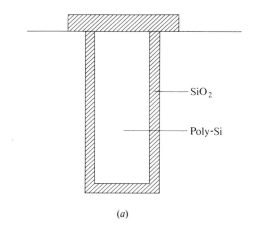

(a)

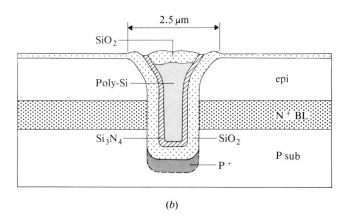

(b)

FIGURE 14.12
(a) Trench isolation; (b) U-groove isolation. (*After Hayasaka et al.* [20]. *Reproduced with permission* © *1982 IEEE.*)

Base   Emitter   Collector

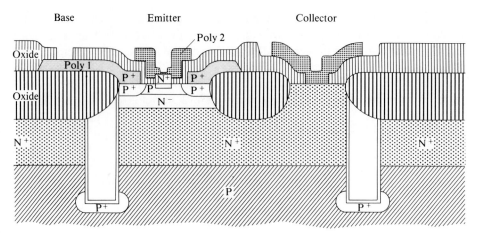

FIGURE 14.13
Cross section of an advanced 1 μm industrial bipolar transistor design. (*After Wilson et al.* [22]. *Reproduced with permission* © *1988 IEE.*)

junction depth of 0.415 μm, are $C_{je} = 16$ fF, $C_{jc} = 21$ fF, $C_{sub} = 25$ fF, at zero bias and the f_t peaks at 5.5 GHz.

Figure 14.13 shows a cross section of an advanced industrial transistor [22] using 0.5 μm *low temperature oxide* (LTO) isolation plus a 5.5 μm trench. The trench sidewall has thermal oxide, CVD nitride, and a layer of LTO, and the trench is filled with polysilicon. The transistor has a polysilicon emitter and the emitter-base isolation uses an L-shaped composite dielectric sidewall spacer. The published f_t of this structure with a 1 μm × 5 μm emitter peaks at 14 to 22 GHz, and the capacitance values are given as $C_{je} = 19$ fF, $C_{jc} = 17$ fF, $C_{sub} = 30$ fF.

14.6   SCHOTTKY TRANSISTOR LOGIC AND THE EFFECT OF OXIDE ISOLATION

This section serves to illustrate both a novel bipolar logic family called STL for *Schottky transistor logic* [34], and also the effect of varying the spacing between the oxide "wall" and the emitter diffusion.

Figure 14.14 shows the circuit and cross section view of an STL inverter. The inverter may be thought of as deriving from IIL. The logic swing is the difference in threshold voltages, V_{SB1}, V_{SB2}, of the two Schottky diodes (function of areas and barrier heights). This may be seen by considering a logic high V_H at the input to the base. For a collector voltage $V_{CE\,sat}$ the logic low output voltage is

$$V_L = V_{CE\,sat} + V_{SB2}$$

But the collector voltage is equal to

$$V_{CE\,sat} = V_H - V_{SB1}$$

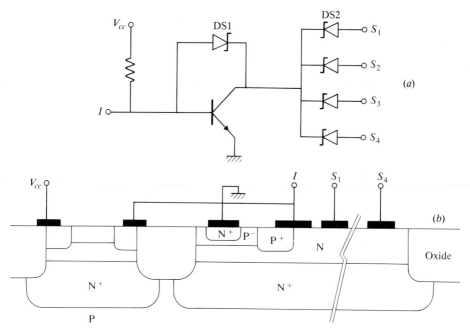

FIGURE 14.14
(a) Circuit of a Schottky transistor logic inverter; (b) cross section diagram of structure. (*After Roulston and Depey* [23]. *Reproduced with permission* © *1982 IEE.*)

Hence it is clear that the logic swing is given by

$$V_H - V_L = V_{SB1} - V_{SB2}$$

The Schottky barriers are chosen by using suitable alloys, for example, PtSi for the clamp, TiW for the output diodes. The advantage of a low logic swing is, as we saw in IIL, a low power delay product. The transistor is in this case a normal (downward operating) NPN giving optimum performance. The pull-up resistor must be incorporated with each inverter, unlike IIL. This means a larger area and also necessitates the choice of operating point (bias current) to be made before fabrication. If the performance is calculated for various resistance values, corresponding to different current levels, a curve of delay versus power similar to that for IIL is obtained.

Effect of Oxide Wall on an STL Logic Gate

Figure 14.15(a) shows a structure used in an advanced Schottky transistor logic (STL) gate. The spacing between the emitter oxide wall and the emitter diffusion edge was varied in a computer simulation and the propagation delay time t_{pd}, determined numerically in each case. Figure 14.15(b) shows the resulting variation of delay time versus spacing [23]. The initial rise in t_{pd} as the spacing increases from zero is due to the introduction of the sidewall capacitance on three

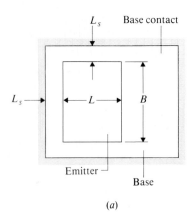

(a)

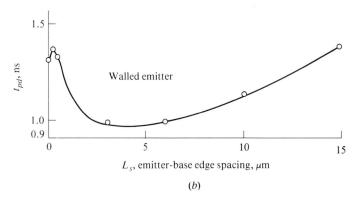

(b)

FIGURE 14.15
(a) Walled emitter transistor with variable spacing between emitter diffusion and oxide region; (b) computed propagation delay time versus spacing. (*After Roulston and Depey* [23]. *Reproduced with permission* © *1983 IEE.*)

sides. Subsequent reduction of the delay occurs as the spacing around the emitter (in the extrinsic base region) increases; this is due entirely to the reduction in total base resistance. For a spacing above about 5 microns the delay gradually starts to increase again. This is because there is no longer a significant reduction in total base resistance, but there is a steady increase in base-collector capacitance due to the increased junction area. The same general trends would be observed for maximum oscillation frequency, in which r_{bb} and C_{jc} are two of the three determining parameters.

14.7 THE P^{++} EXTRINSIC BASE REGION

The use of an extra heavily doped P^{++} base implant/diffusion becomes essential for very shallow junction devices. This is because the intrinsic base is formed by a shallow implant and its sheet resistance is quite high (of order 1000 Ω/sq). Figure 14.16 illustrates the structure. The inclusion of the P^{++} extrinsic implant step

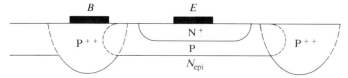

FIGURE 14.16
The P^{++} extrinsic base structure.

introduces a new set of problems [24]. Firstly, in order to keep the extrinsic base resistance to a minimum, the P^{++} region should overlap the normal N^+ emitter. This has two drawbacks, however: firstly an inevitable increase in sidewall capacitance, secondly a decrease in emitter-base breakdown voltage and increase in leakage currents. In fact, in extreme cases, with very abrupt transitions between the P^{++} and N^+ regions, tunneling currents have been observed [25]. The presence of such base current degrades the performance at low bias. Ideally, therefore, the extrinsic base implanted layer should almost touch, but not overlap, the normal emitter region.

14.8 GALLIUM ARSENIDE AND HETEROJUNCTION BIPOLAR TRANSISTORS

Although most commercial bipolar transistors are currently made using silicon, a significant effort over the past several years has been put into the use of gallium arsenide, initially as a substitute for silicon and more recently to make GaAlAs-GaAs heterojunction devices. This latter has become possible with the advent of molecular beam epitaxy (MBE) and metal organic chemical vapor deposition (MOCVD) techniques, which allow very fine control of the material deposited in successive layers of the structure.

The use of GaAs as a substitute for silicon for otherwise "conventional" NPN transistors originated with the idea that the higher electron mobility (five to seven times that of silicon) and peak drift velocity (twice that of silicon) would provide lower neutral base and base-collector space-charge layer transit times, with a consequent improvement in f_t. This fact is indisputable; however, the hole mobility of GaAs is slightly lower than that for Si for doping levels up to near 10^{18} cm^{-3} and the relative dielectric constant is 10 percent higher [26], so the vastly increased cost of manufacturing GaAs devices as opposed to silicon made it unattractive economically for the rather slight overall performance advantage.

The availability of layer-by-layer control of deposited material created the possibility of altering the composition in such a way that the energy bandgap could be altered from emitter to base. In particular, the use of GaAlAs with a carefully controlled fraction of aluminum enables the bandgap in the emitter to be made several tenths of an electronvolt higher than in the GaAs base. The actual values of band-edge changes are given for GaAl$_x$Ga$_{1-x}$As by [27]

$$\Delta E_g = 1.25x \qquad \text{where } x < 0.4 \qquad (14.1)$$

with the band-edge changes being

$$\Delta E_c = 0.85 \ \Delta E_g$$

$$\Delta E_v = 0.15 \ \Delta E_g \qquad (14.2)$$

Figure 14.17(a) shows the band diagrams for a typical wide-gap GaAlAs layer and a GaAs layer. χ_e and χ_b are the electron affinities in the GaAlAs and GaAs, respectively; ϕ_e, ϕ_b are the respective work-functions; E_{ge}, E_{gb} are the respective bandgaps; and E_f is the Fermi level. Figure 14.17(b) shows the combined band diagram in thermal equilibrium. Since the Fermi levels align, the built-in barrier is the difference in the work-function levels. Note that for the typical case shown, a discontinuity in the form of a spike and a notch are formed in the conduction band edge. Although there is necessarily a corresponding discontinuity in the valence band edge, it appears simply as a sudden "jump" in E_v. The barrier to electrons qV_{bin} is shown on the diagram.

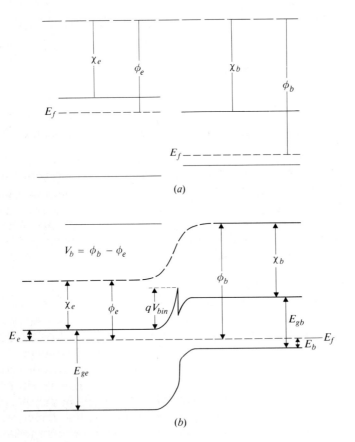

FIGURE 14.17
Energy band diagram for a GaAlAs/GaAs emitter-base heterojunction. (a) Separate diagrams; (b) combined diagram.

The magnitude of the total built-in barrier can be deduced from the diagram as follows: the difference in work functions

$$qV_{bi} = \phi_b - \phi_e$$
$$= E_{cb} - E_{ce} + \Delta E_c$$

where E_{cb}, E_{ce} are the conduction and valence band edges shown in the diagram and ΔE_c is the shift in the conduction band edge. If we define

$$E_e = E_f - E_{ce}$$
$$E_b = E_{cb} - E_f$$

the above may be rearranged thus:

$$qV_{bi} = (E_{cb} - E_f) - (E_{ce} - E_f) + \Delta E_c$$
$$= (E_{gb} - E_b) - E_e + \Delta E_c$$
$$= E_{ge} - \left(\frac{kT}{q}\right) \ln \left(\frac{N_{Ab}}{N_{vb}}\right) - \left(\frac{kT}{q}\right) \ln \left(\frac{N_{De}}{N_{ce}}\right) \qquad (14.3)$$

where N_{vb}, N_{ce} are the effective densities of states for the base and emitter materials, and N_{Ab}, N_{De} are the respective doping levels.

Although quantum mechanical tunneling effects would be present in such an abrupt heterojunction, if a slight gradient is introduced as shown in Fig. 14.18, these discontinuous features tend to disappear. The basic performance in the ideal case can be understood as follows. The high bandgap emitter gives a low

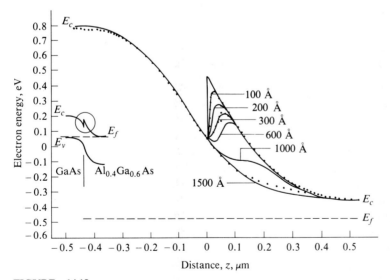

FIGURE 14.18
Effect of grading on the conduction band edge, for a GaAs-Al$_{0.4}$Ga$_{0.6}$As heterojunction; the doping level on both sides of the junction is 5×10^{15} cm^{-3}. (*After Cheung et al.* [28]. *Solid State Electronics, reprinted with permission, Pergamon Press PLC.*)

intrinsic carrier concentration [recall from Chap. 1 that $n_i^2 \propto \exp\left(-E_g/kT\right)$]. The injected carrier (hole) concentration is thus decreased since $p(0) \propto n_i^2/N_D$. The gain is thus (ideally) increased by a factor $\exp\left(\Delta E_g/kT\right)$ compared to a homojunction with the same doping levels.

For example, a 0.35 eV increase in bandgap gives a reduction of order 10^{-6} in hole current injected into the emitter. This means that the emitter can be doped at a fairly low level and still maintain negligible hole injection. The advantage of this is a lower emitter-base depletion-layer capacitance and in several devices reported in the literature the emitter is doped at around 10^{17} cm^{-3} for a base doping of order 10^{18} cm^{-3} (high in order to obtain a low base resistance value).

The capacitance of the emitter-base junction may be derived from the solution to Poisson's equation, as in Chap. 2. However, since the doping levels on each side of a typical heterojunction may be comparable, but not equal, the general formula must be used, as opposed to that for a one-sided abrupt junction or a symmetrical junction. Furthermore, the use of different bandgap materials means that the relative dielectric constant will not be the same in both regions. This leads to a slightly modified result for depletion-layer thickness on the two sides. Following the derivation contained in Eqs. (2.13) to (2.17), with the electric field at the junction now being given by Eq. (2.16) and a corresponding expression on the P side, but with different values of dielectric constant, it is a simple matter, by equating the charge on either side, to derive the depletion-layer thickness for the N type emitter

$$d_n = \sqrt{\frac{2N_{Ab}\,\varepsilon_p\,\varepsilon_n\,V_{jt}}{qN_{De}(\varepsilon_p\,N_{Ab} + \varepsilon_n\,N_{De})}} \qquad (14.4)$$

and for the P type base

$$d_p = \sqrt{\frac{2N_{De}\,\varepsilon_p\,\varepsilon_n\,V_{jt}}{qN_{Ab}(\varepsilon_p\,N_{Ab} + \varepsilon_n\,N_{De})}} \qquad (14.5)$$

where ε_n, ε_p are the dielectric constants of the material on the N and P sides, respectively, and V_{jt} is the total junction voltage (determined by the built-in barrier voltage and the applied bias as discussed in Chap. 2). The value of the dielectric constant for the wide bandgap emitter may be obtained from the empirical expression [37]:

$$\varepsilon_n = (13.1 - 3.0x)\varepsilon_p/13.1 \qquad (14.6)$$

where $\varepsilon_p = \varepsilon_{GaAs} = 1.16 \times 10^{-12}$ F/cm and x is defined as for Eq. (14.1).

The ratio of the total voltage on either side of the junction is now given by

$$\frac{V_p}{V_n} = \frac{N_D\,\varepsilon_n}{N_A\,\varepsilon_p} \qquad (14.7)$$

where the total junction voltage is given by

$$V_{jt} = V_p + V_n = V_{bi} - V_a \qquad (14.8)$$

The junction capacitance is most readily derived by considering the charge in one side of the depletion layer, for example, $Q_p = q d_p N_{Ab} A$ and taking the derivative $C_j = dQ_p/dV_{jt}$. This gives the result

$$C_j = A \sqrt{\frac{q \varepsilon_p \varepsilon_n N_{Ab} N_{De}}{2(\varepsilon_p N_{Ab} + \varepsilon_n N_{De}) V_t}} \qquad (14.9)$$

The major electrical properties of the bipolar heterojunction transistor are thus obtained. For the (common) case of uniform doping in each region, it is a straightforward matter to calculate the other parameters such as vertical delay times and base resistance, using the results obtained in Chaps. 7 and 9. High current gain and f_t falloff are determined by the effects already dealt with in Chap. 10.

In practice, the collector current is often dominated by the effect of the conduction band spike. Marty et al. [29] have shown that this has the effect over part of the current-voltage range of giving an ideality factor somewhat greater than unity. $m \sim 1.1$ has been quoted in [29] over quite a wide range. Furthermore, a large recombination current is associated with the transition region thus degrading the current gain. Nevertheless, quite acceptable values of current gain have been obtained in practical devices. The high electron mobility in the base referred to above ensures a base transit time about one-fifth that for a comparable silicon structure, and high f_t values are therefore obtainable.

Figure 14.19 shows typical GaAlAs/GaAs transistors [30]. The discrete case (a) uses simple etch-processing for large area devices. For integrated circuit applications, the semi-insulating properties of GaAs can be used for the substrate, with an N^+ "buried layer" for the collector low resistance path. Note that the possibility exists of having either a wide-gap collector or a GaAs collector. The choice has implications from a circuit point of view, specially in logic inverters where the V_{CE} offset voltage is determined by the differences between the e-b and c-b junctions, particularly in respect of their built-in barriers.

Some excellent high frequency performance has been reported on experimental GaAlAs/GaAs heterojunction transistors, with f_t values in excess of 40 GHz for base widths of order 0.1 μm [31].

Figure 14.20 shows the impurity profile of an MBE GaAlAs/GaAs transistor with junction depths of 0.15 and 0.30 microns. The doping levels are 5×10^{17} cm^{-3} for the emitter and 5×10^{18} cm^{-3} for the base. The epitaxial layer is doped 2×10^{16} cm^{-3} to a depth of 1 micron, where the N^+ substrate starts. The emitter is a 1 micron wide stripe, 10 microns long. This is similar to the experimental device reported in [31] and uses self-aligned base contacts, a graded Al fraction, and proton bombardment for isolation under the base contact. Reported values are $f_t = 50$ GHz, $f_{m\,osc} = 70$ GHz.

A BIPOLE computer simulation of this device indicates zero bias capacitances of 14 and 50 fF for the e-b and c-b junctions, respectively. The computed neutral base and base-collector space-charge layer delay times are 2.0 and 0.9 ps, respectively. Figure 14.21 shows the computed f_t and $f_{m\,osc}$ values versus collector

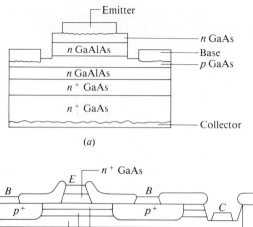

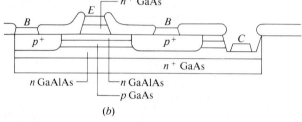

FIGURE 14.19
Cross sections of a gallium arsenide heterojunction transistor. (*a*) Discrete large area device; (*b*) integrated device. (*After Tiwari et al.* [30]. *Reproduced with permission © 1987 IEEE.*)

current. Note that although the f_t is excellent, the maximum oscillation frequency only increases as the square root of f_t. This, coupled with the rather mediocre hole mobility in the base, detracts from some of the performance potential of such a device.

It is also worth noting that although the injection efficiency limited current gain of this particular device is of order 10^4, the actual gain is limited in practice by the low recombination lifetime of GaAs devices. In this case, the gain is 150.

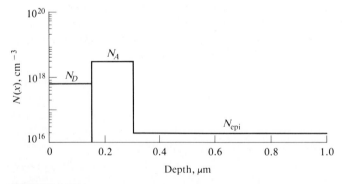

FIGURE 14.20
An MBE grown GaAlAs/GaAs heterojunction transistor impurity profile used in computer simulations.

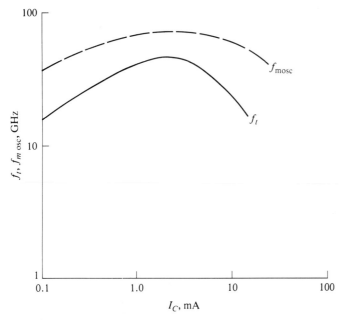

FIGURE 14.21

f_t and $f_{m\,osc}$ versus I_C for the heterojunction transistor of Fig. 14.20, computed using the BIPOLE program.

Other MBE Structures

Experiments with silicon emitters and silicon-germanium bases have been reported recently [32]. This combination has the bandgap shift in the right direction for good injection efficiency (high gain) and the emitter can again be doped at a level equal to or lower than the base, with a consequent reduction in junction capacitance. This structure could therefore be expected to yield excellent high frequency characteristics. An added advantage of the Si-Ge heterojunction is that conventional silicon processing could be used in other parts of the wafer.

Other heterojunction structures are being developed for different applications. For example, the use of InGaAsP/InP has been reported [33] in an integrated device so that a 1.3 μm wavelength laser diode could be incorporated on the same chip.

14.9 CONCLUSIONS

In this chapter we have attempted the (difficult) task of giving the reader an overview of various aspects of advanced technology. Some aspects of this technology are already incorporated in commercial devices. While the focus has been on silicon, we have also indicated the progress in gallium arsenide and other heterojunction bipolar transistors. Published performance of these devices is impressive, with f_t and $f_{m\,osc}$ values approaching 100 GHz.

As a closing note, it is worth pointing out that for an equal complexity in technology between GaAs and Si bipolar transistors (an equality which is inherently difficult to define, since many of the techniques are not interchangeable, e.g., the ease of growing oxide layers on silicon), the GaAs bipolar high speed performance can only be a factor 2 to 5 times better than silicon *at the most*. This is due to the high electron mobility of GaAs. In fact, since the maximum oscillation frequency $f_{m\,osc}$ is a better performance criterion than the unity gain frequency f_t, the performance advantage of GaAs is probably only between 1.4 and 2.2 times that of Si. Bearing in mind the simplicity (and low cost) of many silicon processes, it would appear that its future is assured.

PROBLEMS

14.1. If a lithographic process has a minimum feature size and contact to emitter spacing of 2 μm, draw the mask diagrams and calculate the emitter area and capacitance for two cases: (i) conventional emitter with contact mask and diffusion mask (assume a 0.5 μm deep emitter with a capacitance per unit area of 10^{-7} F/cm²), (ii) shallow junction polysilicon contacted emitter. Note the ratio reduction possible in area and capacitance.

14.2. Refer to Fig. 14.11 for a conventional process in which the emitter is oxide-walled on three sides and the base on four sides. Assuming the same emitter area in both cases, calculate the values of C_{je} and C_{jc} for both the walled structure shown and for a conventional structure. Use the design rules of Prob. 14.1, with emitter-base capacitance 10^{-7} F/cm², collector-base capacitance 10^{-8} F/cm², and make reasonable assumptions for the sidewall regions and assuming e-b and c-b junction depths of 0.5 and 1.0 μm respectively.

14.3. Consider a device with a separate P^{++} extrinsic base region as shown in Fig. 14.16. Compare the sidewall capacitance per unit length of perimeter and breakdown voltage for two cases: (i) no encroachment of extrinsic base on emitter, (ii) substantial overlap so that in the sidewall region the extrinsic layer can be considered as a vertical diffusion. Use Eqs. (2.48) etc. and (9.1) etc. and $N_e = 10^{21}$ cm⁻³, $X_{jeb} = 0.2$ μm, $X_{jbc} = 0.5$ μm, for case (i) with $N_b = 10^{18}$ cm⁻³. For case (ii) use $N_b = 10^{20}$ cm⁻³, with $X_{jbc} = 0.5$ μm. In both cases $N_{epi} = 10^{16}$ cm⁻³. For breakdown calculations, you can use Eq. (3.126) with $E_{br} = 10^6$ V/cm. Discuss the implications.

14.4. Consider a GaAlAs/GaAs heterojunction transistor with $\Delta E_g = 0.25$ eV. The base width is 0.1 μm doped 10^{18} cm⁻³, with a collector-doped 10^{16} cm⁻³, and $V_{cb} = 2$ V. For a 0.1 μm emitter depth, estimate the maximum possible gain and f_t plus the emitter-base junction capacitance for a 2 μm × 2 μm emitter, for emitter doping levels of 10^{16}, 10^{17}, and 10^{18} cm⁻³. Discuss the implications of low emitter doping on performance (what are the trade-offs?).

14.5. Use of MBE or MOCVD gives scope for creating devices with very thin bases. Assuming that the base dopant integral (the Gummel number) is kept at 10 times the integral of the collector-base depletion layer $N_{epi} d_n$ product to maintain high output resistance (see Chap. 9), and that the value of V_{CB} is one-quarter the maximum (plane) breakdown value (i.e., of order BV_{CEO}), use the results from Fig. 7.12(a) to determine for a silicon transistor designed to operate at $V_{CB} = 3$ V, the collector doping level and thickness. Calculate the output resistance at $I_C = 1$ mA. Determine the corresponding base widths for a base doping of 10^{17}, 10^{18}, and 10^{19} cm^{-3}. Calculate the maximum limiting f_t values for these three cases. If an emitter stripe width of 2 μm is used, calculate, for a symmetrical two-base contact transistor, the maximum oscillation frequency $f_{m\,osc}$ for the three cases. Discuss the implications with reference to the "optimum" transistor design discussed in Sec. 9.6.

14.6. Repeat Prob. 14.5 for the case of a GaAs base assuming the results of Fig. 7.12 still apply.

14.7. Calculate by hand the f_t and $f_{m\,osc}$ curves for the GaAs heterojunction transistor used in Fig. 14.20. Use appropriate estimates for electron mobility and drift velocity.

14.8. Using the approach of Chap. 2, derive Eqs. (14.4), (14.7), and (14.9).

14.9. Consider a heterojunction transistor with a silicon emitter and a silicon-germanium base. For a 2 μm wide emitter, with two base contacts, and a collector-doped 10^{17} cm^{-3}, biased at 3 V, a base doping of 10^{18} cm^{-3}, e-b and b-c junction depths of 0.1 and 0.2 μm, respectively, calculate the transit times t_{bb}, t_{scl}, the maximum f_t and $f_{m\,osc}$. Assume mobilities in the base region one-half those for Ge.

REFERENCES

1. A. B. Glaser and G. E. Subak-Sharpe, *Integrated Circuit Engineering*, Addison-Wesley, Reading, Massachusetts (1977).
2. S. M. Sze, *VLSI Technology*, McGraw-Hill, New York (1983).
3. S. B. Ghandhi, *VLSI Fabrication Principles*, John Wiley and Sons, New York (1983).
4. A. C. Adams in *VLSI Technology*, ed. S. M. Sze, McGraw-Hill, New York, p. 104 (1983).
5. P. Ashburn, D. J. Roulston, and C. R. Selvakumar, "Comparison of experimental and computed results on arsenic- and phosphorus-doped polysilicon emitter bipolar transistors," *IEEE Trans. Electron Devices*, vol. ED-34, pp. 1346–1353 (June 1987).
6. G. R. Wolstenhome, N. Jorgensen, P. Ashburn, and G. R. Booker, "An investigation of the thermal stability of interfacial oxide in polycrystalline silicon emitter bipolar transistors by comparing device results with High Resolution Electron Microscopy observation," *Jnl. Appl. Phys.*, vol. 61, pp. 225–233 (January 1987).
7. T. Sakai, S. Konaka, Y. Yamamoto, and M. Suzuki, "Prospects of SST technology for high speed LSI," *IEDM Tech. Digest*, pp. 18–21 (1985).
8. T. Sakai, S. Konaka, Y. Kobayashi, M. Suzuki, and Y. Kawai, "Gigabit logic bipolar technology: advanced super self-aligned process technology," *Electronics Letters*, vol. 19, pp. 283–284 (April 1983).
9. D. D. Tang, T. H. Ning, R. D. Isaac, G. C. Feth, S. K. Wiedmann, and H-N. Yu, "Subnanosecond self-aligned I^2L/MTL circuits," *IEEE Trans. Electron Devices*, vol. ED-27, pp. 1379–1384 (August 1980).
10. T. H. Ning, R. D. Isaac, P. M. Solomon, D. D-L. Tang, H-N. Yu, C. Feth, and S. K. Wiedmann, "Self-aligned bipolar transistors for high-performance and low-power-delay VLSI," *IEEE Trans. Electron Devices*, vol. ED-28, pp. 1010–1014 (September 1981).
11. A. Cuthbertson and P. Ashburn, "Self-aligned transistors with polysilicon emitters for bipolar VLSI," *IEEE Jnl. Solid State Circuits*, vol. SC-20, pp. 162–167 (February 1985).
12. F. Hébert, A. Solheim, and D. J. Roulston, "A V-groove emitter self-aligned bipolar technology," *Solid State Electronics*, vol. 31, pp. 1558–1560 (October 1988).
13. A. Solheim and D. J. Roulston, "Base link-up in the base-etched self-aligned transistor," 4th Canadian Semiconductor Technology Conference, Ottawa (August 1988) and A. Solheim, "Design and optimization of novel transistor structures with applications to monolithic optical preamplifiers," Ph.D. thesis, University of Waterloo (1988).
14. T. Sugii, T. Yamazaki, T. Fukano, and T. Ito, "Epitaxially grown base transistor for high-speed operation," *IEEE Electron Device Letters*, vol. EDL-8, pp. 528–530 (November 1987).

15. M. Tamura, N. Natsuaki, and S. Aoki, "Epitaxial transformation of ion-implanted polycrystalline Si films on $\langle 100 \rangle$ Si substrates by rapid thermal annealing," *Japan Jnl. Appl. Physics*, vol. 24, L151–L154 (February 1985).

16. F. Hébert and D. J. Roulston, "Pedestal bipolar transistor with polysilicon active base and emitter which achieves minimized capacitances," *Journal de Physique*, colloque C4, suppl. to no. 9, pp. 375–378 (September 1988).

17. T. E. Seidel in *VLSI Technology*, ed. S. M. Sze, McGraw-Hill, New York, pp. 251–252 (1983).

18. D. D. Tang, P. M. Solomon, T. H. Ning, R. D. Isaac, and R. E. Burger, "1.25 µm deep-groove-isolated self-aligned bipolar circuits," *IEEE Jnl. Solid State Circuits*, vol. SC-17, pp. 925–931 (October 1982).

19. H. Goto, T. Takada, R. Abe, Y. Kawabe, K. Oami, and M. Tanaka, "An isolation technology for high performance bipolar memories—IOP-II," *IEEE Intl. Electron Devices Meeting Digest*, pp. 58–61 (1982).

20. A. Hayasaka, Y. Tanaki, M. Kawamura, K. Ogiue, and S. Ohwaki, "U-groove isolation technique for high speed bipolar VLSI's," *IEEE Intl. Electron Devices Meeting Digest*, pp. 62–66 (1982).

21. D. J. Roulston and F. Hébert, "Study of delay times contributing to f_t of bipolar transistors," *IEEE Electron Device Letters*, vol. EDL-7, pp. 461–462 (August 1986).

22. M. C. Wilson, P. C. Hunt, S. Duncan, and D. J. Bazley, "10.7 GHz frequency divider using double layer silicon bipolar process technology," *Electronics Letters*, vol. 24, pp. 920–922 (July 1988).

23. D. J. Roulston and M. Depey, "Computer simulation of an oxide-walled emitter STL gate," *Electronics Letters*, vol. 19, pp. 21–22 (January 1983).

24. C. T. Chuang, "The effect of extrinsic base encroachment on the switch-on transient of advanced narrow emitter bipolar transistors," *IEEE Trans. Electron Devices*, vol. ED-35, pp. 309–313 (March 1988).

25. J. M. C. Stork and R. D. Isaac, "Tunneling in base-emitter junctions," *IEEE Trans. Electron Devices*, vol. ED-30, pp. 1527–1534 (November 1983).

26. S. M. Sze, *Physics of Semiconductor Devices*, 2d ed., John Wiley, New York (1981).

27. B. L. Sharma and R. K. Purohit, *Semiconductor Heterojunctions*, Pergamon Press (1974).

28. D. T. Cheung, S. Y. Chiang, and G. I. Pearson, "A simplified model for graded-gap hetero-junctions," *Solid State Electronics*, vol. 18, pp. 263–266 (1975).

29. A. Marty, G. Rey, and J. P. Bailbe, "Electrical behavior of an NPN GaAlAs/GaAs heterojunction transistor," *Solid State Electronics*, vol. 22, pp. 549–557 (1979).

30. S. Tiwari, S. L. Wright, and A. W. Kleinsasser, "Transport and related properties of (Ga, Al)As/GaAs double heterostructure junction transistors," *IEEE Trans. Electron Devices*, vol. ED-34, pp. 185–187 (February 1987).

31. O. Nakajina, K. Nagata, Y. Yamauchi, H. Ito, and T. Ishibashi, "High performance AlGaAs/GaAs HBT's utilizing proton-implanted buried layers and highly doped base layers," *IEEE Trans. Electron Devices*, vol. ED-34, pp. 2393–2404 (December 1987).

32. H. Temkin, J. C. Bean, A. Antreasyan, and R. Leibenguth, "Ge_xSi_{1-x} strained-layer hetero-structure bipolar transistors," *Appl. Phys. Lett.*, vol. 52, pp. 1089–1091 (March 1988).

33. H. Tsujii, K. Ohnaka, Y. Sasai, and J. Shibata, "Monolithic integration of InGaAsP/InP HBTs with a 1.3 µm laser diode for lightwave telecommunications," *IEEE Bipolar Circuits and Technology Meeting Digest*, pp. 68–72 (September 1987).

34. H. H. Berger and S. K. Wiedman, "Schottky transistor logic," *ISSCC*, p. 172 (1985).

35. A. Solheim, F. Hébert, and D. J. Roulston, "Self-aligned V-groove etched devices," *Solid State Electronics*, vol. 32, pp. 235–242 (March 1989).

36. M. J. Howes and D. V. Morgan (eds.), *Gallium Arsenide—Materials, Devices, and Circuits*, John Wiley and Sons, Chichester (1985).

37. H. C. Casey and M. B. Panish, *Heterojunction Lasers*, Academic Press, New York, 1978.

CHAPTER
15

NUMERICAL
ANALYSIS
AND CAD
MODELS
OF BIPOLAR
TRANSISTORS

15.1 INTRODUCTION

Computer-aided design means different things to different people. The circuit designer will normally be interested in a simulation program which is capable of predicting frequency response, propagation delay time, and so on as a function of various circuit topologies and component values, and for given CAD models of active devices (diodes, transistors). For the active devices such a designer will normally use a fairly simple CAD model such as the Ebers-Moll, or Gummel-Poon model [1]. These models describe the terminal characteristics for both static and high frequency or dynamic operation in a manner which is suitable, provided the intricate variations of the device and their dependence on internal processing are not of concern. Indeed, such a designer will not normally be interested in the internal operation of the transistor or diode. For the engineer who wishes to design, or improve the performance of, a new bipolar transistor, on the other hand, a simulation which is capable of relating the fabrication data (mask layout and dimensions, vertical impurity profile data) to terminal (i, v, t) characteristics is essential.

In this chapter we will outline some of the numerical analysis methods currently used to obtain terminal characteristics of bipolar transistors (and diodes) from fabrication data; we will also discuss briefly the commonly used CAD models for circuit analysis programs. Figure 15.1 shows a flowchart of the sequence used in developing an integrated circuit involving design of both the transistor and the circuit. Impurity profile data can be obtained either from measurements using, for example, the spreading resistance technique or the secondary ion mass spectroscopy (SIMS) method; this information can also be obtained prior to fabrication from numerical simulation, the most widely used program today being that developed at Stanford University—SUPREM [2]. Whichever method is used to obtain the profile data, a second and usually totally separate simulation package must then be used to analyze the operation of the device. We will be focusing our attention on one particular package in subsequent parts of this chapter, namely, BIPOLE [3], but there are a number of other simulation methods which will be discussed briefly. The output of the numerical analysis device simulation package can be used to observe terminal characteristics directly, but a very useful system is to obtain the CAD model parameters for the circuit simulator (e.g., SPICE [4], WATAND [5]) directly from the device analysis program. The complete loop may then be run on the computer with, as shown on Fig. 15.1, intervention of the engineer at various stages, either iterating around the processing loop or around the device design loop or around the circuit analysis loop or, without looking at intermediate outputs, the design engineer could in certain cases study the effect of varying process conditions (such as furnace temperature, diffusion times, implant doses) on the circuit performance (for example, frequency response, gain). These techniques have been in use at the author's university for a number of years [28, 44] and provide a very powerful tool for adapting device design to specific circuit performance, both in the case of single transistor amplifiers for use in microwave applications, for example, and for small sections of integrated circuits (up to several tens of transistors to simu-

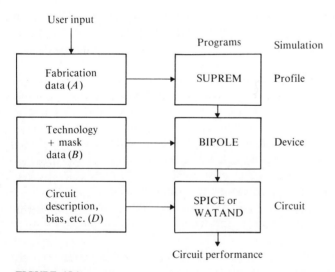

FIGURE 15.1
Flowchart showing the sequence process simulation, device simulation, circuit simulation.

late several coupled logic inverters, or a flip-flop circuit). The technique has also been applied successfully to the study of high voltage power switching transistors, including Darlington pairs; in this case the conventional analytic CAD model has been replaced by a tabular model which better represents the severe nonlinearities encountered in such structures. This model contains tables of base and collector current, charge, resistance, versus both V_{BE} and V_{CB} [6].

15.2 "EXACT" NUMERICAL ANALYSIS SCHEMES

The semiconductor equations which are normally solved in so-called exact numerical analysis of semiconductor devices are derived from the time-independent Boltzmann transport equations, assuming quasi-static carrier transport, zero electron temperature gradient, electron temperature equal to lattice temperature [39]. These simplifications lead to the following equations:

$$J_n = q\mu_n nE + qD_n \text{ grad } n \tag{15.1}$$

$$J_p = q\mu_p pE - qD_p \text{ grad } p \tag{15.2}$$

where $E = \text{grad } \psi$ and ψ is the electrostatic potential.

$$\frac{\partial n}{\partial t} = G_n - U_n + \frac{1}{q} \text{ div } J_n \tag{15.3}$$

$$\frac{\partial p}{\partial t} = G_P - U_P - \frac{1}{q} \text{ div } J_P \tag{15.4}$$

$$\text{div grad } \psi = \frac{q}{\varepsilon} [(N_D - N_A) + p - n] \tag{15.5}$$

In two-dimensional form the reader can easily appreciate that solution of such a system of coupled nonlinear partial differential equations will involve a substantial amount of computing power. The schemes invariably employed involve either finite element or finite difference methods.

In the finite difference method, the derivatives of the above semiconductor equations are replaced by finite difference formulas at each of the nodes in a finite difference mesh. Figure 15.2 shows a typical nonuniform mesh arrangement for a semiconductor device. The boundary conditions are normally Dirichlet (values fixed) at the contacts, and Neumann (gradients fixed) at other surfaces. The system of equations may be solved either by direct methods or by iterative schemes. A commonly employed method is to use *successive over-relaxation* (SOR). One advantage of the iterative method is smaller memory requirements.

The finite difference method is relatively easy to set up. In semiconductor problems, it is important to employ small mesh spacing in the vicinity of junctions where the potential ψ is changing rapidly. This is shown in the example of Fig. 15.2. The simplest schemes lead to "wasted" areas where closely spaced nodes are not required. The finite box method [39] partially overcomes this drawback. It is difficult, however, to use a finite difference grid to cope with

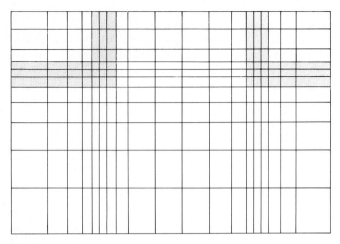

FIGURE 15.2
Typical finite difference nonuniform mesh arrangement for a semiconductor device. (*After Snowden* [39]. *With permission, IEE, 1988.*)

geometries (in cross section) which are not close to rectangular, although some algorithms can handle regions such as beveled edges quite effectively [38].

The finite element method lends itself to structures which are far from rectangular. Since they employ elements which can be quite different in size for different regions, the method is also ideal for having small elements in the vicinity of junctions where the potential is rapidly changing. Figure 15.3 shows a typical finite element arrangement.

The solution method approximates the solution of the differential equations with solutions which follow simple functions for each element. The overall solution is found by combining the solutions of all elements. The boundary conditions are incorporated as integrals in a function which is minimized, and the scheme is independent of the specific boundary conditions; this provides a good

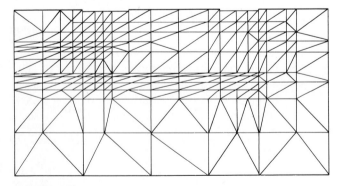

FIGURE 15.3
Typical finite element arrangement for a semiconductor device. (*After Snowden* [39]. *With permission, IEE, 1988.*)

degree of flexibility, particularly since elements of different sizes may be added without increasing the complexity. One main drawback to the finite element method is the complexity of setting up the algorithms; also, stability and convergence criteria are not as well understood as for finite difference methods. The interested reader is referred to one of several texts on numerical methods such as that by Snowden [39], which gives an excellent presentation of both finite difference and finite element methods.

The above two methods make use of the drift-diffusion equations already presented. For devices where very short distances are involved for transport, the concepts of constant (or doping-dependent) mobility and velocity may no longer be valid. For example, when region widths are less than about 0.1 μm the distance becomes comparable to the mean free path length between collisions.

The Monte-Carlo method [39] finds increasing use for these very small-dimensioned high frequency devices, specially for simulating quantum mechanical effects [40, 41]. In this method, the motion of carriers is followed through sequences of free flight between collisions. The method uses the generation of sequences of random numbers with given distribution probabilities; these are used to describe processes such as scattering events (collisions). A detailed description of the physical system is required including energy band structures, scattering rates, lattice temperature. The Monte-Carlo method has been used to calculate the velocity versus electric field law for various semiconductors [39] and to predict velocity overshoot in devices with regions less than 0.1 μm [42]. Baccarani et al. [43] reported a study on silicon bipolar transistors using Monte-Carlo simulations. They concluded that even for base widths considerably less than 0.1 μm the error in computation of base transit time was not too significant; they reported a relative error of 20 percent for a base width of 0.02 μm and noted that the base transit time was then so small that the base-collector transit time dominated the electrical behavior.

Numerical simulation can be performed for one- or two-dimensional dc or transient solutions. The order of complexity (and execution time) of the software increases drastically between one- and two-dimensional solutions and between static and dynamic solutions. A relatively simple (and extremely useful) two-dimensional solution can be obtained for Poisson's equation [Eq. (15.5)] with zero current flow (see, for example, Ref. [7]). Incorporation of the ionization integral can provide invaluable information about avalanche multiplication and breakdown voltages for the case where field plates, guard rings, or beveled structures are used [8]. These "Poisson solvers" have direct engineering applicability in aiding in the optimization of high voltage structures before fabrication.

One of the earliest attempts at a one-dimensional static solution of the complete set of equations was that of H. Gummel [9], and his algorithm still forms the basis of some schemes. A. Demari [10] and V. Arandjelovic [11] published detailed results for a one-dimensional solution of the PN junction. A widely used one-dimensional program for bipolar transistors is SEDAN from Stanford [12].

SEDAN is a general one-dimensional device simulation program which solves for electrostatic potential, electron, and hole concentrations as functions of space and time for the one-dimensional structure. Various one-dimensional device parameters and one-dimensional I-V characteristics can be obtained. It can handle both silicon (including polysilicon emitters) and GaAs-GaAlAs structures. The five coupled semiconductor equations are solved including the option of using Fermi-Dirac statistics (instead of Boltzmann statistics). Grid spacing is determined in such a way that fine spacing exists in the vicinity of junctions and coarse spacing where profiles are changing slowly with distance. The actual numerical analysis uses a modified finite difference method (to cope with the nonuniform grid spacing). The discretization method proposed by Scharfetter and Gummel [13] is used. SEDAN takes typically between five and ten minutes on a VAX-11 785 minicomputer for a complete 1D static analysis of a bipolar transistor. Output includes J_n and J_b versus V_{BE}, f_t, and junction capacitances. Convergence is usually obtained and is only a problem with some unusual impurity profiles.

Two-dimensional semiconductor analysis programs include PISCES from Stanford University [14, 15], BAMBI, from The Technical University of Vienna [16, 17], and GALATEA [18] from the Institut für Theoretische Elektrotechnik in Aachen.

The PISCES program performs a full two-dimensional static and transient analysis for semiconductor devices. Both Gummel and Newton numerical solution techniques are included [15] and the program is based on finite element methods. One of the major problems with two-dimensional analysis concerns the establishment of the mesh structure. PISCES-IIB contains a sophisticated grid generator with a new triangulation algorithm [19].

The BAMBI program is also aimed at both MOS and bipolar devices, and again solves the five semiconductor equations in two dimensions and under transient conditions. Models for parameters such as mobility and recombination rates are supplied by the user. Mixed Newton and Gummel methods are used.

Convergence is often a problem with such complex two-dimensional analyses, and some experimentation is often necessary to achieve the final result. Execution times are necessarily very long, typically of an hour or more on a mainframe computer. These programs are extremely powerful and provide much information about internal carrier concentration distributions and terminal characteristics. The reader interested in studying the mathematical techniques used in such solutions can study the above references, or texts such as Kurata [20] or the NASECODE conference proceedings [21].

Apart from the considerable computer requirements, software of this type often involves sorting out problems of grid spacing and convergence. While a wealth of data is thus eventually produced concerning the detailed internal operation of the device, in many real life design situations what is really required is a knowledge of how the main terminal characteristics depend upon the mask dimensions and impurity profiles, plus an indication of how to improve the device performance. Sometimes it is not easy to obtain this information from two-dimensional simulation packages. Figure 15.4 shows a three-dimensional representation of recombination rate in the emitter of a bipolar transistor [22].

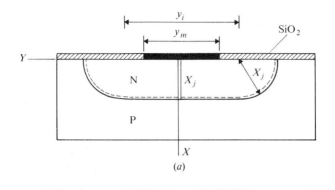

(a)

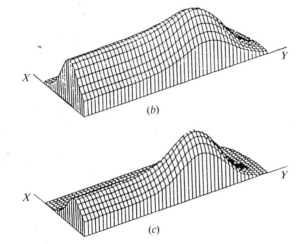

FIGURE 15.4
Two-dimensional solution of carrier concentration for injection into the emitter of a bipolar transistor. (*After Eltoukhy and Roulston* [22].) (a) Structure, surface doping 5×10^{20} cm^{-3}. $y_m = 0.8$ μm, $t_i = 1.0$ μm, $X_j = 0.3$ μm; (b) using bandgap reduction data from [22]; (c) using bandgap reduction data from [24]. (*Reproduced with permission* © *1981 IEE.*)

While pictures of this sort may appear quite artistic, they do not in themselves necessarily provide a very clear insight about how to go about altering either the mask dimensions or the impurity profile data in order to improve the performance of the device.

Figure 15.5 shows the two-dimensional solution for the sidewall region of the emitter of a bipolar transistor [25]. Once again the diagrams in themselves do not provide a lot of useful information; in this particular case the program was established in order to compute the recombination properties of the sidewall region as mentioned and described in Chap. 9, Fig. 9.7. This is one region where one-dimensional solutions cannot be applied.

Figure 15.6 shows the significant difference between peak and high current f_t computed from a one-dimensional solution and from a complete simulation. Differences are due both to the emitter current-crowding effect discussed in Chap. 10 and to sidewall effects where significant capacitance can exist, as we discussed from an analytical point of view in Chap. 9. It is perhaps worth noting at this point that advanced technology is tending to produce bipolar transistors which are more and more one-dimensional, i.e., devices which approach the ideal

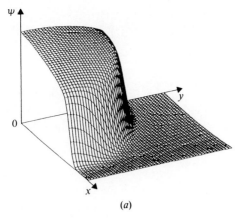

(a)

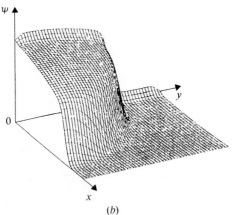

(b)

FIGURE 15.5
Two-dimensional simulation of the electro-static potential in the sidewall space-charge region of the emitter base junction of a bipolar transistor for an applied forward bias of 0.4 V. (a) $Q_{ss} = 0$; (b) $Q_{ss} = 4 \times 10^{11}$ cm^{-2}. (*After Roulston and Eltoukhy* [25]. *Reproduced with permission* © *1985 IEE.*)

behavior. Such techniques as oxide isolation and double polysilicon which were discussed in the previous chapter, using design rules of the order of one or two microns, are used for this type of device. Reference to Eqs. (10.1) through (10.4) shows that for typical pinched-base sheet resistance values of order 1 to 5 kΩ/sq, emitter current crowding becomes almost, but not quite, negligible for $L = 2$ μm.

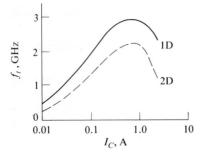

FIGURE 15.6
Variation of transition frequency f_t versus collector current from one-dimensional and complete analysis using the BIPOLE program for an RF interdigitated power transistor: emitter = 4 μm × 1 cm (total), $N_{epi} = 3 \times 10^{15}$ cm^{-3}.

Even the one-dimensional exact solution provided by SEDAN while giving valuable insight to the internal operation of the device requires a considerable amount of computing time. As a design tool, and also for attaining substantial physical insight into the physical operation of the transistor, the quasi-two-dimensional variable boundary regional approach used in the BIPOLE program is extremely powerful.

Before examining this "approximate" numerical analysis approach, let us refer again to the so-called two-dimensional "exact" method. Solutions of Eqs. (15.1) through (15.5) involve approximations at several levels: (a) the discretization is necessarily rather coarse, in order to obtain manageable execution times; (b) there are a number of physical parameters which themselves depend on "models": mobility, SRH, and Auger recombination rates, bandgap narrowing are the most important (in the case of recombination rates, the model parameters are process-dependent which complicates their determination); (c) the equations themselves are a simplification of the Boltzmann transport equations and are only an approximation to the more physical world; (d) isothermal conditions are implicitly assumed; this is often not valid at high currents.

Finally, one must bear in mind the precision with which the input data to the program is available—particularly the impurity profile. There are two alternate techniques for obtaining this data. Numerical simulation using a program such as SUPREM [2]; in this case one is relying on ion implant data, furnace temperature data, and models for the diffusion of the atoms. As in solving the semiconductor device equations, the models are at best an approximation to physical reality and some degree of experimentation is invariably required to obtain perfect agreement between computed and simulated impurity profiles. Use of measured information from test wafers is often used directly as input to device simulation programs. Impurity profiles are often obtained using the SIMS method, which gives the total number of impurities versus depth, or using spreading resistance methods, which gives the electrically active doping concentration versus depth. In either case, the results must be interpreted correctly before being used as input, and the user must always bear in mind the ultimate resolution capabilities of the method being used. We conclude therefore that *exact* input impurity profile data is seldom available; this must be taken into account when weighing the advantages and disadvantages of the various software available. This author has found that an essential cross-check before making extensive use of any simulation results, and assuming that at least the nominal values of junction depths are known, is to compare computed and measured (on test chips) intrinsic (pinched) base sheet resistance with the computed values. Since this parameter is fairly sensitive to process conditions it is a very good indication of whether or not the correct impurity profile data is being used.

It is important to bear the above points in mind when considering the usefulness of simulation work for a particular application. We repeat that while very valuable physical insight may often be obtained by exact two-dimensional simulation programs, the potential user of software must reflect carefully on the above factors before selecting a particular method or software package.

15.3 THE QUASI-2D VARIABLE BOUNDARY REGIONAL SIMULATION METHOD

We will give an overview in this section of the essential features of the BIPOLE variable boundary regional approach and how it can be very simply applied to give two-dimensional characteristics of bipolar transistors [32, 26, 27]. The BIPOLE program is a widely used industrial and research tool for studying bipolar device characteristics. It has the advantage of being very fast (typically tens of seconds on a VAX-11 785 or a few seconds on an IBM 4341) and exists also in an IBM PC-AT version. The BIPOLE program is also used to supply CAD model parameters for input to programs such as SPICE [4] or WATAND [5]. The purpose of the following is to serve both as a review of the bipolar transistor physical equations and as a starting point for readers wishing to write their own numerical programs.

Figure 15.7 shows a cross section of the transistor and the impurity profile. In the impurity profile diagram we have shown the neutral and the space-charge regions. Clearly, if we are to solve Poisson's equation neglecting free carriers, the solution is very simple, and this may be done for both the emitter base and the collector base junctions for given bias conditions. Boundaries may thus be established between the space-charge layers and the neutral regions; application of the Boltzmann relations (which as we know, strictly imply zero current conditions) provide a relationship between the injected carrier concentrations for holes and electrons and the applied bias. It is then a relatively simple matter to solve the transport equations for diffusion and drift in the presence of arbitrary recombi-

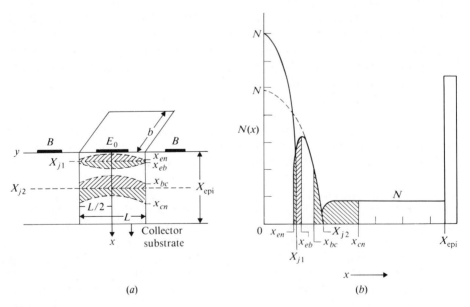

(a)   (b)

FIGURE 15.7
Cross section of the transistor and the impurity profile. (*After Hajj et al.* [27].)

nation in both the quasi-neutral emitter and quasi-neutral base regions. This scheme is extremely rapid in computing time because we are not solving the coupled set of semiconductor equations. Let us now examine more closely how such a scheme can be established.

15.3.1 Solution of Poisson's Equation for the Emitter Base Junction

Starting at an arbitrary point x_{en} on the emitter side of the emitter base junction, Poisson's equation can be integrated to supply electric field versus distance and total voltage across the space-charge layer

$$\frac{dE}{dx} = \frac{q}{\varepsilon} N(x) \tag{15.6}$$

$$E(x) = \int_{x_{en}}^{x} \frac{q}{\varepsilon} N(x) \, dx \tag{15.7}$$

$$V(x) = -\int_{x_{en}}^{x} E(x) \, dx \tag{15.8}$$

Having obtained the solution for electric field and voltage versus distance, it is straightforward to now apply the Boltzmann relations and obtain the distribution of free carriers inside the space-charge layer, including particularly the values at the two boundaries for a particular bias solution. To relate the applied bias V_{BE} to the total junction voltage, the solution in thermal equilibrium must first be obtained. This may be defined as the solution for which the voltage obtained from Poisson's equation is equal to the built-in barrier voltage obtained from the Boltzmann equations as discussed in Chap. 2. In the BIPOLE scheme, the emitter base space-charge layer is thus solved for a number of bias points ranging from 10 to 50 and the results stored in an array.

It is important to write the equations relating injected carrier concentration versus voltage in such a way that high level injection is included and so that large forward bias is allowed, with the total junction voltage tending to zero. Applying space charge neutrality at the depletion-layer boundaries gives the Fletcher boundary conditions which may be rearranged thus:

$$n(x_{eb}) = \frac{N_{A\,\text{eff}}(x_{eb}) + N_{D\,\text{eff}}(x_{en}) \exp\left(V_{jeb}/V_t\right)}{\exp\left(2V_{jeb}/V_t\right) - 1} \tag{15.9}$$

and

$$p(x_{en}) = \frac{N_{D\,\text{eff}}(x_{en})}{2} \left\{ \left[1 + \frac{4n(x_{eb})[N_{A\,\text{eff}}(x_{eb}) + n(x_{eb})]}{N_{D\,\text{eff}}(x_{en})^2} \right]^{1/2} - 1 \right\} \tag{15.10}$$

Note that in all equations pertaining to minority carrier concentration, the effective doping level (including bandgap narrowing) must be used.

15.3.2 Solution for the Base and Collector Regions

A valid simplification for the bipolar transistor in the vertical direction (emitter to collector) consists in setting the hole current density approximately equal to zero, thus giving an equation for electric field

$$J_p = 0 = -qD_p \frac{dp}{dx} + \mu_p qpE_x \tag{15.11}$$

This may be combined with the equation for electron current density

$$J_n = qD_n \frac{dn}{dx} + \mu_n qnE_x \tag{15.12}$$

to give

$$\frac{dn}{dx} = \frac{N_{eff} + n}{N_{eff} + 2n} \left[\frac{J_n(x)}{qD_n} - \frac{n}{N_{eff} + n} \left(\frac{dN_{eff}}{dx} \right) \right] \tag{15.13}$$

Doping-dependent mobility must be used and recombination can be included, in which case J_n is a function of x. Clearly for a given impurity profile and given mobility versus doping level, it is a simple matter, using well-established numerical integration methods, to solve this equation starting at the emitter base space-charge layer boundary x_{eb} and proceeding for increasing x values toward the base collector space-charge layer boundary. In order to determine when the boundary has been reached, it is necessary to monitor the value of electron concentration n and the normalized space charge R_p [29]. The electron concentration must not fall below the value n_c discussed in Chap. 7

$$n_c = \frac{J_n}{qv_d} \tag{15.14}$$

The normalized space charge may be defined in the p type base as

$$R_p = \frac{p - n - N_A(x)}{p} = \frac{\varepsilon}{q} \frac{dE}{dx} \frac{1}{p} \tag{15.15}$$

with the gradient of $E(x)$ evaluated from successive points using Eq. (15.11). When either of the above conditions is encountered, the equation for electron current flow in the quasi-neutral base region is changed to a solution of Poisson's equation in the presence of the electron current and the electron concentration, given by

$$\frac{dE}{dx} = \frac{q}{\varepsilon} \left[N(x) - \frac{J_n}{qv(E)} \right] \tag{15.16}$$

By monitoring the value of field it is a simple matter to determine the end of the base collector space-charge layer, i.e., the start of the collector neutral region, beyond which the ohmic drop due to current in the collector epitaxial layer (including conductivity modulation) must be added. The actual BIPOLE program uses these tests, together with a solution of the potential in the gradual transition from quasi-neutrality to space-charge conditions. For a given injected carrier concentration $n(0)$ and electron current density J_n, the above scheme with double integration of Eq. (15.16) will give a solution for a particular value of collector base voltage V_{CB}. Altering the value of $n(0)$ or J_n will give a different voltage as shown in Fig. 15.8. It is of interest to note that this dependence of V_{CB} on the chosen $n(0)$ or J_n is simply the Early effect, discussed in Chap. 9. It is thus a simple matter to introduce an iterative scheme to solve for given terminal conditions. At the same time as carrying out the base integral it is also possible to define a conductivity parameter

$$X_i = \frac{1}{[N(x_{eb}) + n(x_{eb})]\mu_p(x_{eb})} \int_{x_{eb}}^{x_{bc}} [N(x) + n(x)]\mu_p(x)\, dx \qquad (15.17)$$

This will be used subsequently in computing lateral base current flow.

The emitter region as we have already discussed in Chap. 8 is of fundamental importance in the bipolar transistor. The equation to be solved in this case consists of minority carrier hole transport

$$\frac{dp}{dx} = \left[\frac{N_{eff} + p}{N_{eff} + 2p}\right]\left[\frac{-J_p}{qD_p} + \left(\frac{J_n}{qD_n}\right)\left(\frac{p}{N_D + p}\right) - \left(\frac{p}{N_{eff} + p}\right)\frac{dN_{eff}}{dx}\right] \qquad (15.18)$$

$$\frac{dJ_p}{dx} = \frac{-qp'}{\tau} \qquad (15.19)$$

Recombination is very important in the emitter and must in general include both Shockley-Read-Hall recombination [24]

$$\tau_{SRH} = \frac{\tau_{PO}}{[1 + N(x)/N_0]} \qquad (15.20)$$

and Auger recombination

$$\tau_A = \frac{1}{C_N N(x)^2} \qquad (15.21)$$

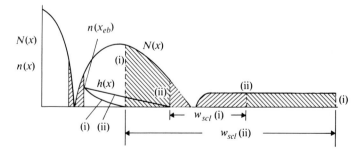

FIGURE 15.8
Illustration of successive solutions varying J_n. (i) J_n high, W_b low, V_{CB} high; (ii) J_n low, W_b high, V_{CB} low.

The combined lifetime may be defined as

$$\frac{1}{\tau} = \frac{1}{\tau_A} + \frac{1}{\tau_{SRH}} \tag{15.22}$$

A similar iterative scheme may be employed here for a given injected hole concentration $p(0)$ at $x = x_{en}$. A value of J_p for hole current is guessed and iterated upon until the surface boundary conditions defined by the surface recombination velocity are satisfied, as shown in Fig. 15.9, curve (ii).

We have thus obtained effectively a solution for a current flow in the transistor for particular base-emitter and base-collector voltages. However, this is only a one-dimensional solution; in order to obtain real terminal characteristics it is necessary to include horizontal base current flow and sidewall effects. Horizontal current flow in the base region may be analyzed as follows:

The hole current density in the lateral (y) direction is given by

$$J_p(y) = -qD_p \frac{dp}{dy} + q\mu_p p E_y \tag{15.23}$$

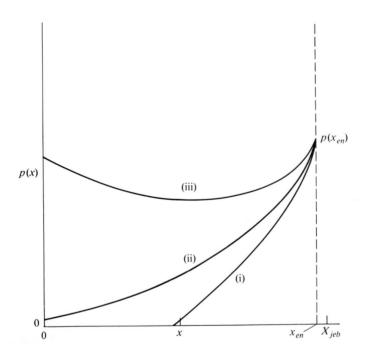

FIGURE 15.9
Emitter hole concentration for various guesses of J_p. (i) J_p too high; (ii) correct solution; (iii) J_p too low.

Since the electron current is essentially zero, the field is given by

$$E_y = -\frac{V_t}{n}\frac{dn}{dy}$$ (15.24)

Combining Eqs. (15.23) and (15.24) gives

$$J_p(y) = -qD_p\left(2 + \frac{N_A}{n}\right)\frac{dn}{dy}$$ (15.25)

For a given value of x (constant for all y), both D_p and N_A are constant in Eq. (15.25). If we now integrate over the base region $x_{eb} < x < x_{bc}$, we can write the total base current at a given y in the form

$$I_p(y) = -qBX_i\mu_p(0, y)V_t[2n(0, y) + N_A(0, y)]\left[\frac{1}{n(0, y)}\frac{dn(0, y)}{dy}\right]$$ (15.26)

where X_i is defined by Eq. (15.17) and where, for ease in reading, we refer to the point (x_{eb}, y) as $(0, y)$. The base current can also be expressed in terms of the components that determine the gain, h_{FES}, where

$$h_{FES} = \frac{J_n(x_{bc})}{J_p(x_{eb})}$$ (15.27)

to give

$$\frac{dI_p(y)}{dy} = -\frac{BJ_n(0, y)}{h_{FES}(y)}$$ (15.28)

Since X_i, $J_n(0)$, and h_{FES} are available from the vertical analysis for a given value of $n(0)$, Eqs. (15.26) and (15.28) may be solved directly, starting from the center of the emitter stripe and using, as an initial condition, an arbitrary value of $n(0)$.

Integration in the base region from the center to the edge of the emitter (assuming a symmetrical two-base contact transistor) is straightforward. The voltage and current at the edge of the emitter will, for most practical purposes, be the same as the terminal current and voltage; if not, a simple $I_b R_b$ drop for the extrinsic region may be added.

The delay times may all be calculated directly from the static solutions. Junction capacitances are obtained from the Poisson equation solution (with an additional solution in cylindrical coordinates for the sidewall regions), emitter and base delays from the neutral region minority carrier charge integrals, base collector space-charge layer delay from the integral of free carrier charge; in addition the small component of delay due to the free carriers within the emitter-base space-charge layer is computed from the Boltzmann relations at each bias and included with the base and emitter charges. The collector RC time constant (including the substrate capacitance term for integrated devices) is also accessible.

The above is a condensed and simplified summary of the variable boundary regional method as implemented as a quasi-two-dimensional analysis in the

BIPOLE program. The actual program includes all major documented effects such as mobility versus doping and temperature, bandgap narrowing (several models), lifetime versus doping, deionization of impurity atoms, the ionization integral for breakdown studies; in addition both tunneling and thermionic emission are included in the polysilicon emitter model and the basic GaAs data is incorporated for heterojunction devices.

15.3.3   Major Approximations of the Variable Boundary Regional Method

Let us summarize the major approximations inherent in the *variable boundary regional analysis.*

Initial neglect of the free carriers in solving the emitter-base space-charge layer may appear to be a serious approximation. However, the free carrier distribution is computed using the Boltzmann relations for each Poisson solution of the potential distribution. The error is negligible at low forward bias but the error in determination of the *positions* of the space-charge layer boundaries will increase for large forward bias. This may be considered significant in a precise examination of the physical meaning of "space-charge layer." For determination of terminal characteristics, however, the likely error is small. This is because the *charge distribution* is continuous (free carriers within the space-charge layer being included) and only the "depletion layer" capacitance will be affected by the exact positions of the boundaries. Since this capacitance has a negligible effect at high forward bias (being masked by the diffusion capacitance terms due to the free carrier charge), errors in overall delay time are likely to be small. It may be noted furthermore that at large forward bias, the "depletion layer" becomes very narrow and the formulation for $n(0)$ and $p(0)$, using Eqs. (15.9) and (15.10), ensures that these two quantities tend to the same (high) value for high forward bias; this is consistent with reality. There will also be an error in determination of the applied forward voltage at high currents. This error will in practice be masked by the additional lateral voltage drop (included in the analysis of horizontal base current flow).

Errors associated with the determination of the boundary between the neutral base and the base-collector space-charge layer are due to the artificial division into two regions. However, since the charge due to free carriers in both regions is included, any error in determination of the boundary position will have a self-canceling tendency on the magnitude of the free carrier charge.

There are some two-dimensional effects present in the bipolar transistor, other than the horizontal base current flow which is included in the analysis, see Fig. 15.10. Firstly, at the emitter edge or sidewall region, electrons are injected and will be collected outside the collector area directly under the emitter. This lateral charge spreading is negligible in all transistors studied, at normal operating current [because of the large ratio of emitter stripe width L to base width W_b (typically 10 to 1)]. It becomes significant in the presence of Kirk effect base

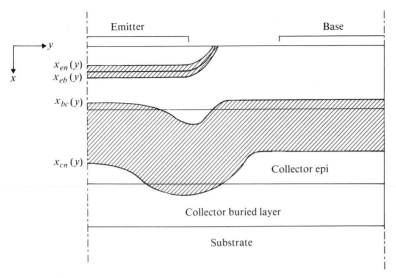

FIGURE 15.10
Two-dimensional effects in the bipolar transistor.

stretching [30]. This can be modeled empirically and the BIPOLE results appear to be satisfactory, but could contain significant errors in unusual device geometries. Two-dimensional injection of holes into the emitter could also be a problem, but two-dimensional studies [22] tend to indicate that, in most devices, the errors are slight.

Two-dimensional base current flow around the emitter (see Chap. 13) is a difficult problem unless one resorts to a two-dimensional solution. In the BIPOLE formulation, empirical formulas based on a detailed two-dimensional analysis [31] are used. By the same technique, the two-dimensional current flow vertically within the collector resistive region can be handled with empirical formulas, giving quite good accuracy in most cases.

The above very simple overview of the variable boundary regional approach extended to include quasi-two-dimensional analysis has been shown (on devices ranging from 600 V power switching transistors with f_t values of the order of 6 MHz to 2 μm VLSI structures with f_t values of 7 GHz) to give very good engineering agreement between computed and measured terminal characteristics. In addition, because the regions have already been defined, it is a trivial matter to print out the delays corresponding to each of the regions of the device. Thus the designer may see at a glance which regions are contributing most to the delay terms limiting the f_t of the transistor. Because terminal characteristics are available, for example, f_t versus I_C and h_{fe} versus I_C it is evident that the CAD model parameters for Ebers-Moll or Gummel-Poon type models can also be extracted from such a simulation package, as is done in the BIPOLE program.

Figure 15.11 shows measured and computed f_t versus collector current for a

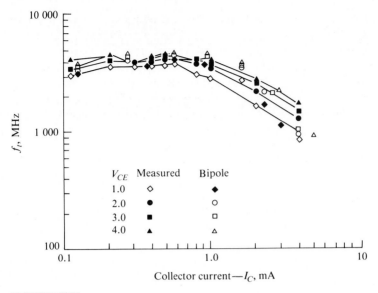

FIGURE 15.11

Measured and BIPOLE computed f_t versus I_C for a small VLSI transistor. (*After Roulston and Hébert* [32]. *Reproduced with permission* © *1986 IEEE.*)

small VLSI transistor [32]. The agreement is seen to be quite acceptable given the degree of uncertainty that always exists in impurity profile data. Table 15.1 gives the delay times in the different regions for various values of collector current density. In this case, the profile data was obtained from spreading resistance measurements. Indeed, when considering any numerical analysis scheme for devices, it is of fundamental importance to realize that input data to the software is *never* available with 100 percent precision. Whether it is obtained from computer simulation using software such as SUPREM or by measurements, the data is subject to error in impurity concentration value at a given depth.

A second rather important factor concerning the comparative merits of exact two-dimensional numerical analysis schemes and the above variable

TABLE 15.1

Computed delay times in picoseconds for various collector current densities (A/cm²) for $V_{CB} = 1.2$ V

J_n	t_{poly}	t_{em}	t_{re}	t_{bb}	t_{scl}	t_{RC}	t_{qbe}	t_{tot}
29	1.0	4.1	290	8.6	2.4	0.4	35	342
465	1.0	4.4	21	8.6	2.4	0.4	12	50
3 720	1.3	5.5	4	9.0	2.5	0.4	5	27.7
14 900	1.8	7.2	1.8	10.0	2.8	0.3	1.7	25.6
29 800	2.7	9.3	1.7	12.0	2.7	0.3	0.7	32.1

boundary regional approach with its quasi-two-dimensional solution concerns grid size. In the two-dimensional solutions, for reasons of computer memory and reasonable execution time, grids tend to be rather coarse with total number of points in general not exceeding the order of 10 000. In the BIPOLE variable boundary regional approach just described, it is easy to have several hundred points in the vertical analysis discretization; in many cases, up to 800 points are used to obtain very high precision. Since the second-dimensional effect in bipolar transistors is very much of secondary importance, the above described scheme where only the orthogonal current in the base is taken into account, with typically 10 to 20 discretization points in the lateral dimension, is quite adequate. Moreover, the execution time on a mainframe such as the IBM 4341 is of the order of 10 seconds for one V_{CB} voltage covering the complete range of collector currents. This is very fast compared to existing exact solutions and enables multiple runs to be made to study the effect of parameter changes in impurity profile or mask data.

15.4 CAD MODELS FOR USE IN CIRCUIT ANALYSIS PROGRAMS

In Chap. 7 we introduced the Ebers-Moll model for the bipolar transistor. In Chap. 9 we discussed the Gummel integral. The most widely used model today in network analysis programs such as SPICE and WATAND is the Gummel-Poon model. Let us examine this briefly and discuss its characteristics. For an excellent detailed explanation of Ebers-Moll and Gummel-Poon models the reader is referred to the work by Getreu [1]. The equation for collector current is

$$I_C = \frac{1 - V_{BE}/V_B - V_{BC}/V_A}{1/2 + \sqrt{1/4 + (I_C/I_k)^2}} \, I_{cs}\left[\exp\left(\frac{V_{BE}}{V_t}\right) - \exp\left(\frac{V_{BC}}{V_t}\right)\right] \qquad (15.29)$$

V_A, V_B are the normal and inverse Early voltages (see Chap. 9). One of the major weaknesses of the widely used Gummel-Poon model is the fact that in a modern VLSI transistor with quasi-neutral base widths of the order 0.1 to 0.2 microns, these voltages are by no means constant. The reader should refer back to Chap. 9 where the voltage dependence of the Early voltage was discussed in some detail. This dependence has been incorporated in a modified Gummel-Poon model [34, 35]. In the denominator of the equation for collector current is a term involving I_k. This is intended to represent the change in the I_C versus V_{BE} law at high currents. I_k may be used to represent either high level injection or base widening (Kirk effect). The former is accurately modeled by the expression in the above equation; the Kirk effect may be modeled approximately inasmuch as the quantity of prime concern is the variation of gain with collector current. The denominator is equivalent to introducing a variable base majority carrier charge Q_B.

In order to model f_t falloff at high collector currents, the Gummel-Poon model normally employs an equation based on the Van der Ziel effect discussed

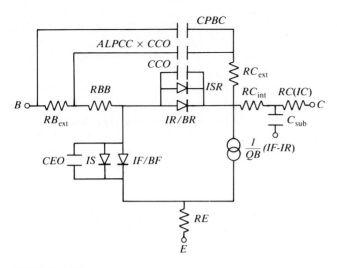

FIGURE 15.12
Circuit representation of the Gummel-Poon model.

in Chap. 10 [1]

$$\tau_F = \tau_{F0}\left\{1 + \frac{1}{4}\left(\frac{L_E}{W_b}\right)^2\left[\left(\frac{I_c}{I_{KF}}\right)^2 - 1\right]\right\} \tag{15.30}$$

for $I_c \geq I_{KF}$ and is the collector current given by Eq. (15.29) [45].

The above equation is based on Eq. (10.29) and represents the increase in lateral base width when the collector current exceeds the Kirk current value. As we already discussed in Chap. 10, since both the Van der Ziel and the Kirk effects occur at precisely the same collector current, for modeling the falloff of f_t this equation is usually adequate; this is particularly true since normally a very accurate model for the transistor operation under these bias conditions is not necessary. Operation would normally be confined to a collector current at or only slightly beyond the value which gives peak f_t.

As can be seen by looking at the circuit representation of the Gummel-Poon model (Fig. 15.12), there are two other parameters of importance in predicting accurately the performance of transistors. These concern the base and collector resistances. In the SPICE model a simple current dependence of R_b is assumed; this is based on Hauser's formulation referred to in Chap. 10 on emitter current crowding. Unfortunately it is clear from looking at Fig. 15.13 that *high*

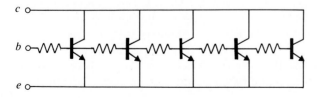

FIGURE 15.13
Sectional representation of the bipolar transistor.

frequency emitter current crowding can occur even at relatively low dc bias. This is due to the fact that the base resistance is distributed along the transistor and base current will flow into the capacitive components as we proceed from the edge of the base toward the center of the base under the emitter. The simplest way to represent this effect is in fact to use a number of sectionalized Ebers-Moll or Gummel-Poon type models each coupled by a fraction of the total (possibly current-dependent) base resistance, as shown in Fig. 15.13. If this sectional modeling technique is used, then quite accurate expressions can be used to model the dependence of base resistance, incorporating both high level injection and base widening effects [34]:

$$R_{bb} = R_{bbi0}\left[\left(\frac{N_{av}}{N_{av}+N_{bm}}\right)\left(\frac{W_b}{W_b+W_k}\right)\left(\frac{W_b}{W_b+W_{sat}}\right)\left(\frac{Q_{b0}}{Q_{b0}+Q_{bc0}-Q_{bc}}\right)\right]+R_{bbx}^{Early}$$

(15.31)

where R_{bb} is the total base resistance, R_{bbi0} is the low current intrinsic base resistance (computed from the device dimensions and sheet resistances for zero bias collector-base junction), N_{av} is the average base doping, N_{bm} is the average free carrier concentration in the base resulting from high level injection (HLI) or saturation, W_b is the low current neutral base width, W_k is the increase in base width due to Kirk effect, W_{sat} is the increase in base width due to saturation, Q_{b0} is the zero bias intrinsic base charge, Q_{bc0} and Q_{bc} are the charges in the base-collector depletion layer (base side) at zero bias and V_{bc}, respectively, to take into account the Early effect.

Figure 15.14 shows the excellent agreement between numerical simulation of base resistance and values computed using Eq. (15.31).

The second extension to the Gummel-Poon formulation concerns the collector resistance. In the WATAND version it is divided into two components, one in series with the intrinsic or active region of the transistor, the other in series with the extrinsic base area as shown in Fig. 15.12. However, it should be recognized that this collector resistance can seldom be treated as a constant value for accurate circuit simulation purposes. It varies due to collector base voltage V_{CB} because of the variation in space-charge layer width extending into the epitaxial layer; it also varies due to the quasi-saturation effect discussed in Chap. 10. In quasi-saturation the collector resistance is a complicated function of both collector current and base collector voltage. A convenient one-dimensional model for this is given by Hébert [37]:

$$R_{cm} = \left[\frac{x_0 - W_b}{qA_e'\mu_{nc}N_{cm}}\right]+\left[\frac{W_{epi}-(x_0-W_b)}{qA_e'\mu_{nc}N_{epi}}\right]$$

(15.32)

where x_0, W_b, W_{epi}, N_{epi} are as shown in Fig. 15.15, A_e' is the effective emitter area, μ_{nc} the electron mobility in the collector, N_{cm} the "average" carrier concentration in the region $W_b < x < x_0$, determined from the collector current. This

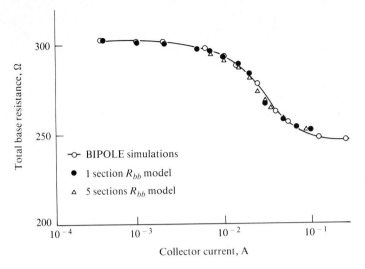

FIGURE 15.14
Base resistance versus collector current using (i) BIPOLE simulation, (ii) Eq. (15.31) for 1- and 5-section models. (*After Hébert and Roulston* [34, 37].)

has been derived by considering the injection of carriers into the collector region as shown in Fig. 15.15. The reader is referred back to Chap. 10 for a more detailed discussion of quasi-saturation effects.

Figure 15.16 shows the measured I_C versus V_{CE} characteristic of two transistors with the curves obtained using the standard Gummel-Poon model and the enhanced model incorporating nonconstant Early voltage, and nonlinear base and collector resistance. It is seen that excellent agreement can be obtained with the improved model.

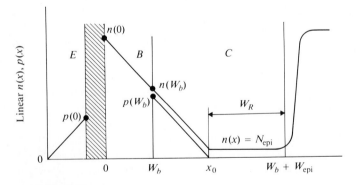

FIGURE 15.15
Carrier concentrations in the base and collector under quasi-saturation conditions. (*After Hébert and Roulston* [34, 37].)

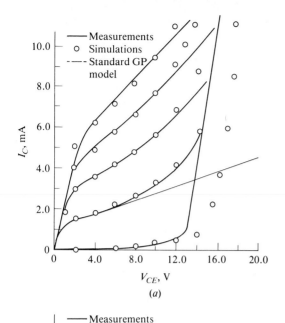

Measurements
Simulations
Standard GP model

V_{CE}, V

(a)

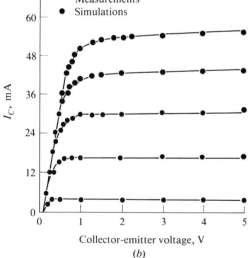

Measurements
Simulations

I_C, mA

Collector-emitter voltage, V

(b)

FIGURE 15.16
Measured and calculated I_C-V_{CE} characteristics of a narrow base transistor using conventional and Gummel-Poon models. (*After Hébert and Roulston* [35, 37].) (a) A narrow base transistor exhibiting punchthrough; (b) a poly emitter device with quasi-saturation at high current. (*Reproduced with permission* © *1987 IEEE.*)

15.4.1   Two-Dimensional Effects in CAD Circuit Models

There are two important areas which necessitate further refinement to the Gummel-Poon type CAD models, particularly in integrated devices. These concern the two- (or three-) dimensional nature of collector and base current flow.

Figure 15.17(a) shows the base current path in a typical integrated transistor. For accurate predictions where base resistance is important (e.g., in high

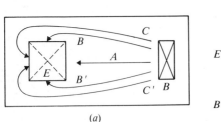

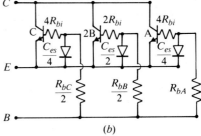

(a)   (b)

FIGURE 15.17
Modeling of two-dimensional base current. (a) Base current paths; (b) possible sectional representation with each transistor described by a Gummel-Poon model. (*After Hébert and Roulston* [36, 37].)

frequency power gain analysis or in some high speed switching applications) it is necessary to take this two-dimensional current flow into account. This can be done with a usually adequate degree of precision by suitably dividing the device into several Gummel-Poon type cells as shown in Fig. 15.17(b) [36].

Figure 15.18 shows the two- or three-dimensional collector current flow in an integrated device. Current flows down from the collector diffusion to the buried layer, spreads laterally in the buried layer and spreads vertically under the emitter (active) area. The three components may be calculated separately and added for dc analysis. Note that both R_{cv} and R_{cs} should incorporate conductivity modulation effects when the transistor enters the quasi-saturation region, whereas normally the buried layer component R_{bL} remains constant (due to the relatively high doping of the buried layer).

15.4.2 Tabular Current Charge Voltage Resistance (CCVR) Models

Another CAD modeling technique for circuit analysis programs, in particular WATAND, uses tabular data as opposed to analytic equations. The advantage of tabular models is that no analytic approximations are necessary in going from the device simulation program to the circuit simulation program. The tabular model approach as implemented in the **BIPOLE WATAND** piecewise linear scheme consists in describing one-dimensional collector current versus base-emitter voltage and base-collector voltage by tabular data [6]. The CCVR tabular model also includes emitter-base charge and collector-base charge, and base resistance and collector resistance. These tables are then used in the sectional model discussed in the previous section. Since each section is described by a tabular model and the sections are coupled by nonlinear base resistance, and since each contains a nonlinear current and voltage-dependent collector resistance, this model gives extremely detailed representation of all the nonlinear and high frequency effects commonly encountered in bipolar transistors. Provided enough values of V_{BE} and V_{CB} are used and provided an adequate number of sections are employed in the circuit analysis program, the major remaining inac-

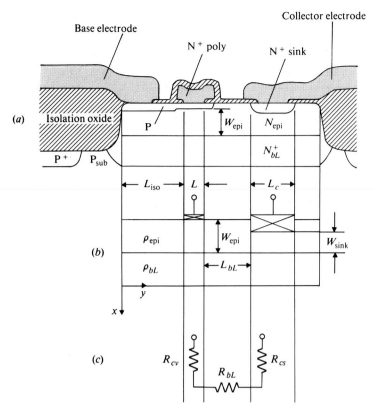

FIGURE 15.18
Two/three-dimensional collector current flow in an integrated device. (*After Hébert* [37].)

curacy concerns the time-dependence of the collector resistance. As we saw in
Chap. 11, under transient conditions, a substantial amount of charge is injected
from the base into the collector epitaxial region; because of the relatively thick
collector layer, this charge is not injected or extracted instantaneously but
requires a diffusion transit time of order $W_{epi}^2/2D_p$. Apart from this one
restriction, the tabular current charge voltage resistance model has been used
with considerable success in predicting such behavior as reverse base current
drive with inductive load conditions for very high voltage power switching transistors.

15.5   CONCLUSIONS

In this chapter we have presented an overview of some of the numerical analysis
and modeling aspects of bipolar transistors. Each approach has its place in the
study and design of devices. At one extreme, there exists the full two-dimensional
solution of the coupled semiconductor equations (Poisson, transport, continuity);

the intermediate approach of the variable boundary regional method with a coupled solution for the lateral base region has the advantage of speed and easy identification of the contributions of the different regions; finally for circuit analysis programs, the Ebers-Moll or Gummel-Poon analytic CAD models are universally used, but as we pointed out, these require some care and/or extensions for reliable circuit models.

Parameters for the CAD models are frequently obtained from measurements on fabricated devices. While this is satisfactory for the circuit designer restricted to one or several already developed transistor structures, it does not enable the effect of process changes (intended or random) on the circuit performance to be determined. This can be done using the combined scheme of a process simulator (like SUPREM), a device simulator (like BIPOLE), and a circuit simulation program (for example, SPICE or WATAND).

Of course, it must be recognized that even in the most favorable situation, there are usually *some* process or device parameters which are not accurately known. This applies also to some of the physical modeling data such as recombination lifetime, minority carrier mobility, bandgap narrowing versus doping level. It is thus clear that no advanced development on bipolar devices can be carried out independently of extensive fabrication of test structures and extensive measurements. Comparison of measured and computed data on test structures enables discrepancies to be taken into account; then the computer simulation of device and circuit can be used as a predictive design tool; this enables the development engineer to see and to understand device and circuit performance, in terms of absolute values if the simulations and models have been sufficiently refined with fitted values where necessary, otherwise performance trends can at least be predicted.

PROBLEMS

15.1. For the solution of Eqs. (15.6) through (15.8), write a self-contained Fortran computer program using first-order integration methods: $E_{new} = E_{old} + (q/\varepsilon)N(x) \, dx$ and $V_{new} = V_{old} - E_{av} \, dx$. The program should use a double gaussian impurity profile as in Eq. (2.47), with the quantities N_e, N_b, N_{epi}, x_e, x_b as input data. It is particularly convenient to supply also a junction depth as input and to start the integration at some value $x = X_j - d_n$ where d_n is a depletion-layer thickness estimated by hand calculation. By monitoring the sign of the electric field, a test can be used to stop when the far side of the depletion layer is reached. Use the program for the impurity profile given by: $N_e = 10^{21}$, $N_e = 10^{19}$, $N_{epi} = 10^{16}$ cm^{-3}, $x_e = 0.3$ μm, $x_b = 0.6$ μm. As a first run, use $d_n = 0.02$ μm, with the value of the emitter-base junction depth (0.744 μm) as input. You can use the program to study the shape of the space charge, electric field, and potential distributions for a few values of d_n. You can also estimate the breakdown voltage by adjusting d_n until the peak value of electric field attains a value of 5×10^5 V/cm. The program should contain about 8 lines of code, plus input and initialization data and write statements.

15.2. Derive the expression given by Eq. (15.13) for electron concentration gradient in the base of an NPN transistor.

15.3. Write a self-contained computer Fortran code for the base region based on Eq. (15.13) using first-order integration methods, and an impurity profile as for Prob. 15.1. Use a constant $\mu_n = 500$ cm^2/s. If integration is started at the collector side of the neutral base ($x = W_b$), the integration for a given J_n gives the injected electron concentration $n(x_{eb})$ of Fig. 15.8. The value of x_{eb} may be supplied as input to determine a stop condition (a convenient value is available from the result of Prob. 15.1). The value of x_{bc} (W_b) may be estimated by hand for a given V_{CB} (for example, 10 V) or computed using the program of Prob. 15.1 but for the base-collector junction. The initial value of $n(x_{bc})$ may be set to zero, or to the value given by Eq. (7.1). Use the program to compute $n(x)$ and the base delay time from the integral of $n(x)$ and the current density (use 100 A/cm^2). The electric field may also be computed and studied from $(V_t/N_{eff})\, dN_{eff}/dx$. Try also using a high current density (for example, 10^4 A/cm^2) so that the base enters partial high level injection, and observe the modified shape of $n(x)$ and the new value of t_{bb}.

15.4. Derive Eq. (15.18) for hole concentration gradient in the emitter of an NPN transistor.

15.5. Write a Fortran computer program for solving the hole distribution in the emitter, based on similar ideas to those used in the above questions (first-order integration, gaussian impurity profile). Include the current continuity equation (15.19) with values of SRH lifetime parameters and Auger coefficient discussed in Chap. 1. Use constant mobilities, $\mu_p = 100$, $\mu_n = 500$ cm^2/V $\cdot$ s. The program may be used, starting the integration at $x = 0$ (the emitter surface) $p = p_{n0}$, and J_p chosen at some arbitrary small value (for example, 0.01 A/cm^2), terminating at the emitter-base depletion-layer boundary x_{en} of Fig. 15.9 (guessed or found from Prob. 15.1 for a forward bias of about 0.6 V). The value of injected hole concentration $p(x_{en})$ is the result of the integration. Print a table of x, $N(x)$, N_{eff}, p, J_p, τ_p. The program should contain less than 20 lines of code plus input, initialization, and write statements.

It may be noted that the integration of the base region equation and the emitter region equation results in two injected concentrations. A meaningful physical result will be that for which these two concentrations are obtained for the same applied voltage V_{BE} as per Eqs. (15.9) and (15.10). The complete solution would require an iterative procedure.

15.6. Plot the SPICE type Gummel-Poon equation (15.29) in the form log (I_C) versus linear V_{BE} and observe the shape of the curve at high currents; assume very large values of V_A, V_B using the data of Prob. 10.3. Compare this to a hand calculation based on Kirk effect using the data of Prob. 10.3. Note that Eq. (15.29) models high level injection in the base perfectly, but is only an approximation to Kirk base widening.

REFERENCES

1. I. Getreu, *Modeling the Bipolar Transistor*, Elsevier, New York (1978).
2. D. A. Antoniadis, S. E. Hansen, and R. W. Dutton, "SUPREM-II. A program for IC process modeling and simulation," Stanford Electronics Laboratories Technical Report no. 5019-2 (June 1978).
3. D. J. Roulston, "Discrete and integrated bipolar device analysis using the BIPOLE fast computer program," *Proc. IEEE Custom Integrated Circuits Conf., Rochester, New York*, pp. 2–6 (May 1980).

4. L. W. Nagle and D. O. Pederson, "Simulation Program with Integrated Circuit Emphasis (SPICE)," *Proc. 16th Midwest Symposium on Circuit Theory, Waterloo, Ontario* (April 1973).

5. I. Hajj, K. Singhal, J. Vlach, and P. Bryant, "WATAND—a program for the analysis and design of linear and piecewise linear networks," *Proc. 16th Midwestern Symposium on Circuit Theory, Waterloo, Ontario* (April 1973).

6. I. N. Hajj, D. J. Roulston, and P. R. Bryant, "Generation of transient response of nonlinear bipolar transistor circuits from device fabrication data," *IEEE Jnl. Solid State Circuits*, vol. SC-12, pp. 29–83 (February 1977).

7. R. Kumar, D. J. Roulston, and S. G. Chamberlain, "Two-dimensional simulation of a high voltage PIN diode with overhanging metallization," *IEEE Trans. Electron Devices*, vol. ED-28, pp. 534–540 (May 1981).

8. M. S. Adler and V. A. K. Temple, "A general method for predicting the avalanche breakdown voltage of negative beveled devices," *IEEE Trans. Electron Devices*, vol. ED-23, pp. 956–960 (August 1976).

9. H. Gummel, "A self-consistent iterative scheme for one-dimensional steady state transistor calculations," *IEEE Trans. Electron Devices*, vol. 11, pp. 455–465 (October 1964).

10. A. Demari, "An accurate numerical steady state one-dimensional solution for the PN junction," *Solid State Electronics*, vol. 11, pp. 33–58 (January 1968).

11. V. Arandjelovic, "Accurate numerical steady state solution for a diffused one-dimensional junction diode," *Solid State Electronics*, vol. 13, pp. 865–871 (June 1970).

12. Z. Yu and R. W. Dutton, "SEDAN III—A generalized electronic material device analysis program," Integrated Circuits Laboratory Technical Report, Stanford University (July 1985).

13. D. L. Scharfetter and H. K. Gummel, "Large-signal analysis of a silicon Read diode oscillator," *IEEE Trans. Electron Devices*, vol. ED-16, pp. 64–77 (January 1969).

14. M. R. Pinto, C. S. Rafferty, H. R. Yeager, and R. W. Dutton, "PISCES-IIB, Supplementary Report," Stanford Electronics Laboratories, Dept. of Electrical Engineering, Stanford University (1985).

15. M. R. Pinto, C. S. Rafferty, and R. W. Dutton, "PISCES-II—Poisson and continuity equation solver," Stanford Electronic Laboratory Technical Report, Stanford Unversity (September 1984).

16. W. Kausel, G. Nanz, S. Selberherr, and H. Potzl, "BAMBI—a transient two-dimensional device simulator using implicit backward Euler's method and a totally self-adaptive grid," NUPAD II Workshop, San Diego, California, Digest no. 105/106 (May 9–10, 1988).

17. S. Selberherr, *Analysis and Simulation of Semiconductor Devices*, Springer, Wien, New York (1984).

18. W. L. Engl, H. K. Dirks, and B. Meinerzhagen, "Device modeling," *Proc. IEEE*, vol. 71, pp. 10–33 (January 1983).

19. R. E. Bank and W. Fichtner, "PLTMG User's Guide," Bell Laboratories Technical Memorandum 82-52111-9.

20. M. Kurata, *Numerical Analysis for Semiconductor Devices*, Lexington Books, Lexington, Massachusetts (1982).

21. "Numerical analysis of semiconductor devices," *NASECODE Conference Proceedings*, 1979–1985, Boole Press, Dublin.

22. A. A. Eltoukhy and D. J. Roulston, "Sidewall effects in shallow emitter of small bipolar transistor," *Electronics Letters*, vol. 17, pp. 845–846 (August 1981).

23. J. W. Slotboom and H. C. de Graaf, "Measurements of bandgap narrowing in Si bipolar transistors," *Solid State Electronics*, vol. 19, pp. 857–862 (1976).

24. D. J. Roulston, N. D. Arora, and S. G. Chamberlain, "Modeling and measurement of minority carrier lifetime in heavily doped N diffused silicon diodes," *IEEE Trans. Electron Devices*, vol. ED-29, pp. 284–291 (February 1982).

25. D. J. Roulston and A. A. Eltoukhy, "Modeling bulk and surface recombination in the sidewall space-charge layer of an emitter-base junction," *IEE Proc. Part I, Solid State and Electron Devices*, vol. 132, pp. 205–209 (October 1985).

26. D. J. Roulston, S. G. Chamberlain, and J. Sehgal, "A simplified computer aided analysis of double-diffused transistors including two-dimensional high level effects," *IEEE Trans. Electron Devices*, vol. ED-19, pp. 809–820 (July 1972).

27. I. N. Hajj, D. J. Roulston, and P. R. Bryant, "A three-terminal piecewise-linear modeling approach to dc analysis of transistor circuits," *Int. Jnl. Circuit Theory and Applications*, vol. 2, pp. 133–147 (1974).

28. D. J. Roulston, "CAD of bipolar custom chips using a coupled process-device-circuit simulation package," *Proc. IEEE Custom Integrated Circuits Conf., Rochester, New York*, pp. 229–232 (May 1983).

29. R. B. Schilling, "The bipolar transistor—an analytic solution obtained via regional approximation and computer monitoring techniques," in *Semiconductor Device Modeling for Computer Aided Design*, G. J. Herskowitz and R. B. Schilling (eds.), McGraw-Hill, New York (1972).

30. C. T. Kirk, "A theory of transistor cutoff frequency (f_t) falloff at high current densities," *IRE Trans. Electron Devices*, vol. ED-9, pp. 164–174 (March 1962).

31. F. Hébert and D. J. Roulston, "Base resistance of bipolar transistors from layout details including two-dimensional effects at low currents and low frequencies," *Solid State Electronics*, vol. 31, pp. 283–290 (February 1988).

32. D. J. Roulston and F. Hébert, "Study of delay times contributing to the f_t of bipolar transistors," *IEEE Electron Device Letters*, EDL-7, pp. 461–462 (August 1986).

33. H. K. Gummel and H. C. Poon, "An integral charge control model of bipolar transistors," *Bell Syst. Tech. Jnl.*, vol. 49, pp. 827–851 (1970).

34. F. Hébert and D. J. Roulston, "Unified model for bipolar transistors including the voltage and current dependence of the base and collector resistances as well as the breakdown limits," 18th ESSDERC Conference, Montpellier, France, *Journal de Physique*, C4, suppl. to no. 9, vol. 49, pp. 371–374 (September 1988).

35. F. Hébert and D. J. Roulston, "Modeling of narrow-base bipolar transistors including variable-base-charge and avalanche effects," *IEEE Trans. Electron Devices*, vol. ED-34, pp. 2323–2327 (November 1987).

36. F. Hébert and D. J. Roulston, "High frequency base resistance and the representation of two- and three-dimensional ac and dc emitter and base current flow of bipolar transistors," 18th ESSDERC Conference, *Journal de Physique*, C4, suppl. to no. 9, vol. 49, pp. 379–382 (September 1988).

37. F. Hébert, "Advanced modeling and fabrication techniques for bipolar transistors," Ph.D. thesis. University of Waterloo, 1988.

38. R. Kumar, S. G. Chamberlain, and D. J. Roulston, "An algorithm for two-dimensional simulation of reverse-biased beveled PN junctions," *Solid State Electronics*, vol. 24, pp. 309–311 (April 1981).

39. C. M. Snowden, *Semiconductor Device Modeling*, Peter Peregrinus, London (1988).

40. T. Kurosawa, Proc. Int. Conf. on Physics of Semiconductors, Kyoto, *J. Phys. Soc.*, Japan, vol. 21, suppl., p. 527 (1966).

41. R. W. Hockney and J. W. Eastwood, *Computer Simulation Using Particles*, McGraw-Hill, New York (1981).

42. C. Moglestue, "Monte-Carlo particle simulation of the hole-electron plasma formed in a PN junction," *Electronics Letters*, vol. 22, pp. 397–398 (1986).

43. G. Baccarani, C. Jacoboni, and A. M. Mazzone, "Current transport in narrow-base transistors," *Solid State Electronics*, vol. 20, pp. 5–10 (January 1977).

44. D. J. Roulston, P. R. Bryant, and I. N. Hajj, "Computer-aided design and evaluation of bipolar transistors and circuits for high-efficiency microwave power conversion," European Space Agency (ESA) SPACECAD '79 International Symposium on Computer Aided Design of Electronics for Space Applications, Bologna, Italy, Conference Proceedings, pp. 177–187 (September 1979).

45. F. Hébert and D. J. Roulston, "Current dependence of forward delay time of bipolar transistors," *Electronic Letters*, vol. 22, p. 126 (January 1986).

DENSITY
OF STATES
AND CARRIER
CONCENTRATIONS

The kinetic energy (KE) of an electron in the conduction band $E - E_c$, (see Fig. 1.2) is related to the momentum as follows:

$$\frac{1}{2} m_e^* v^2 = \frac{p^2}{2m_e^*} = E - E_c \qquad (A1.1)$$

where $p = m_e^* v =$ momentum, that is,

$$p = \sqrt{2m_e^*(E - E_c)} \qquad (A1.2)$$

In three-dimensional momentum coordinates, a constant KE corresponds to a spherical surface centered at the origin. The radius of the sphere is p, the magnitude of the momentum. The differential momentum volume dV_p is then

$$dV_p = 4\pi p^2 \, dp = 2\pi p d(p^2) \qquad (A1.3)$$

$$= 2\pi \sqrt{2m_e^*(E - E_c)} \, d(2m_e(E - E_c))$$

$$= 4\sqrt{2} \, \pi m_e^{*3/2} \sqrt{(E - E_c)} \, dE \qquad (A1.4)$$

399

Minimum Momentum Volume

From Heisenberg's uncertainty principle, we have the minimum value of momentum for given physical dimensions of the material

$$\Delta P_x = \frac{\hbar}{L_x}$$

$$\Delta P_y = \frac{\hbar}{L_y}$$

$$\Delta P_z = \frac{\hbar}{L_z}$$

where L_x, L_y, and L_z are physical dimensions of a "cube" of material, and ΔP_x, ΔP_y, and ΔP_z are the components of momentum in each direction.

$$\Delta P_x \, \Delta P_y \, \Delta P_z = \Delta V_p = \frac{\hbar^3}{L_x L_y L_z} = \frac{\hbar^3}{V} \tag{A1.5}$$

where V is the volume of the material.

ΔV_p is the smallest possible momentum volume, and corresponds to one energy level. The number of energy levels is thus given by

$$\frac{d(\text{energy levels})}{dV_p} = \frac{V}{\hbar^3} \tag{A1.6}$$

or, since there are two states per energy level,

$$d\left(\frac{\text{states}}{\text{cm}^3}\right) = \frac{2dV_p}{\hbar^3}$$

$$= 8\sqrt{2}\,\pi\, \frac{m_e^{*3/2}}{\hbar^3}\, \sqrt{E - E_c}\; dE \tag{A1.7}$$

This gives us the density of states per unit energy.

The carrier concentration per unity energy, n', is thus given by

$$n' \, dE = f(E) \times d\left(\frac{\text{states}}{\text{cm}^3}\right) \tag{A.1.8}$$

where $f(E)$, the Fermi factor (which we may approximate by the Boltzmann factor in the conduction band, for nondegenerate material) is simply the probability of occupancy of a state.

The total carrier concentration is obtained by integrating over all energy levels above E_c.

$$\bar{n} = \int_{E_c}^{\infty} n' \, dE \tag{A1.9}$$

$$\bar{n} = \frac{8\sqrt{2}\,\pi m_e^{*3/2}}{h^3}\sqrt{kT}\,e^{(E_F - E_c)/kT}\int_{E_c}^{\infty}\sqrt{\frac{E - E_c}{kT}}\,e^{-(E - E_c)/kT}\,dE \quad (A1.10)$$

By substituting $x = (E - E_c)/kT$

$$dx = \frac{1}{kT}\,dE \qquad \text{and} \qquad \int_0^{\infty}\sqrt{x}\,e^{-x}\,dx = \frac{\pi}{2} \qquad (A1.11)$$

we obtain

$$n = \frac{4\sqrt{2}}{h^3}(\pi m_e^* kT)^{3/2}e^{(E_F - E_c)/kT} \qquad (A1.12)$$

A similar analysis for holes gives

$$\bar{p} = \frac{4\sqrt{2}}{h^3}(\pi m_h^* kT)^{3/2}e^{(E_v - E_F)/kT} \qquad (A1.13)$$

These two relationships are usually written in the form

$$\bar{n} = N_c\,e^{(E_F - E_c)/kT} \qquad (A1.14)$$

$$\bar{p} = N_v\,e^{(E_v - E_F)/kT} \qquad (A1.15)$$

where N_c, N_v are constants for a given material at a given temperature.

APPENDIX

2

BOLTZMANN AND EINSTEIN RELATIONS

A. Consider a semiconductor (with nonuniform doping) in thermal equilibrium, no current flowing ($J_n = J_p = 0$).

$$qD_n \frac{dn}{dx} + \mu_n qnE = 0 \tag{A2.1}$$

hence

$$\frac{1}{n}\frac{dn}{dx} = -\frac{\mu_n}{D_n} E$$

where $E = -dV/dx$.

Integrating from one point in the semiconductor to another point gives

$$\frac{n(x_1)}{n(x_2)} = \exp\left[\frac{\mu_n(V_1 - V_2)}{D_n}\right] \tag{A2.2}$$

where V_1 is the potential at the point x_1 where $n = n(x_1)$ and
V_2 is the potential at the point x_2 where $n = n(x_2)$

B. Recall the solution for carrier concentration $\bar{n} = N_c\, e^{(E_F - E_C)/kT}$. In thermal equilibrium with no current flowing, the Fermi levels must align, i.e., be the same at every part in the crystal

$$\frac{n(x_1)}{n(x_2)} = \exp\left[-\frac{E_{c1} - E_{c2}}{kT}\right] \tag{A2.3}$$

Now an energy difference $E_{c1} - E_{c2}$ is simply $-q(V_1 - V_2)$ where $V_1 - V_2$ is the potential difference between points x_1 and x_2, that is,

$$\frac{n(x_1)}{n(x_2)} = \exp\left[\frac{q(V_1 - V_2)}{kT}\right] \tag{A2.4}$$

Comparing (1) with (2) we see that

$$\boxed{\frac{D_n}{\mu_n} = \frac{kT}{q}} \tag{A2.5}$$

This is the Einstein relation.

APPENDIX
3

THE SHOCKLEY-READ RECOMBINATION MODEL

Consider the one-level recombination model shown in Fig. 1.7. The rate of trapping, R, of electrons can be written

$$R(E) = n(E)v(E)A(E) \tag{A3.1}$$

where $n(E)$ is the electron concentration, $v(E)$ the thermal velocity of the electrons, and $A(E)$ the effective "capture cross section." Note that each of these parameters is a function of the energy E of the particle in the conduction band. The rate of decrease U_{nc} of the electron concentration due to trapping is, therefore,

$$U_{nc}(E) = -n(E)v(E)A(E)p_t \tag{A3.2}$$

where p_t is the concentration of empty traps and N_t the total concentration of traps at energy level E_t. It is customary to write this equation in the form

$$U_{nc}(E) = -c_n(E)n(E)p_t \tag{A3.3}$$

where $c_n(E) = v(E)A(E)$ and is called the "capture probability per unit time." Note that the number of occupied traps n_t is given by the relation

$$N_t = n_t + p_t \tag{A3.4}$$

i.e., a trap must either be filled (by an electron) or be empty (occupied by a hole).

404

Since the rate of trapping of electrons from the conduction band given by Eq. (A3.3) must be equal to the rate of emission from the traps into the conduction band, we have

$$c_n(E)n(E)p_t = e_n(E)[N(E) - n(E)]n_t \qquad (A3.5)$$

where $e_n(E)$ is the "emission probability per unit time." Note that Eq. (A3.5) is valid even in nonequilibrium conditions. Using the relation

$$n(E) = N(E)F(E) \qquad (A3.6)$$

and

$$n_t = N_t[F(E_t)] \qquad (A3.7)$$

where $F(E)$ is the Fermi-Dirac probability function for electrons, we obtain (using the fact that $F(E) \ll 1$, for $E \geq E_c$ and $F(E) \to 1$, for $E \approx E_t$)

$$e_n(E) = c_n(E) \exp\left[-\frac{E - E_t}{kT}\right] \qquad (A3.8)$$

A similar expression relates hole capture and emission probabilities per unit time

$$e_p(E) = c_p(E) \exp\left[-\frac{E_t - E}{kT}\right] \qquad (A3.9)$$

In general, one is interested in the total electron concentration n in the conduction band. Integrating Eq. (A3.3) over all energy levels E greater than E_c gives

$$U_{nc} = p_t n c_n \qquad (A3.10)$$

where

$$c_n = n^{-1} \int_{E_c}^{\infty} c_n(E)n(E)\, dE \qquad (A3.11)$$

where c_n is the capture probability per unit time between the trap level and the whole of the conduction band. Similarly, we may define U_{ne} and e_n for emission from the trap level to the whole of the conduction band, assuming a nearly empty conduction band, i.e., assuming $n \ll N_c$

$$U_{ne} = n_t N_c e_n \qquad (A3.12)$$

where

$$e_n = N_c^{-1} \int_{E_c}^{\infty} e_n(E)[N(E) - n(E)]\, dE \qquad (A3.13)$$

and N_c is the effective density of states for the conduction band.

In order to derive a relationship between e_n amd c_n, we take Eq. (A3.13) and substitute Eq. (A3.8) for $e_n(E)$. This gives

$$e_n = c_n \exp\left[-\frac{E_c - E_t}{kT}\right] \tag{A3.14}$$

A similar derivation gives

$$e_p = c_p \exp\left[-\frac{E_t - E_v}{kT}\right] \tag{A3.15}$$

The exponential terms may be replaced by defining

$$n_1 = N_c \exp\left[-\frac{E_c - E_t}{kT}\right] \tag{A3.16}$$

$$p_1 = N_v \exp\left[-\frac{E_t - E_v}{kT}\right] \tag{A3.17}$$

These are simply convenient constants (it is obvious that they represent the electron and hole concentrations, respectively, for the case where the Fermi level lies at the trap level E_t. Since E_t is near the middle of the bandgap, it is apparent that n_1 and p_1 will be comparable in magnitude to n_i). In other words, we have

$$e_n N_c = c_n n_1 \tag{A3.18}$$

and

$$e_p N_v = c_p p_1 \tag{A3.19}$$

We can thus write

$$U_n = U_{nc} - U_{ne} = p_t n c_n - n_t N_c e_n \tag{A3.20}$$

$$= (p_t n - n_t n_1)c_n \tag{A3.21}$$

where U_n is the net capture rate per unit volume of electrons from the conduction band. Similarly

$$U_p = U_{pc} - U_{pe} = (n_t p - p_t p_1)c_p \tag{A3.22}$$

where U_p is the net capture rate per unit volume of holes from the valence band.

Under steady state nonequilibrium conditions, we must have no build-up of carriers in the trap levels. Therefore

$$U_n = U_p \tag{A3.23}$$

Equating (A3.21) and (A3.22), putting p_t in terms of N_t and n_t [from (A3.4)] we obtain

$$U = \frac{(np - n_1 p_1)N_t}{(n + n_1)/c_p + (p + p_1)/c_n} \tag{A3.24}$$

This may be rewritten in the form

$$U = \frac{np - n_i^2}{(n + n_1)\tau_{p0} + (p + p_1)\tau_{n0}} \tag{A3.25}$$

where

$$\tau_{p0} = (N_t c_p)^{-1} \tag{A.3.26}$$

$$\tau_{n0} = (N_t c_n)^{-1} \tag{A3.27}$$

It is obvious that τ_{p0} and τ_{n0} are the minority lifetimes for moderately to heavily doped N-type and P-type material, respectively, e.g., for N-type material with $N_d \gg n_1$ and $N_d \gg p_1$, we can write

$$\frac{dp}{dt} = U = \frac{p - n_i^2/N_d}{\tau_{p0}} = \frac{p'}{\tau_{p0}}$$

APPENDIX
4

QUASI-FERMI LEVELS

By rearranging Eq. (1.9), we may express the free carrier concentration n as a function of the intrinsic energy level E_i

$$n = n_i \exp\left[\frac{E_F - E_i}{kT}\right] = n_i \exp\left[\frac{\psi - \phi}{V_t}\right] \tag{A4.1}$$

where ψ = intrinsic level of potential = $-E_i/q$
   ϕ = Fermi level = $-E_F/q$
   It follows that we may define

$$n \triangleq n_i \exp\left[\frac{\psi - \phi_n}{V_t}\right] \tag{A4.2}$$

where ϕ_n is called the "quasi-Fermi level," which in the general case is position-dependent. The concept of a quasi-Fermi level is frequently used in analysis of semiconductor devices. It enables current flow to be expressed in terms of one derivative instead of two as is the case when current flow is written as a combination of drift and diffusion. This may be seen as follows by differentiating the above expression for n with respect to distance:

$$\begin{aligned} \frac{dn}{dx} &= \frac{n_i}{V_t} e^{(\psi - \phi_n)/V_t} \left(\frac{d\psi}{dx} - \frac{d\phi_n}{dx}\right) \\ &= \frac{n}{V_t}\left(\frac{d\psi}{dx} - \frac{d\phi_n}{dx}\right) \end{aligned} \tag{A4.3}$$

408

hence, rearranging terms

$$\frac{V_t}{n}\frac{dn}{dx} - \frac{d\psi}{dx} = \frac{d\phi_n}{dx} \tag{A4.4}$$

or, using the Einstein relation,

$$qD_n \frac{dn}{dx} - q\mu_n n \frac{d\psi}{dx} = -q\mu_n n \frac{d\phi_n}{dx} \tag{A4.5}$$

Since $d\psi/dx$ is the electric field E, this equation is equivalent to writing

$$J_n = qD_n \frac{dn}{dx} + q\mu_e nE = -q\mu_n n \frac{d\phi_n}{dx} \tag{A4.6}$$

In other words, the electron current density is given directly by the electron concentration and the derivative of the quasi-Fermi level

$$J_n = -q\mu_n n \frac{d\phi_n}{dx} \tag{A4.7}$$

APPENDIX
5

HOLE DIFFUSION CURRENT AT CENTER OF SPACE-CHARGE LAYER

For the case of a forward biased junction, the result [Eq. (2.44)] for hole concentration versus distance can be used with n_c replacing n_i

$$p(x) = n_c \exp\left(-\frac{x}{x_c}\right) \tag{A5.1}$$

where, from Eq. (2.70), $n_c = n_i \exp(V_a/2V_t)$.

Equation (2.46) thus gives (for an assumed linearly graded junction in the space-charge layer region)

$$\frac{x_c}{d_n} = \frac{4}{3} \frac{V_t}{V_{bi} - V_a} \tag{A5.2}$$

The hole diffusion current $J_p = -qD_p \, dp/dx$ can therefore be obtained at the metallurgical junction $x = 0$:

$$J_{\text{diff scr}} = \frac{qD_p n_c}{(4/3)d_n} \frac{V_{bi} - V_a}{V_t} \tag{A5.3}$$

For a narrow base neutral region, using Eq. (3.8), the hole diffusion current is

$$J_{\text{diff neut}} = \frac{qD_p\, p_n(0)}{W_n}$$

with $p(0) = p_{n0} \exp(V_a/V_t)$ and $p_{n0} = n_i^2/N_d$. The ratio of the two currents r_{curr} is thus

$$r_{\text{curr}} = \frac{J_{\text{diff scr}}}{J_{\text{diff neut}}} = \frac{3}{4}\frac{D_p}{D_n}\frac{n_c}{p_n(0)}\frac{W_n}{d_n}\frac{V_{bi} - V_a}{V_t} \tag{A5.4}$$

$$= \frac{3}{4}\frac{D_p}{D_n}\frac{N_d}{n_i}\frac{W_n}{d_n}\frac{V_{bi} - V_a}{V_t} \exp\left(-\frac{V_a}{2V_t}\right) \tag{A5.5}$$

Typical values are $V_a = 0.7$ V, $V_{bi} = 0.8$ V, $N_d = 10^{16}$ cm^{-3}, $d_n = 0.1$ μm, $W_n = 10$ μm. The ratio r_{curr} is thus of order $10^8 \exp(-V_a/2V_t)$.

APPENDIX

6

CHARGE CONTINUITY EQUATION

Consider the time-dependent continuity equation for holes

$$-\frac{\partial I_p}{\partial x} = qA\left(\frac{p'}{\tau_p} + \frac{\partial p'}{\partial t}\right) \tag{A6.1}$$

If we substitute the expression for hole diffusion current, we obtain

$$D_p\left(\frac{\partial^2 p'}{\partial x^2}\right) = \frac{p'}{\tau_p} + \frac{\partial p'}{\partial t} \tag{A6.2}$$

If we now integrate this equation from some point $x = 0$ (a space-charge layer boundary, for example) to some point $x = w$, we obtain

$$qAD_p\left[\left(\frac{\partial p'}{\partial x}\right)_{x=w} - \left(\frac{\partial p'}{\partial x}\right)_{x=0}\right] = \frac{d}{dt}\int_0^w qAp'\,dx + \frac{1}{\tau_p}\int_0^w qAp'\,dx \tag{A6.3}$$

This can be written as

$$I_{p(x=0)} - I_{p(x=w)} = \frac{dQ}{dt} + \frac{Q}{\tau_p} \tag{A6.4}$$

where Q is the total instantaneous stored charge corresponding to the excess holes.

For any structure in which the total minority carrier current is zero at some point distance w from the point of injection of the excess carriers, we can write

$$I_p = \frac{dQ}{dt} + \frac{Q}{\tau_p} \qquad \text{(A6.5)}$$

This is the charge continuity equation. It is extremely useful in evaluating the characteristics of a wide range of devices. Notice that it is much simpler than the complete form of the continuity equation simply because we have lost all information about the *shape* of the minority carrier distribution.

APPENDIX
7

SOME USEFUL PHYSICAL CONSTANTS

Permittivity in vacuum	ε_0	8.85×10^{-14} F/cm
Electron charge	q	1.60×10^{-19} C
Boltzmann's constant	k	1.38×10^{-23} J/K
Thermal voltage at 300 K	kT/q	0.0259 V
Electron rest mass	m_0	9.11×10^{-31} kg
Planck's constant	h	6.63×10^{-34} J·s

APPENDIX
8

SOME
USEFUL
MATERIAL
PROPERTIES
FOR
SEMICONDUCTORS

Property and units	Ge	Si	GaAs
Bandgap at 300 K E_g (eV)	0.66	1.12	1.42
Maximum electron mobility μ_e (cm^2/V s)	3900	1500	8500
Maximum hole mobility μ_h (cm^2/V s)	1900	450	400
Dielectric constant ε	16.0	11.9	13.1
Effective density of states in conduction band, N_c (cm^{-3})	1.04×10^{19}	2.8×10^{19}	4.7×10^{17}
Effective density of states in valence band, N_v (cm^{-3})	6.0×10^{18}	1.04×10^{19}	7.0×10^{18}
Thermal conductivity at 300 K (W/cm °C)	0.6	1.5	0.46

INDEX

417